THE WORLD AT RISK: NATURAL HAZARDS AND CLIMATE CHANGE

AIP CONFERENCE PROCEEDINGS 277

THE WORLD AT RISK: NATURAL HAZARDS AND CLIMATE CHANGE

CAMBRIDGE, MA 1992

EDITOR:
RAFAEL BRAS
MIT
CAMBRIDGE, MA

American Institute of Physics   New York

Authorization to photocopy items for internal or personal use, beyond the free copying permitted under the 1978 U.S. Copyright Law (see statement below), is granted by the American Institute of Physics for users registered with the Copyright Clearance Center (CCC) Transactional Reporting Service, provided that the base fee of $2.00 per copy is paid directly to CCC, 27 Congress St., Salem, MA 01970. For those organizations that have been granted a photocopy license by CCC, a separate system of payment has been arranged. The fee code for users of the Transactional Reporting Service is: 0094-243X/87 $2.00.

© 1993 American Institute of Physics.

Individual readers of this volume and nonprofit libraries, acting for them, are permitted to make fair use of the material in it, such as copying an article for use in teaching or research. Permission is granted to quote from this volume in scientific work with the customary acknowledgment of the source. To reprint a figure, table, or other excerpt requires the consent of one of the original authors and notification to AIP. Republication or systematic or multiple reproduction of any material in this volume is permitted only under license from AIP. Address inquiries to Series Editor, AIP Conference Proceedings, AIP, 335 East 45th Street, New York, NY 10017-3483.

L.C. Catalog Card No. 93-71333
ISBN 1-56396-066-4
DOE CONF-9201138

Printed in the United States of America.

Contents

Preface ix
    Rafael L. Bras

INTRODUCTION—NATURAL DISASTERS AND CLIMATE CHANGE

Challenges of the Decade: Natural Disasters and Global Change 3
    James P. Bruce
Climate Change and Natural Disasters: Where are the Links? 13
    James C. I. Dooge

IMPACTS OF CLIMATE CHANGE ON NATURAL PHENOMENA

The Dependence of Hurricane Intensity on Climate 25
    Kerry A. Emanuel
Tropical Cyclone Frequency and Global Warming 34
    Richard E. Peterson and Thomas E. Warner
The Response of Sea Level to Global Warming 38
    Andrew R. Solow
The Effect of Rising Sea Level on the Hydrology of Coastal Watersheds 43
    William K. Nuttle
Glacier-Related Hazards and Climatic Change 48
    Steven G. Evans and John J. Clague
Changes in Water Supply in Alpine Regions Due to Glacier Retreat 61
    Mauri S. Pelto
Lightning and Forest Fires in a Changing Climate 68
    Colin Price and David Rind
The Global Electrical Circuit as Global Thermometer 77
    Earle Williams and Stan Heckman
The Relationship Between ENSO Events and California Streamflows 86
    Ercan Kahya and John A. Dracup
Droughts and Climate Change 96
    Ignacio Rodríguez-Iturbe
An Approach for Assessing the Sensitivity of Floods to Regional Climate Change 112
    James P. Hughes, Dennis P. Lettenmaier, and Eric F. Wood
Blowing Dust and Climate Change 125
    Richard E. Peterson and James M. Gregory
Biomass Burning and Global Change 131
    Joel S. Levine, Wesley R. Cofer III, Donald R. Cahoon, Jr.,
    Edward L. Winsted, and Brian J. Stocks

MODELING AND MEASUREMENTS

Forecast Cloudy: The Limits of Global Warming Models 143
    Peter H. Stone

Oceanic Aspects and Global Change 150
    Jochem Marotzke
Bridging the Gap Between Microscale Landsurface Processes
and Land–Atmosphere Interactions at the Scale of GCM's 156
    Roni Avissar
The Representation of Landsurface–Atmosphere Interaction in
Atmospheric General Circulation Models 168
    Dara Entekhabi and Peter S. Eagleson
Spatial Distribution of Precipitation Recycling in the Amazon Basin 174
    Elfatih A. B. Eltahir and Rafael L. Bras
Global Estimation of Rainfall: Certain Methodological Issues 180
    Witold F. Krajewski
Methods and Problems in Assessing the Impacts of Accelerated Sea-
Level Rise 193
    Robert J. Nicholls, Karen C. Dennis, Claudio R. Volonte,
    and Stephen P. Leatherman
Autonomous Aerosondes for Meteorological Soundings in Remote Areas 206
    Tad McGeer

INDUSTRY, CLIMATE CHANGE, AND NATURAL DISASTERS

Global Warming and the Insurance Industry 217
    G. A. Berz
Natural Hazards and Climate Change: The Response of the Energy
Industry 224
    David C. White
Adapting to Agricultural Hazards Created by Climate Change 254
    Paul E. Waggoner
Global Trends in Motor Vehicles and Their Implications for Climate
Change 262
    James J. MacKenzie
Global Warming, Natural Hazards, and Transportation in Europe 275
    Allen Perry
Potential Impact of Climate-Induced Natural Disasters on the
Construction Industry 280
    Ahsan Kareem

SOCIETAL RESPONSES TO GLOBAL CLIMATE AND NATURAL HAZARDS

Societal Response to Global Climate Change: Prospects for Natural
Hazard Reduction 289
    Joanne M. Nigg
Societal Response to Chronic Environmental Change: The Role of Evolving
Scientific and Technical Information in State Coastal Erosion
Management 295
    Mark Meo, Thomas E. James, and Robert E. Deyle
A Consortium Approach for Disaster Relief and Technology
Research and Development: Fire Station EARTH 305
    Douglas C. Ling

NATIONAL RESEARCH ACTIVITIES

Engineering Response to the Dual Risk of Natural Hazards and Global Change 317
    Joseph Bordogna and J. Eleonora Sabadell

Italian National Response to the International Decade for Natural Disaster Reduction 320
    Lucio Ubertini

United States Response to the International Decade for Natural Disaster Reduction 328
    J. Eleonora Sabadell

RESPONSE OF NATIONS

Natural Hazards and the U.S. Global Change Research Program 337
    Dallas L. Peck

Venezuelan Policies and Responses on Climate Change and Natural Hazards 343
    C. Caponi and A. Rosales

Concluding Remarks 347
    Peter S. Thacher

Author Index 351

Preface

There is no question that for the first time in history humanity has altered the chemical composition of the atmosphere. Carbon dioxide, largely the product of fossil fuels burning, has increased from about 270 parts per million at the beginning of the Industrial Revolution to about 355 ppm in 1992. Not since 130 000 years ago, between two major ice ages, has the Earth seen such levels of carbon dioxide. Then it was naturally produced as the result of little understood biochemical and physical phenomena that spanned thousands of years in the Earth's evolution. It is predicted that in 40 years the concentration of carbon dioxide could reach 600 ppm. No such levels have occurred in the past as far as we can tell from paleoclimate records.

It is commonly predicted that as a result of such chemical changes in the atmosphere (i.e., doubling of CO_2 concentration), the global mean temperature of the Earth will increase 1.5 to 4 °C. These predictions are fraught with uncertainties. Nevertheless, many, if not most, scientists believe that our knowledge is sufficient to warrant serious concern. The evidence of changes in the past 100 years is tenuous—about 0.5 °C global mean temperature increase, accelerated retreat of glaciers, thinning of ocean ice, etc.—but sufficiently consistent to encourage study and possibly action to prevent further changes in atmospheric composition.

The big uncertainties and concerns, though, lie not on mean temperature predictions, but on related effects. How much will oceans rise, mostly due to thermal expansion, flooding the already marginal lands like Bangladesh? Will hurricanes, fed by ocean temperatures, increase in magnitude and frequency? Are droughts going to be more frequent? How will agricultural production change? Who will suffer and who will benefit? What demographic and political changes will follow? In other words, what natural hazards will become disasters with major sociopolitical implications at the regional and national levels?

The potential for human-induced climate change and other global environmental issues have captured the attention of everybody, from the child in the local elementary school to the heads of state that attended the United Nations Conference on Environment and Development. But the linkage between global and regional environmental damage and natural disasters is too commonly glossed over. Ironically, this great movement for global environmental protection is occurring during the International Decade for Natural Disaster Reduction (IDNDR). The commonalities of these two international efforts must be emphasized and the synergism between those involved encouraged. With those two goals in mind the Center of Global Change Science of the Massachusetts Institute of Technology hosted a symposium on The World at Risk: Natural Hazards and Climate Change, held January 14–16, 1992, at the MIT campus. The meeting was sponsored by the U.S. National Science Foundation, Engineering Directorate, Natural Hazards and Man-Made Hazard Mitigation Program. About 200 engineers and scientists dealt with issues of predicting hazards impacted by climate change; discussed impacts on various industries and sectors; outlined strategies of prevention and/or adaptation to such hazards; and debated appropriate responses of society, professions, and nations. This was not only a meeting to discuss the science of global climate change. The discussions concentrated on the possible impacts of climate change and how society can respond to those possible threats.

As commonly is the case, any collection of selected papers will necessarily fail in capturing, in its entirety, the nuances, exchanges, and excitement of the meeting. Nevertheless, the *Proceedings* of this Conference include a good portion of the formal presentations.

The material is introduced with articles by James P. Bruce, Chair of the U.N. Scientific and Technical Committee for the International Decade for Natural Disaster Reduction, and by J. C. Dooge, Chairman of the Organizing Committee for the Second World Climate Conference, President-elect of the International Council of Scientific Unions, and advisor to the Secretary

General of the U.N. on the proposal for IDNDR. Dr. Bruce discusses the paradigm of environmentally sustainable development and argues that it is necessary to view efforts in natural disaster reduction in a similar way. National disaster preparedness must be part of any successful development strategy. IDNDR is the framework for international cooperation to achieve that goal. Professor Dooge explores the common scientific, institutional, and social grounds of climate change and natural disasters. He argues for new partnerships between scientists and engineers and between the scientific-technical community and other groups including international governmental organizations, national governments, private business, and the public at large.

The next series of articles deals with the possible *Impacts of Climate Change on Natural Phenomena*. Professor Kerry A. Emanuel estimates the changes in frequency and intensity of tropical cyclones that may result from human-induced climate change. He concludes that significant increase in the destructive potential of hurricanes is possible. Dr. Richard E. Peterson and Dr. Thomas F. Warner discuss potential data sets that may be used to detect any changes in tropical cyclone behavior as a result of ocean warming. Sea-level rise and its potential impact are discussed in articles by Dr. Andrew R. Solow and Dr. William K. Nuttle. Solow provides a critical review of evidence of sea-level rise. Nuttle deviates from the traditional warnings of coastal inundation and argues that sea-level rise poses a serious threat to the land hydrology. Twenty centimeters of sea-level rise could double the mean surface runoff and reduce the groundwater discharge by half in a small watershed in Cape Cod, Massachusetts. Articles by Dr. Steven G. Evans and Dr. John J. Clague, and by Dr. Mauri S. Pelto focus on the observed significant glacier ice loss around the world. Loss of glacier ice induces instabilities that result in avalanches and floods commonly resulting in disasters of very large magnitude. Evans and Clague catalog and discuss these impacts very well. Pelto argues that glaciers act as reservoirs of water, damping periods of high flow and providing large volumes of meltwater during otherwise dry times. The retreat of glaciers increases flood risk and endangers water supply during late summer periods.

Lightning kills a very large number of individuals annually as well as significantly disrupts many key activities of modern-day society. Dr. Colin Price and Dr. David Rind show results indicating that doubling CO_2 concentrations in the atmosphere may lead to a 32% increase in lightning activity. This combined with the amazing possibility that the present-day 100-year drought may become a common two-year drought by the year 2060 in parts of the U.S. could result in increased forest fires, among other impacts. Dr. Earle Williams and Mr. Stan Heckman use the relationship between lightning and temperature in a suggestive way to detect climate change. It is argued that the Schumann resonance, a global electromagnetic phenomenon driven by global lightning activity, will respond quickly and uniformly to temperature changes. This could provide a means for detection of global temperature changes with local observations.

Mr. Ercan Kahya and Dr. John A. Dracup highlight the interdependence of climate processes over large global scales. They study the relationship between streamflows in southern California and El Niño/Southern Oscillation signals in the equatorial Pacific. Dr. Ignacio Rodríguez-Iturbe also dwells on interdependence, this time between surface hydrologic processes and the atmosphere. He couches the behavior of droughts and soil moisture on the nonlinear feedbacks between the atmosphere and the continental land masses. Dr. James P. Hughes *et al.* outline a possible procedure to assess the impact of climate change on extreme floods. The technique is designed to properly transfer information from the large scale of global climate models to the subgrid, regional, scale. Dr. Richard E. Peterson and Dr. James M. Gregory look at records from Lubbock, Texas to understand the relationship between weather, climate, and blowing dust. In the last paper of this section Dr. Joel S. Levine *et al.* illustrate the magnitude of human-initiated biomass burning and argue that this may be the source of as much as 40% of the global annual production of CO_2 and a similarly large proportion of other greenhouse gases.

Predictions of the potential impact of greenhouse gases on climate change are generally based on models. These models are approximations to reality and in many ways flawed. Ultimately model results must be verified with appropriate global scale information. The next section (*Modeling and Measurements*) concentrates on models and their limitations, particularly the issue of parametrization of processes that occur at scales smaller than the common model resolutions. Furthermore, several papers discuss the problem and ideas to measure key phenomena at global scales. Dr. Peter H. Stone puts global circulation models in perspective. He explains how they operate and where are some of their known flaws. Dr. Jochem Marotzke makes a strong argument for the need to integrate models of oceans and atmospheric circulation. Presently this link is at best weak, and without it predictions of climate change are seriously compromised. The next three papers deal with the representation of processes that occur at scales smaller than the common grid size of a global circulation model. This is the problem of subgrid parametrization. Dr. Roni Avissar suggests a prognostic equation for mesoscale kinetic energy to relate subgrid-scale landscape heterogeneity to subgrid-scale convective clouds. Dr. Dara Entekhabi and Dr. Peter S. Eagleson focus on the interaction of subgrid-scale soil hydrology and the atmosphere. They outline the influence of landsurface processes on climate and describe ways of modeling those processes within the context of GCM's. Dr. Elfatih A. B. Eltahir and Dr. Rafael L. Bras accent the landsurface–atmosphere interactions by estimating the level of moisture recirculation at mesoscales over the Amazon basin. About 25% of all rain that falls within the Amazon basin is contributed by evaporation within the basin. Dr. Witold F. Krajewski and Dr. Robert J. Nicholls *et al.* discuss the large scale measurement of precipitation and sea-level rise, respectively. Dr. Tad McGeer advocates the use of small, remote-controlled aircraft to obtain global scale environmental data.

The next section is on *Industry, Climate Change, and Natural Disasters.* Dr. G. A. Berz opens by stating that the international insurance industry has been confronted with a drastic increase in the scope and frequency of great natural disasters. Although this increase cannot be attributed to any climate change, the prospects of some of the predictions coming true are sobering. He warns about national and international insurance industries running into capacity problems. Dr. David C. White provides a thorough view of the relationship between the electric industry and climate change, particularly the production of greenhouse gases. He concludes that the only feasible strategy to reduce CO_2 emissions is to increase the efficiency with which energy is used and to shift the energy systems toward electric energy that is produced by more efficient generating systems using lower carbon-based fuels and eventually to noncarbon-based energy sources. Dr. Paul E. Waggoner calls for adaptation as a valid and reasonable strategy for the agricultural industry. Historically, changing technology and flexible strategies by farmers have overcome threats of similar nature. Waggoner bets that they will again succeed in facing this new challenge. Dr. James J. MacKenzie and Dr. Allen Perry offer two views of climate change from the perspective of the transportation industry, particularly the automobile. MacKenzie argues that growth in motor vehicle use will overwhelm gains in fuel efficiency, hence not reversing the trend of increasing greenhouse gas production. He believes that ultimately electric and hydrogen vehicles, charged by nonfossil electric sources, are the only long-term technical options that can reverse the increasing trends of greenhouse gas production and air pollution. This position contrasts with White's opinion that this nonfossil technology for vehicles is not in the cards. Finally Dr. Ahsan Kareem outlines the impact of climate-induced natural hazards on the construction industry. Some of the hazards of concern are extreme winds, storm surges, hurricanes, erosion, etc. The industry has to be concerned about new construction as well as about retrofitting the infrastructure that supports us.

Whether an extreme geophysical event becomes a disaster depends largely on how society responds to the menace. The next three papers in the proceedings deal with societal responses to environmental change (*Societal Responses to Global Change and Natural Hazards*). Dr. Joanne M. Nigg identifies several factors that affect the way societies respond to hazards: societal resources, realistic expectations, vulnerability, and past experience. Climate change-induced natural hazards add an element of uncertainty not necessarily present in other societal experiences.

The question is how well society responds to possible future disaster events, due to a changing, but still not characterized, physical environment. Dr. Mark Meo *et al.* chronicle the contrasting experiences of Florida, Massachusetts, and North Carolina in handling coastal erosion. Successful policies involve institutions and mechanisms that facilitate the communication between key decision makers and the scientific/technical community. Dr. Douglas C. Ling calls for an international consortium involving governments, academia, industry, and businesses as a way to effectively respond to natural and man-made disasters. He calls his concept Fire Station Earth.

Research is an integral element of any approach to the dual issues of climate change and natural hazards. Several countries presented their research initiatives, particularly within the framework of the International Decade for Natural Hazards Reduction (*National Research Activities*). Here three such contributions are presented. Dr. Joseph Bordogna (Assistant Director for Engineering, U.S. National Science Foundation) and Dr. J. Eleonora Sabadell (Program Director, Natural and Man-Made Hazards Mitigation, U.S. NSF) outline the initiatives of the Engineering Directorate of the U.S. National Science Foundation. Dr. Lucio Ubertini, Director of the Research Institute for Hydrogeological Protection in Central Italy, details activities in his country in response to IDNDR. Similarly, Dr. J. Eleonora Sabadell discusses U.S. activities.

The Symposium ended with a panel discussion of how governments, at the agency and policy-making levels, are responding to the new challenges and questions posed by the prospect of climate change and associated natural hazards (*Response of Nations*). The discussion centered around the positions of developed and developing countries. It was evident that although the objectives of sound environmental management are prevalent, the approach and concerns differ significantly. The papers by Dr. Dallas L. Peck, Director of the U.S. Geological Survey and former Chairman of the Committee on Earth and Environmental Sciences, and that of Dr. C. Caponi and A. Rosales, from the Ministry of Environment and Natural Resources of Venezuela, present both positions. Dr. Peck emphasizes the need for increased knowledge in order to base policy on sound scientific foundations. Caponi and Rosales struggle with the sometimes apparently conflicting needs of development and environmental protection. They wonder about who is going to pay for the alternative, and presumably more expensive, strategies for sustainable development. Mr. Peter S. Thacher, Senior Advisor to the Secretary General of the U.N. Conference on Environment and Development, and Senior Counselor at the World Resources Institute, moderated the panel. He ends this volume by putting in perspective the north–south debate evident in the previous two statements. He correctly points out that "Both the North and the South are preoccupied with immediate problems (even the rich feel poor) and find it convenient to hide behind uncertainty," and calls for more attention to the legitimate questions related to the costs of reducing climate risk and committing our economies to providing goods and services on a sustainable basis.

Although considerable uncertainties about the nature, and the magnitude, of human-induced climate change exist, the consequences are potentially too large to ignore. U.S. Vice President Al Gore in his recent book, *Earth in the Balance*, refers to global environmental changes as strategic threats to civilization. Many of these threats are related to potentially devastating natural hazards. All too often natural hazards and society's ability to deal with them are ignored in the debate of global environmental change and sustainable development. A society cannot develop if it cannot appropriately respond to nature's unavoidable extremes. Hopefully the Symposium and *Proceedings* of The World at Risk: Natural Hazards and Climate Change will help raise the consciousness of engineers, scientists, and policy makers to these issues.

Rafael L. Bras

ACKNOWLEDGMENTS

This book and the Symposia on which it is based were sponsored by the National Science Foundation, Natural and Man-Made Hazards Mitigation Program. I want to thank Dr. Eleonora Sabadell for her support of this and many other initiatives.

MIT's Industrial Liaison Program was also instrumental in organizing the meeting and generously providing complementary funds.

I would like to acknowledge Professor Ronald Prinn, colleague and Director of the Center for Global Change Science at MIT. Similarly, I acknowledge the help of Ms. Ann Slinn.

Finally, Ms. Elaine Healy and Mr. Read Schusky were responsible for all the publishing details, including reformatting and retyping all the papers. I thank them for their effort and dedication.

INTRODUCTION—NATURAL DISASTERS AND CLIMATE CHANGE

CHALLENGES OF THE DECADE:
NATURAL DISASTERS AND GLOBAL CHANGE

James P. Bruce
Chair, U.N. Scientific and Technical Committee for
International Decade for Natural Disaster Reduction
Chair, Canadian Climate Program Board

ABSTRACT

The 1990s can be seen, in many ways, as a transitional decade. The concept of environmentally sustainable development is becoming accepted as a paradigm for the 21st century, gradually overtaking the concepts of environmental protection and resource management. It will be increasingly unacceptable to view environmental protection as an add-on or afterthought to economic development activities. The U.N. Conference on Environment and Development, the "Earth Summit" of June 1992, sought to strengthen economic development in ways that reinforce environmental values.

Two of the major, inter-related, issues driving these changes in approach to both environment and development, are climate change as influenced by human-generated greenhouse gases, and the reduction of losses, human and economic, due to natural environmental hazards. The Second World Climate Conference in late 1990, lead to opening of international negotiation of a global convention on climate change. This convention, signed at the Earth Summit, ensures that energy, forest management and agricultural policies, in all countries, are being closely re-examined for their environmental effects. The climate negotiations also addressed adaptation to changes in risks of natural hazards, due to changing climate.

The International Decade for Natural Disaster Reduction should be viewed in a similar developmental context. It is clear that losses during natural disasters can seriously set back economic development, especially in poorer countries and regions. In many vulnerable countries, periodic declines in national economic output are highly correlated with occurrences of major disasters due to tropical cyclones, floods, earthquakes or volcanos. It is also evident that by adequate measures for prevention, warning and preparedness, these disaster losses can be substantially reduced. It is, thus, essential that all national development plans incorporate disaster preparedness activities, and that these activities be seen as one major step towards ensuring development that is sustainable. The International Decade for Natural Disaster Reduction provides the framework for international cooperation to achieve this goal.

INCREASING DISASTER LOSSES AND GLOBAL CHANGE

The rising toll of losses from *natural disasters* is an important, but often neglected manifestation of *global change*. The Munich Re-insurance Company estimates that natural disaster losses increased three times from the 1960s to the 1980s and insured losses nearly five times, adjusted for inflation.[1] Total global losses for 1990 were estimated to be of the order of $47 billion.[2] There are tricky problems involved in estimating the costs of disasters. While initial damage estimates pose serious difficulties, secondary effects, and the longer term impacts on economic development are even more elusive. Thus, while total damages may well be in considerable error, and probably understate the case, the trends, using consistent methodology, are probably more reliable, although the counting may be better now than in previous decades.

Accepting that there has been something like a three-fold increase since the 1960s, what are the global changes that have brought this about? Are the natural hazards, that turn into disasters, more severe and frequent than in earlier decades? Or has the growing human population and its activities become increasingly exposed to the hazards, turning hazard into disaster?

The casual observer would suspect that the hazards themselves are increasing in severity and frequency. Since the beginning of the IDNDR in 1990 we have seen a sequence of major disasters. These include the tropical cyclones and floods of 1990 and 1991 in Bangladesh and the S.W. Pacific affecting 15–20 million people and resulting in hundreds of thousands of lost lives; earthquakes in Iran, California, and elsewhere; the disastrous floods of 1991 in China with losses estimated at $12 billion; the Mt. Pinatubo eruption in the Philippines, affecting 80,000 people; and renewed drought and famine in eastern and southern Africa.[2]

In 1978, Burton et al.[3] estimated that 90% of the world's major natural disasters are due to four types of hazards—floods, tropical cyclones, earthquakes and droughts. More recent estimates for 1964–89 based on data from the Office of U.S. Foreign Disaster Assistance (OFDA), indicate that some 93% of people affected in disasters including civil strife, are affected by those due to hydrometeorological (including drought and famine) and geophysical causes.[4] Heavily populated Asia and the Pacific Islands are by far the most disaster prone areas.

But there are difficulties even in trying to compare the relative magnitude of disasters caused by various environmental hazards. For example, in the U.S.-OFDA compilation of statistics on disaster history since 1900, "declared" disasters are only the more serious events but there are differing criteria for different hazards. For example, earthquakes and volcanos qualify if 25 or more people are injured or killed. Weather and flood disasters (excluding droughts) are "declared" only if 50 or more people are injured or die, and drought disasters are included only if "the number affected is substantial."[5] Given these difficulties and the fact that hazard occurrences are only well reported when they produce disasters, i.e., when they affect human settlements, it is difficult at this stage to say whether natural environmental hazards have been increasing in intensity or frequency over the period from the '60s to the '80s. There are, however, some factors that lend credence to the idea that hazards may be increasing.

HUMAN INFLUENCES ON NATURAL HAZARDS

We used to think of natural hazards as the forces of Nature or "acts of God." We now recognize that there are a number of ways in which global change due to human actions have been increasing the incidence and severity of such hazards. The removal of forests from the slopes of river basins, the erosion of agriculture soils leading to sedimentation of river mouths, and obstructions placed in channels, all contribute to increased flood levels and more damage. Droughts are often made more prolonged and severe by human actions which remove woody vegetation, and unwise agricultural and water management practices, changing the albedo, and the roughness characteristics of the land surface. There are some hints that a warming climate may also be at work in some types of disasters, but this must be considered mainly as a problem for the future rather than a process that has been documented.

There is the other side of this coin as well, as has been pointed out by Rattien.[6] That is, some kinds of natural disasters, for example, volcanic eruptions, have significantly changed atmospheric composition on a geological time scale. Even now, the eruption at Mount Pinatubo in the Philippines has injected large quantities of SO_2 and particles into the atmosphere, affecting significantly the atmosphere's radiation balance leading to a temporary cooling effect, for at least a year or so.

HAZARDS BECOME DISASTERS

But there is little doubt that the main factor causing the increased natural disaster losses is not the changes in natural hazards, but the increasing vulnerability of human communities to these environmental hazards. The major global change affecting the damage statistics is the rapidly increasing population and human activity in coastal zones and in earthquake, drought and flood-prone areas of the world. The limited efforts in disaster prevention, mitigation, warning and preparedness have simply not kept pace with the increasing exposure of populations to natural hazards.

Even advanced industrialized countries have much work to do in order to reduce potential losses. The Japanese National Land Agency estimates that an earthquake similar to that of 1923 would kill or injure 350,000 people in Tokyo if it occurred tomorrow.[7] The greatest tragedy is that the less developed countries, which are the most vulnerable to natural hazards, are least able to mount effective disaster reduction programs without substantial technical assistance. For example, of those developing countries with a high risk of floods, 55 do not have adequate warning systems.[8] Yet experience shows that reliable flood forecasts and warnings can reduce damages by between 6 and 40%. These are among the major challenges of the International Decade for Natural Disaster Reduction in the 1990s.

THE IDNDR–PROGRAM FRAMEWORK

The 25-member Scientific and Technical Committee for the IDNDR has established three main targets to be achieved by all countries by the year 2000.[2]

As part of their plans to achieve sustainable development, each country should have in place:

1. Comprehensive national assessments of risks from natural hazards, with these risks taken into account in development plans,
2. Mitigation plans at national and/or local levels involving long term prevention and preparedness and community awareness, and
3. Ready access to global, regional, national and local warning systems and broad dissemination of warnings.

To achieve these three targets, the Scientific and Technical Committee (STC) identified a program framework consisting of seven main activities:

1. Identifying hazard zones and hazard assessment
2. Assessing vulnerability and risks
3. Raising awareness of decision and policy makers
4. Monitoring, predicting and warning of natural hazards
5. Undertaking long-term preventive measures, both non-structural (e.g., land use planning), and structural (e.g., civil works and building codes)
6. Undertaking short-term protective measures and preparedness
7. Planning and executing early intervention measures (at the time of or immediately after hazards strike).

It is recognized that to achieve the targets within the program framework, a number of supporting activities are required. These activities cut across all program elements and include:

- Education and training,
- Public information programs,
- Transfer of appropriate technologies both between and to, developing countries, and
- Application of existing knowledge and techniques.

In addition, there are a number of important research needs, to devise better and more accessible techniques and to provide socio-economic input to policy development.

These targets and the program framework elements have been endorsed by the Special High Level Council for the Decade. This Council, which consists of ten prominent members chaired by the Hon. Miguel de la Madrid, past president of Mexico, will promote and advance the work of the IDNDR at the highest political and industrial levels.

The STC could not possibly review and approve all projects that fit into this program framework, since IDNDR projects are required at local, national, regional and global levels and in many fields. However, as an initial step the Committee did consider it important to identify from a number of proposals before it, some international demonstration projects which could make valuable contributions, and would demonstrate the types of activities needed.

These projects include:

1. Tropical cyclone projects: Research and improved warning systems for tropical cyclone risk assessments and preparedness by the International Council of Scientific Unions (ICSU) and World Meteorological Organization (WMO).

2. Volcano hazard reductions, including better risk mapping, instrumentation and mobile early-warning systems through ICSU and the U.N. Educational Scientific and Cultural Organization (UNESCO).

3. <u>Minimizing earthquake vulnerability</u> for improved global hazard assessment, warnings and disaster management involving ICSU and U.N. Disaster Relief Organization (UNDRO).

4. An <u>improved disaster data base</u> and information system—coordinated by UNDRO.

5. <u>Education, research and training</u> activities, including disaster management training by UNDRO and the U.N. Development Program (UNDP), preparation of better training materials (UNESCO), and Roving Seminars (World Federation of Engineering Organizations (WFEO) and the Union of International Technical Associations (UITA).

6. <u>Improved risk assessment</u>, including input from the tropical cyclone project[1] and the development of comprehensive risk assessment techniques from all types of hazards; coordinated by WMO.

7. <u>Preventive projects</u>: Low-cost structural designs for developing countries to minimize casualties and damages (World Health Organization (WHO) and WFEO and UITA, with CERESIS—Regional Centre of Seismology for South America).

8. <u>Technology transfer</u>: Extension of a simple successful PC-based technology transfer scheme for floods and droughts developed by WMO to other types of hazards in a project called System for Technology Exchange for Natural Disasters (STEND).

9. <u>Public Health projects</u> to improve disaster preparedness and management plans and responses (WHO) and disaster mitigation for hospital facilities (Pan American Health Organization (PAHO) part of WHO).

10. <u>International Centers</u> for research and training in disaster reduction: the furthest along in planning is Morocco. Should these be a part of the IGBP–START initiative for global change research centers in developing countries?

11. The <u>instability of mega-cities</u>: led by the International Agency of Engineering Geology (IAEG) and International Union of Geological Science (IUGS).

It should be clear from this group of important projects that the international scientific community and agencies are coming together effectively to advance the objectives of the IDNDR.

National and regional organizations are also urged to participate in these international projects and to initiate projects of their own. The IDNDR Secretariat in Geneva acknowledges and keeps a registry of all projects world-wide which fit within the program framework and wish to use the IDNDR designation.

At last count some 100 countries have designated national committees or focal points for the IDNDR and many are working vigorously to improve disaster management in their own countries and regions. Regional cooperation in Latin America has been stimulated by a regional IDNDR Conference, sponsored by Pan American Health Organization (PAHO) and held in Guatemala in September 1991. Similar regional IDNDR meetings are planned for the Caribbean and Africa in the first half of 1992, all leading to a global IDNDR Conference in 1994, to stimulate both national and international cooperative actions. Taken together, after a rather slow start, there are now heartening signs of progress under the IDNDR banner.

CLIMATE CHANGE—POTENTIAL EFFECTS ON NATURAL HAZARDS

One worry in the minds of many, is the potential impact of climate change on natural hazard frequency and severity. Since hydrometeorological hazards are the most important causes of disasters in much of the world, this is not an unreasonable concern. To assess the potential effects requires a projection or scenario of future climate. The projections which depict the most likely future, are those of the Intergovernmental Panel on Climate Change (IPCC), established by WMO and UNEP in 1988.[9] While intergovernmental in initial structure, many leading scientists from academic and private institutions also participated actively in the scientific assessment and reviews of socio-economic impacts and policy options. More than 1000 specialists from 70 or so countries participated. The IPCC report of 1990 was reviewed at the Second World Climate Conference of Oct.–Nov. 1990. The science assessment was accepted as the best scientific consensus that could be achieved at this time. The IPCC is currently updating its science assessment. In private discussions, Sir John Houghton of U.K. the chair of the Scientific Assessment Working Group, has said that the research and modelling results of the past year and a half have not significantly changed the main conclusions of their 1990 report.

I am sure that many of the readers are familiar with that report, but let me remind you of some of the key conclusions and projections as they might affect disasters world-wide. The IPCC reviewed the rapid increase in atmospheric greenhouse gas concentrations, especially in the last few decades and stated that:

"We are *certain* that these increases will enhance the greenhouse effect, resulting on average in an additional warming of the earth's surface." They go on to predict, under a "business as usual" scenario for greenhouse gas emissions, and on the basis of model outputs from a number of countries, "a rate of increase of global mean temperatures during the next century of 0.3°C per decade with an uncertainty range of 0.2°C to 0.5°C per decade; this is greater than that seen over the past 10,000 years." Their projection for related sea level rise is for the observed rise of the past few decades of 1 to 2 cm/decade to increase to about 6 cm/decade, with an uncertainty range of 3 to 10 cm, giving by the end of the next century an increase of 2/3 ±1/3 meter.

The IPCC noted the uncertainties associated with these global predictions, and that even more surround the regional projections. However, they had sufficient confidence to advance projections for three of the five regions they studied. For Central North America, and the European portion of the Mediterranean basin they indicate summer conditions 2 to 3°C warmer than pre-industrial times by 2030, with 15–20% reductions in soil moisture, i.e., more frequent and severe droughts. In the Southern Asia region (mainly India and Bangladesh) a 1–2°C. temperature increase is projected by 2030, but with a strengthening of summer monsoon circulations, an increase of 5–10% in soil moisture, suggesting more severe river flooding, but fewer droughts. A further robust outcome of the models is more winter snowfall in the Arctic and Sub-Arctic with earlier and more rapid spring melt, suggesting larger and earlier spring floods in Arctic basin rivers.

However, the greatest concern for disasters must center on the potential effects on small island countries and low lying coastal areas of the projected sea level rise, especially in regions where tropical cyclones and accompanying storm surges occur. The IPCC's Response Strategy working group's report concentrated considerable attention on the urgent need for improved coastal zone management to reduce exposure to hazards, as an important adaptation strategy.

One issue of considerable uncertainty is the possible effect of global warming on the frequency and severity of tropical cyclones. Currently an average of 80 of these storms form in tropical regions each year. Conditions required for formation of tropical cyclones are:[10]

1. Ocean surface temperatures greater than 26.5°C, coupled with a relatively deep oceanic mixed layer,
2. Significant values of absolute vorticity in the lower atmosphere,
3. Weak vertical wind shear directly overhead, and
4. Mean upward motion and high mid-level humidities.

These are the conditions that favor deep convection and tropical cyclogenesis. The future climate scenarios from the climate system models indicate higher sea surface temperatures which would imply potentially more severe and widespread tropical cyclones. However, the model outputs also suggest greater atmospheric stability with more warming in the upper troposphere than near the ground and thus a possible inhibiting change in the fourth factor listed above. However, it has been suggested by some workers that there are several factors which would tend to keep tropical sea surface temperatures from rising significantly.

In examining this complex situation for the Australian region, Holland and his colleagues[11] also note the close relationships between regions of tropical cyclone formation in the S.W. Pacific and the state of the El Niño–Southern Oscillation (ENSO) phenomenon. The interaction between greenhouse gas induced warming and ENSO has not been well explored and this makes projections of future tropical cyclone activity of this region very difficult. Early conclusions of Holland et al. are that there could be an increase in cyclone activity in the central South Pacific, which could seriously affect the Cook, Society, and Samoan Islands and possibly Kirabati.

In general, however, the signals are too conflicting to draw conclusions about changes in frequency and severity of tropical cyclones. Whether or not tropical cyclones increase, the frequency of severe storm surges in the islands nations of the Pacific and Indian Oceans, the Caribbean Sea, the Bay of Bengal, the Philippines,

Florida and other regions will be increased by sea level rise. This makes imperative the improvement of coastal zone or island management plans to reduce loss of life, property and human suffering in tropical storms and storm surge flooding.

This has been just a short overview of some of the implications for natural hazards and disasters of a global warming induced by anthropogenic increases in greenhouse gases. Other papers address this theme further.

THE POLICY ISSUES

From the point of view of public policy are the inter-relationships between responses to the issues of climate change, and disaster reduction being recognized? The U.N. Conference on Environment and Development the "Earth Summit" in Rio de Janeiro in June 1992 offered a marvelous opportunity to make the connection. It is evident that natural disasters frequently set back the hard won economic development of developing countries with periodic drops in national economic output highly correlated with occurrence of major natural disasters.[12] It is also evident from experience, especially in developed countries, that human losses and economic impacts due to natural hazards can be significantly reduced. As Mary Anderson[13] has pointed out, "economic losses as a percentage of national wealth are 20% higher in developing countries," and as Burton et al.[3] noted "about 95% of disaster related deaths occur among the two thirds of the world's population that occupy developing countries."

There have been several calls for all countries to build into their plans for sustainable development, the preventive and preparatory measures needed to reduce the human and economic losses from natural disasters. These have come from the High Level Council and the Scientific and Technical Committee for the Decade,[2] from the Latin American countries in their Guatemala Declaration of September 1991,[14] from the Commonwealth Expert Group on Environment and Sustainable Development[15] and from IUCN's prescription for "Caring for the Earth."[16] The IDNDR provides an excellent opportunity to improve international cooperation to reduce losses in the developing world through use of techniques both proven and new.

"The basic argument for integrating disaster awareness into development planning is that it is wasteful not to do so,"[12] and I would add, it is tragic that it has not been done in many vulnerable countries.

The issue of climate change also has a profound North-South dimension. There is widespread realization that the burning of fossil fuels, and other human activities since the beginning of the industrial era, have profoundly changed the chemical composition of the global atmosphere. The changes are continuing at ever-increasing rates, altering the radiation balance at the earth's surface. We have already considered some of the potential consequences. The "Precautionary Principle" in the face of remaining uncertainties, and our obligations to vulnerable countries, especially the island states, makes it abundantly evident that we in the north, especially in North America, must reduce our dependency on fossil fuels. Many of the energy efficiency and fuel switching measures required to reduce greenhouse gas emissions could also save us money, help make us more competitive in the world, while reducing our burden on the global atmosphere.

To address the issue thoroughly requires a global bargain such as that which was sought by the International Negotiating Committee (INC) for a Climate Convention. We cannot expect the developing countries to withhold the benefits of economic progress from their people. We can and must find ways to assist this progress in a manner that minimizes the additional impact on the global environment. This will involve some profound changes—reversing the flow of financial resources which currently goes from South to North, assisting with appropriate energy technologies and with protection of tropical forests. The developing countries must be helped to follow efficient energy pathways in their economic development, "leapfrogging" over the dirty era of our northern industrial development. These issues are central to the climate issue.

However, even with the greatest imaginable commitments to reductions of greenhouse gas emissions and to reforestation, a major increase in greenhouse gas concentrations has already taken place, and will continue for a number of decades. Thus adaptation strategies including disaster mitigation, are not an *alternative* to greenhouse gas limitation strategies, but are an absolutely essential feature of a considered response to climate change.

SUSTAINABLE DEVELOPMENT

The major goal of the UNCED process is to take cooperative steps, world-wide, towards development which is much more sustainable, environmentally and economically. Sustainable development cannot be achieved without addressing these two major issues, two related challenges of the decade, climate change and disaster reduction. The international framework to address these issues is now in place in the form of the International Decade for Natural Disaster Reduction and the Climate Change Convention. Let us work to ensure that these issues are effectively addressed, and that the vital connections are recognized in national and international policies and decisions.

REFERENCES

1. Gerhard A. Berz. The Worldwide Increase of Natural Catastrophes in the Insurance Industry. Proc. UCLA International Conference on Impact on Natural Disasters, Los Angeles, California. July 1991.
2. Scientific and Technical Committee, IDNDR. First annual report to the U.N. Secretary General, U.N. General Assembly, Economic and Social Council. Doc. A/46/266. 16 Oct. 1991.
3. Ian Burton, R. W. Kates, and G. F. White. *The environment as hazard*. Oxford University Press. 1978.
4. B. N. Heyman, C. Davis, and P. F. Krumpe. An Assessment of World Disaster Vulnerability. Internal Paper of AID. Washington. 6 June 1991).
5. Office of U.S. Foreign Disaster Assistance (AID). *Disaster history: Significant data on major disasters worldwide, 1900–present*. Washington. 1991.
6. S. Rattien. In *Managing natural disasters and the environment*, Eds. A. Kreimer and M. Munasinghe., pp 38–39. The World Bank, Washington. 1990.
7. *The Economist*. Waiting for the big one. 7 Dec.,1991, pp. 101–103.

8. World Meteorological Organization. *Natural disaster reduction: How meteorological and hydrological services can help.* D. K. Smith. WMO #722. 1989.
9. Intergovernmental Panel on Climate Change Scientific Assessment. Impacts assessments and response strategies, 3 volumes. WMO, Geneva. 1990.
10. W. M Frank. Chapter 3 in *A global view of tropical cyclones*, Eds. R. L. Elsberry et al. U.S. Office of Naval Research. 1985.
11. G. L. Holland, J. L. McBride, and N. Nicholls. In *Greenhouse: Planning for climate change*, pp 438–55. Ed. G. I. Pearman, CSIRO, and Brill, Leiden and Melbourne. 1988.
12. Asian Development Bank. Disaster mitigation in Asia and the Pacific. Proc. of Bangkok Regional Disaster Mitigation Seminar, Manila. 10–12 Oct. 1990).
13. Mary B. Anderson. Which costs more: Prevention or recovery? In *Managing natural disasters and the environment*, pp 17–27, Eds. A. Kreimer and M. Munasinghe. The World Bank, Washington. 1990.
14. Guatemala Declaration, Report of the Scientific and Technical Committee for IDNDR, Second Session, Guatemala City. 16–20 Sept. 1991, Annex III.
15. Commonwealth Group of Experts. *Sustainable development: An imperative for environmental protection.* 130 pp. Commonwealth Secretariat, Marlborough House, Pall Mall, London, SW1YSHX. 1991.
16. The World Conservation Union (IUCN), UNEP, and WWF. *Caring for the Earth: A strategy for sustainable living.* Gland, Switzerland. 1991.

CLIMATE CHANGE AND NATURAL DISASTERS: WHERE ARE THE LINKS?

James C. I. Dooge
Centre for Water Resources Research
University College Dublin
Earlsfort Terrace
Dublin 2
Ireland

ABSTRACT

The growth of the partnership between the scientific community and intergovernmental organizations is outlined. Reference is made to the interactive nature of the relationship between climate and society and to the development of new methodologies of climate impact assessment to take account of this. It is emphasized that the essence of a disaster is the social disruption rather than the geophysical hazard which causes it. The importance of new approaches and new partnerships to solve the twin challenges of climate change and natural disasters is stressed. The conditions for a meaningful dialogue between partners from differing backgrounds is discussed.

INTERNATIONAL STUDY OF CLIMATE CHANGE

The developments in relation to climate change over the past few decades represent an interesting case study of the interaction between scientific research and policy formulation. The present Intergovernmental Negotiations on Climate are taking place against the background of the scientific assessments prepared by the Intergovernmental Panel on Climate Change and these in turn drew on the work of the World Climate Programme and cognate studies. Even though most of the work during these decades was planned and carried out by the scientific community, the initiative for such work arose initially in the policy arena. The original impetus that led to the Global Atmospheric Research Programme (GARP) and its successor the World Climate Programme may be said to have arisen from the desire of the superpowers to replace rivalry by cooperation in relation to their space programmes. It is interesting to recall that the story has a close connection with the Massachusetts Institute of Technology where the meeting resulting in these proceedings was held.

In April 1961, Jerome Wiesner, the special assistant for science and technology to the newly elected President John F. Kennedy, consulted his M.I.T. colleagues Bruno Rossi and Jule Charney in regard to possible international cooperation in meteorology on the basis of emerging space technology. As a result, a group of scientists, convened by Tom Malone, prepared a position paper on the basis of which President Kennedy introduced a reference to atmospheric sciences in his State of the Union Message to Congress in May 1961. The topic was also part of the President's brief for the somewhat inconclusive Summit Meeting with President Kruschev in Vienna in June 1961 and was referred to by the President in his address to the United Nations General Assembly in September 1961. On the latter occasion, President Kennedy suggested "further cooperative efforts between all nations in weather predictions and eventually in weather control."

As a result of all this activity the U.N. General Assembly at the end of 1961 passed a resolution which included a call:

> To advance the state of atmospheric science and technology so as to provide greater knowledge of basic physical forces affecting climate and the possibility of large-scale weather modification.

This provided a stimulus for action but the response was delayed due to a number of factors including rivalry between WMO (an intergovernmental organization), and ICSU (a non-governmental organization). It took over five years for the conflicting interests to be reconciled by the common-sense solution of launching an international research programme organized jointly by WMO and ICSU. The first objective of that Global Atmospheric Research Programme was to use the latest science and technology to improve weather forecasting, while the second objective was:

> Studying those physical processes in the troposphere and stratosphere that are essential for an understanding of the factors that determine the statistical properties of the general circulation of the atmosphere which would lead to a better understanding of the physical basis of climate.

While the main emphasis in GARP was on the first objective of improved weather forecasting, nevertheless sufficient progress was made on the second objective to interest a number of scientists in climate research and to build up a new body of knowledge.

This new interest and new knowledge formed the basis of the World Climate Conference in February 1979[1] and the launching later that year of the World Climate Programme[2]. The objectives of the latter Programme were:

1. To aid nations in the application of climate data and present knowledge of climate to the planning and management of all aspects of human activities;
2. To improve significantly the present knowledge of climate and to understand more fully the relative roles of the various influences on climate;
3. To provide the means to foresee possible future changes in climate and to warn of potential manmade hazards that might be adverse to the well being of humanity.

During the decade of the 1980s there developed a wide agreement that the scientific evidence in relation to the effect of the build up of greenhouse gases in the atmosphere was such as to warrant political attention and action. Such a process was greatly facilitated by the 1985 Villach Assessment Conference organized by WMO, UNEP and ICSU[3] which formulated a scientific consensus and resulted in an awakening of political interest which lead to the General Assembly of the United Nations passing a resolution in 1988 on the Protection of Global Climate. In the same year the Intergovernmental Panel on Climate Change (IPCC) was established by WMO and UNEP. The first assessment report of the IPCC[4] was presented to the General Assembly in 1990 and was endorsed by the Second World Climate Conference later that year[5].

IMPACT OF CLIMATE ON SOCIETY

Though a scientific methodology for the study of climate impacts was only developed over the past ten years under the World Climate Impacts Studies component of the World Climate Programme, qualitative descriptions of the impact of climate are almost as old as the written word. A good example is the qualitative calibration by Pliny the Elder in A.D. 77 of the annual flood level on the Nile in terms of economic and social effects. Pliny's was a prose calibration but it can be shown graphically as in Figure 1. He identified a nilometer reading of 15 ells as an indication of security of the harvest and characterized the range from 14 ells to 16 ells as giving an effect ranging

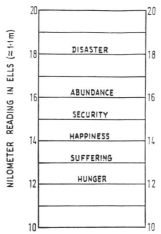

Figure 1. Effect of level of Nile (Pliny).

from happiness to abundance. Below this range, 13 ells meant suffering due to the bad harvest and 12 ells extensive hunger through drought and famine. Above the secure range, a level of 18 ells meant disaster due to catastrophic flooding.

Even in modern times the initial attempts to deal quantitatively with climate impacts were elementary insofar as the impact models used assumed a direct one-directional impact of a given climate variation on a given population or activity in a region producing a single impact. Even when they became more complicated, impact models though multi-stage were still one-directional. Thus meteorological, hydrologic or oceanographic variations in climate were considered to have first order effects on plants, micro-organisms and animals and through these to have second order effects on such sectors as human health, food supplies, transport and communication, wind and water power, etc., and through these to have further higher order socio-economic effects. During the past decade, the realization has grown of the need to use interactive impact models which include the response to the impact, which may have feedback effects on the natural system producing the climate variation or on the organization of the local society, both of which would modify the impact of climate variation or change on activity in the given region[6]. The broad structure of such an interactive model is shown in Figure 2.

An interesting study of such an interactive model is that due to Johnson and Gould[7] who studied the effect of climate fluctuations on human populations through a case study of the Tigris-Euphrates lowland where population data is available for over two millennia. This model, shown on Figure 3, linked up the interactions between annual river flow, food production and consumption, population growth, use of labour and irrigation systems, amount of arable land, etc. The model was used to explore the hypothesis that some of the major collapses of the population in this area might be due to climatic factors and might have given rise to invasions rather than be the direct result of such invasions. Such interactive models bring us well beyond the scope of the physical and biological sciences into the areas covered by the social sciences.

16 Climate Change and Natural Disasters

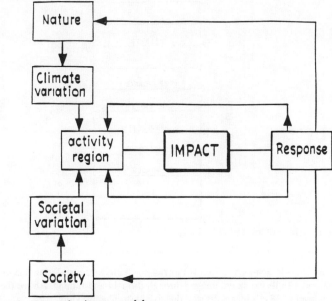

Figure 2. Interactive impact model.

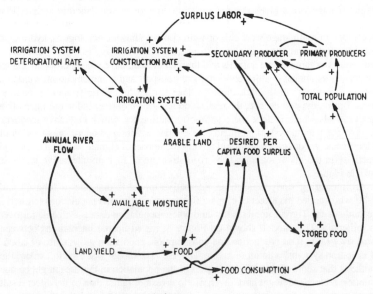

Figure 3. Climate-food-population.

More recently, the world community has become concerned with the possibility of international action aimed at natural disaster reduction. Again, there was an M.I.T. connection since the original proposal for an International Decade on Natural Disasters came from Frank Press[8]. Any meaningful definition or description of a disaster must take account of the fact that the essence of a disaster is the disruption caused rather than the physical or biological stimulus giving rise to the disaster. Thus Lechat[9] placed disasters firmly within the context of human ecology when he wrote in 1984:

> Natural disasters can be defined as ecological disruptions which exceed the community's capacity of adjustment so that outside assistance is necessary.

The same wide concept of the failure of internal resilience is contained in the definition used by the U.N. Disaster Relief Organisation which describes a disaster as:

> A serious disruption of the functioning of a society, normally occurring with little or no warning, causing widespread material and human losses with which the affected society cannot cope only using the resources available to the community in question.

The current argument concerning the need for international action in regard to climate change is essentially a debate as to whether the effects of greenhouse gases and other factors are producing effects so serious for individual regions that they amount to a potential manmade global disaster.

The typical morphology of a disaster, whether natural or man-made, is of successive interactive phases as follows:

The inter-disaster phase
The pre-disaster phase
THE IMPACT
The isolated local response phase
The external relief phase
The rehabilitation phase

The key to successful disaster reduction is to plan and act during the two pre-impact phases in such a way as to improve the effectiveness of the three post-impact phases. This involves making the best use of the inter-disaster phase for preventive measures and for preparedness. This would include such activities relating to risk assessment as archiving the historical record, mapping of hazard levels, evaluation of potential impacts, research to improve understanding of the type of geophysical event involved, study of the nature of the impacts likely to occur, and of the socio-economic factors that would make one community more resilient and another more vulnerable under the same degree of stress. It would also include such disaster preparedness activities as education and training of personnel, land use and building control, design and construction of protective structures, establishment of emergency communications and other infrastructure, and the promotion of public awareness.

It is also necessary to plan and prepare during the inter-disaster phase for activities relating to hazard prediction such as the monitoring of selected indicators, the establishment of early warning systems, the provision of alerts of imminent hazard, and the evaluation of these prediction systems. Finally, during this inter-disaster phase advantage must be taken of the time available to set in place procedures to facilitate the post-impact phases including schemes for evacuation and supplies, training for local self-organization, procedures for national relief, and procedures for the expediting of external aid.

As the world community begins to tackle with seriousness the problems presented by climate change and by natural disasters, the deeper understanding of the complex interactions involved reveals quite a degree of similarity in the general morphology of both processes. Both types of occurrence arise from the operation of geodynamic systems and both have a potential for the disruption of the human environment. While climate change is essentially a slow process compared with the sudden onset of most natural hazards, the deeper study of the two sets of phenomena indicate that in each case there are both slowly varying dynamic parameters and rapidly varying dynamic parameters whose interaction must be taken into account. The interplay of slowly varying and rapidly varying parameters with critical thresholds in a non-linear system constitutes the essence of the new theoretical developments known as catastrophe theory and the theory of chaos[10]. There are also contrasts in spatial scale since most of the impacts of climate change will be regional in nature while those for most natural hazards are local in nature. The appropriate scale provides an initial point of interest but the ultimate solution of the problem will require action and coordination at local, at national and at international level, even if the degree and order of the initiatives is different in the two cases.

In the study of climate processes, the work has become more multi-disciplinary as time progresses. The initial concentration on atmospheric processes has been expanded to include the effects, firstly of the upper ocean and latterly of the deep ocean. Work is now commencing on the probing of climate-vegetation interactions thus involving the biological sciences to a much greater degree than previously. However, there are not only links requiring multi-disciplinary activity within the area of climate studies but also between climate studies as such and the study of natural disasters. The indications are that the climate change due to increased concentration of greenhouse gases will impinge on most of the important types of natural hazard. One representation of the possible linkages is shown on Figure 4. It is important to realize that climate change may affect the variance as well as the mean value of critical factors. If, as many people think, the anticipated climate change will lead to increased frequency of extreme events such as droughts and floods, then the two types of phenomena will be closely linked. An increased frequency of drought, whether intensified by human activity or not, could lead to an increase in both wild fires and wind erosion, thus leading to severe land and vegetation degradation. An increase in floods could lead to landslides and water erosion on slopes, also leading to land and vegetation degradation. An increase in the frequency of storm surges could, in conjunction with the level rise, lead to even more severe coastal impacts. All of these could be accompanied by interactions between natural hazards themselves, such as an increased seismic hazard affecting not only tsunamis and landslides, but also the question of land subsidence with further intensification of the effect of sea level rise on coastal areas.

MEETING THE TWIN CHALLENGES

There are sufficient similarities between the problems posed by climate change and those posed by natural disasters to make it well worthwhile for anyone primarily interested in one of these fields to take due account of developments in the other area. The ultimate objective in each case must be to pass interactively through the successive phases of observation, understanding, prediction, and ultimately control. In order to accomplish this, there would need to be a meaningful partnership between science with its emphasis on prediction based on understanding and engineering with its emphasis on control based on prediction. This is only one of the partnerships that need to be established for the achievement of the objectives of the world community in relation to

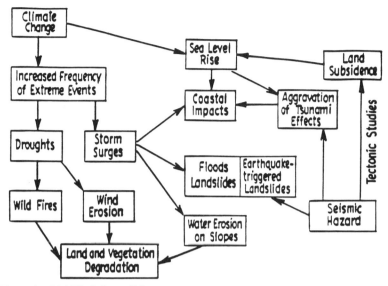

Figure 4. Multidisciplinary links.

the two problems. There must be a continuation of the broadening of the scope of international programs and of multi-disciplinary cooperation.

The history of international cooperation in relation to problems of the geosphere dates back for over a century to the First Polar Year of 1882–1883. Fifty years later, the second Polar Year reflected a much more interdisciplinary approach than on the previous occasion. When a third Polar Year was suggested twenty-five years afterwards, this was broadened substantially to become the International Geophysical Year which marked the coming of age of space science and technology. In the three decades since then, the understanding of the high viscosity dynamics of the solid geosphere has been the subject of international cooperation in successive programmes: in the first decade in the form of the Upper Mantle Project; in the second decade as the International Geodynamics Programme; and in the third decade as the International Lithosphere Programme. During the same three decades we have seen a broadening of the scope of the research on the low viscosity dynamics of the fluid geosphere in the form of the World Weather Watch, established by WMO in the 1960s, the Global Atmospheric Research Programme organized by WMO and ICSU from 1967 to 1979, and the World Climate Programme sponsored by WMO/UNEP/ICSU from 1980 onwards.

The experience of these programmes of international scientific cooperation will be of great use in the planning and the execution of the International Decade for Natural Disaster Reduction (IDNDR) of the United Nations. While the latter is an inter-governmental programme, there has already been a substantial input from the non-governmental science community through ICSU's committee for the IDNDR. This special committee has drawn up a scientific programme based on the principles of

a) concentrating the scientific research on the reduction of the human impact of natural disasters, b) restricting the programme to globally international and inter-disciplinary scientific problems, and, c) covering the various aspects of preparedness involved in research, monitoring, training and awareness.

ICSU's special committee on the IDNDR has in its recent second report identified three spearhead projects which are at a relatively advanced stage of scientific development and hold out hope of early implementation of operational phases designed to bring direct benefit to populations under threat during the first half of the IDNDR. These spearhead projects are:
1. A project aimed at reducing volcanic disasters under the leadership of the International Association for Volcanology and the Chemistry of the Earth's Interior,
2. A project on the global assessment of seismic risk under the leadership of the International Lithosphere Programme, and,
3. A project on tropical ocean disasters under the leadership of the International Association of Meteorology and Atmospheric Physics.

The ICSU special committee for the IDNDR has also identified a number of projects which show marked scientific promise for development during the early years of the decade to a stage where operational benefits can be seen for the later years of the decade. These second wave projects include: 1) intermediate prediction of earthquakes; 2) flood hazard assessment and mitigation; 3) drought hazard assessment, famine disasters, and vulnerable food systems; and, 4) instability of mega-cities. This planning represents an excellent start.

EFFECTIVE IMPLEMENTATION OF NEW PROGRAMS

For real progress in the years ahead, the dialogue between the parties concerned would need to be much more systematic and more effective than the interchange of views which has taken place over the last decade. In this connection, we can perhaps learn by considering what are the requirements for a good conversation between individuals in a social context. It is suggested that four essential elements are:
1. A willingness to talk clearly,
2. An ability to listen patiently,
3. A well-defined focus of interest, and,
4. A language known to all participants.

It is salutary to apply the same criteria to our scientific conversation. The implicit assumptions and the special language of an individual discipline greatly facilitate efficient communication between individuals within that discipline, but frequently are a major barrier to communication with someone outside the particular discipline or craft. The use of the special dialect native to a given discipline can be positively dangerous when the outsider feels that he or she understands clearly what is being said and is unaware of the special meaning attached to a term by individuals within that particular discipline.

Our ability to listen properly is also often defective both in our social and our scientific lives. We often listen, merely because it is not our turn to speak, or listen so that we may receive confirmation of our own preconceived ideas. Scientific mixing and social mixing both require that we should listen patiently, make a real effort to understand the other person's contribution and to identify diverse points with a view to synthesis rather than rebuttal. A real focus of interest is also vital. Interdisciplinary research is rarely effective except when a carefully chosen group tackle as a team a sharply focused problem. Too often in our scientific presentations, we fail to replace the historical order of discovery by the logical order of exposition.

The problems that arise in relation to efficient communication between scientists of different backgrounds occur even more frequently in regard to communication between individual scientists or scientific groups on the one hand, and policy advisers and governments on the other. In the latter case the four precepts indicated above become even more important and their achievement more difficult. A major contribution to dealing both with the climate problem and with the problem of natural disasters would be to make a conscious effort to improve communication in this regard. In this connection particular attention should be paid to the possibility of bridging the communication gap as well as developing meaningful scenarios through regional policy exercises in which some individuals trained as scientists and others trained as policy advisers discuss together scenarios in relation to climate change or in relation to natural disasters with a view to simulating the element of human response to these natural changes and the results of the complex interaction that follows. In this regard both the scientists and the policy makers can learn a good deal from what has been done in the field of military planning.

The basic physics which is of prime importance in the area of climate change and in the area of many natural disasters can be quite disparate. However, the successful prosecution of international scientific programmes and of policy agreements based on scientific understanding requires a number of attitudes and techniques that are relevant to both sets of problems.

REFERENCES

1. WMO. Publication No. 537. Geneva. 1979.
2. WMO. Publication No. 540. Geneva. 1980.
3. WMO. Publication No. 66. Geneva. 1986.
4. J. T. Houghton, G. J. Jenkins, and J. J. Ephraums. *Climate change: The IPCC assessment.* Cambridge University Press, Cambridge, U.K. 1990.
5. J. Jaeger and H. L. Ferguson. *Climate change: Science, impacts and policy.* Cambridge University Press, Cambridge, U.K. 1991.
6. J. C. I. Dooge. *Science, impacts and policy*, pp. 169–175. Edited by J. Jaeger and H. Ferguson. Cambridge University Press. 1991.
7. D. L. Johnson and H. Gould. *Climate and development*, pp. 117–138. Edited by A. K. Biswas. Tycooley, Dublin. 1984.
8. National Research Council. *Confronting natural disasters.* National Academy Press, Washington. 1987.
9. M. Lechat. *Oxford textbook of public health*, Volume 1, pp. 119–132. Edited by W. W. Holland and G. Knox. Oxford University Press. 1984.
10. J. Gleick. *Chaos: Making a new science.* Cardinal, London. 1987.

IMPACTS OF CLIMATE CHANGE ON NATURAL PHENOMENA

THE DEPENDENCE OF HURRICANE INTENSITY ON CLIMATE*

Kerry A. Emanuel
Center for Meteorology and Physical Oceanography
Department of Atmospheric and Planetary Sciences
Massachusetts Institute of Technology
Cambridge, MA 02139

ABSTRACT

Tropical cyclones rank with earthquakes as the major geophysical causes of loss of life and property. It is therefore of practical as well as scientific interest to estimate the changes in tropical cyclone frequency and intensity that might result from short term human-induced alterations of the climate. In this spirit a simple Carnot cycle model is used to estimate the maximum intensity of tropical cyclones under the somewhat warmer conditions expected to result from increased atmospheric CO_2 content. Estimates based on August mean conditions over the tropical oceans predicted by a general circulation model with twice the present CO_2 content yield a 40–50% increase in the destructive potential of hurricanes.

INTRODUCTION

The rising level of carbon dioxide in the earth's atmosphere probably represents the most serious source of human-induced change in the climate, with predicted adverse effects ranging from increases in sea level to significant changes in arable land cover. A National Academy of Sciences report[1] predicts that the total amount of CO_2 in the atmosphere will double sometime in the first half of the 21st century, based on observed changes in CO_2 content over a period of two decades and on projections of the rate of fossil fuel combustion. A number of largely independent means, ranging from simple energy-balance climate models to highly sophisticated three-dimensional general circulation simulations, have been employed to estimate the influence of doubled levels of CO_2 on the environment. The more realistic of these efforts predict a global warming of between 2°C and 4.5°C with greater increases in polar regions. The greatest uncertainties in the size of these increases and the speed with which they will occur derive from poor knowledge of the time scale of heat flow from the near-surface, well-mixed, ocean layer to the deep ocean and from lack of understanding of the response of global cloud cover to climatic change. While the uncertainties are rather large, we may fully expect the model predictions to be empirically tested over the next century. The reader is referred to the National Academy of Sciences report[1] on the subject, for a more detailed review of the various modeling efforts and associated uncertainties.

It is the purpose of this report to point out that relatively modest increases of tropical sea surface temperature can lead to substantial increases in the maximum

---

*This paper is based on a previously published article. It is reprinted by permission from *Nature* 326:483–485. Copyright © 1987, MacMillan Magazines Limited.

intensity of tropical cyclones. This is particularly disturbing since, among national catastrophes, tropical storms rank second only to earthquakes in the loss of life and property they produce[2]. The reader is referred to the latter reference for a general review of the empirical and theoretical understanding of hurricanes.

In the following section we review a simple model[3] of the mature tropical cyclone which is capable of predicting the maximum intensity that can be achieved by such storms as a function of environmental conditions. We shall then apply this model to the estimation of maximum cyclone intensity under the range of predictions of environmental changes resulting from increased atmospheric CO_2, and to a specific scenario predicted by a general circulation model simulation.

A SIMPLE MODEL BASED ON CARNOT'S THEOREM

Virtually all motions in the earth's atmosphere (except for electrically and tidally induced circulations in the very high atmosphere) result from a conversion of thermal to mechanical energy. In the case of tropical cyclones, the conversion is particularly direct. These storms make textbook examples of the operation of a Carnot engine, with the provision that the heat input is largely in the form of the latent heat of vaporization. Figure 1 illustrates the energy cycle[3]. At some radius r_0 air begins to flow inward toward the cyclone center within a frictional boundary layer whose depth is on the order of 1 or 2 km. During its inward trek the air maintains a nearly constant temperature that is very close to the sea surface temperature but acquires water vapor from the ocean, which supplies the latent heat of vaporization. The rate at which it acquires latent heat is a monotonic function of the near-surface wind speed. As the air flows toward lower pressure, heat is added due to isothermal expansion as well. It is

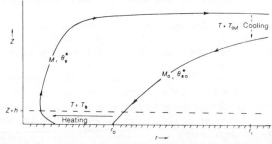

Figure 1. Carnot cycle of the mature tropical cyclone[3]. Air begins with absolute angular momentum per unit mass M_0 and moist entropy θ_{ea}^* at a radius r_0, and flows inward at constant temperature T_B within a thin boundary layer, where it loses angular momentum and gains moist entropy from the sea surface. It then ascends and flows outward to large radii, preserving its angular momentum (M) and moist entropy (θ_e^*). Eventually, at large radii, the air loses moist entropy by radiational cooling to space at a mean temperature T_{out} and acquires angular momentum by interaction with the environment.

acquires latent heat is a monotonic function of the near-surface wind speed. As the air flows toward lower pressure, heat is added due to isothermal expansion as well. It is also during this inflow that air suffers the greatest frictional dissipation. At some much smaller radius (typically from 5 to 100 km) the air abruptly turns upward and ascends through the deep cumulonimbus clouds that constitute the eyewall. During this ascent the total heat content (sensible plus latent) is approximately conserved, though in the process there are large conversions of latent to sensible heat. There is also comparatively little frictional loss of energy in this branch. Finally, the air flows outward at the top of the storm and eventually loses the heat gained from the ocean by longwave radiation to space. The whole process is considered to take place in an atmosphere which is neutral to buoyant and centrifugal convection along angular momentum surfaces.

In the steady state, the mechanical energy available from this thermodynamic cycle balances frictional dissipation. As previously mentioned, the bulk of the latter occurs in the boundary layer inflow where air crosses surfaces of constant absolute angular momentum[4]. The balance may be expressed symbolically

$$W_{BL} = \varepsilon \Delta Q \tag{1}$$

where W_{BL} is the work done against friction in the boundary layer, Q is the total (latent) heat gain from the sea surface, and ε is the thermodynamic efficiency, which is proportional to the difference between the temperature at which the heat is added (the sea surface temperature) and a weighted mean temperature at which it is lost in the upper atmosphere. (See ref. 3 for an exact definition.) The efficiency is largest in the tropics, where the depth of the moist convecting layer is large, and smaller at higher latitudes, approaching zero in many places.

Knowledge of the frictional loss in the boundary layer can be used to calculate the total drop in pressure in towards the eye of the storm through the use of Bernoulli's energy equation, which may be written

$$\tfrac{1}{2}\Delta v^2 + \int \alpha dp + W_{BL} = 0 \tag{2}$$

where $\tfrac{1}{2}\Delta v^2$ is the total change in kinetic energy per unit mass following the air flowing inward in the boundary layer, α is the volume per unit mass, and p is the pressure. Since v is approximately zero at the beginning of the inflow (at r_0) and is exactly zero at r = 0, Bernoulli's equation integrated all the way into the center of the storm may be expressed

$$\int_{r_0}^{0} \alpha dp = -\varepsilon \Delta Q \tag{3}$$

where we have used equation (1) for W_{BL}. Using the ideal gas law to express α as a function of p and T (and water vapor content, which is variable), the author[3] showed that the integral in equation (3) can be evaluated exactly. Specifically, both α and Q can be regarded as functions of temperature, pressure, and relative humidity, so that equation (3) has the functional form

$$\Delta G(p,T,RH) = -\varepsilon \Delta F(p,T,RH) \tag{4}$$

To a very good approximation, the temperature is a constant equal to the sea surface temperature. Moreover, the relative humidity cannot exceed unity. This permits an

thermodynamic efficiency ε, the last of which can be calculated from knowledge of the ambient temperature structure of the atmosphere.

Figure 2 shows the results of a calculation of this kind using September mean sea surface temperatures, relative humidities, and temperature profiles used in calculating ε. Also shown are the measured central surface pressures of several of the most intense tropical cyclones recorded. As storms in these regions tend to move toward the northwest, they often reach maximum intensity on the downstream side of the region of maximum potential intensity. Apparently, the dynamics of tropical storms sometimes permit them to achieve the maximum intensity that is energetically possible. The central pressure (or more precisely the difference between the ambient and central pressures) is a good measure of storm intensity and is also closely related to the maximum wind speeds in tropical cyclones[5].

The pattern of maximum intensity in Figure 2 strongly resembles the September mean sea surface temperature distribution, although the relationship is highly nonlinear. This is because the thermodynamic efficiency, ε, is itself mostly a function of sea surface temperature since the depth of the moist convecting layer is large over relatively warm water and small over colder water. The prediction of maximum cyclone intensity is therefore crucially dependent on estimates of sea water temperature. This will be a main focus of the estimates presented in the next section.

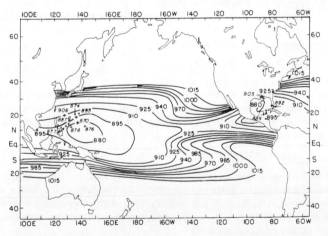

Figure 2. Minimum sustainable central pressures (mb) under September mean climatological conditions, assuming an ambient pressure of 1015 mb[3]. Crosses mark the positions and central pressures of several of the most intense tropical cyclones on record[2].

DEPENDENCE OF MAXIMUM CYCLONE INTENSITY ON CLIMATE

The Carnot cycle model described in the previous section ascribes changes in the maximum intensity of tropical cyclones to changes in three essential parameters: 1) the thermodynamic efficiency, ε, 2) the ambient near-surface relative humidity, and 3) the sea surface temperature. We here argue that the bulk of the change in maximum storm intensity resulting from relatively small changes in climate would be attributable to the third of these parameters.

Fractional changes in ε result from changes in the air temperature near the sea surface (T_B) and changes in the mean temperature of the ambient atmosphere at the levels where air flows out of the top of the storms ($\overline{T}_{out}$). From the definition of ε

$$\frac{d\varepsilon}{\varepsilon} = \frac{1}{T_B - \overline{T}_{out}} \left[\frac{\overline{T}_{out}}{T_B} dT_B - d\overline{T}_{out} \right] \qquad (5)$$

We are principally concerned about regions of the earth where tropical cyclones are particulary frequent and intense; these are characterized by relatively large values of $T_B - \overline{T}_{out}$. Since this quantity has a typical magnitude of about 100°C, the percentage change in ε will be nearly equal to the quantity in bracketts in equation (5). As an upper bound on this change, consider a scenario in which $\overline{T}_{out}$ changes but T_B remains fixed. It follows that a change of 4°C in $\overline{T}_{out}$ would result in only about a 4% change in ε. In fact, dT_B and $d\overline{T}_{out}$ are positively correlated in virtually all climate simulations performed to date. In the particular simulation discussed later in this section, dT_B and $d\overline{T}_{out}$ are about equal. This makes fractional changes in ε, according to equation (5), more on the order of 1%. We conclude that changes in the thermodynamic efficiency resulting from a doubling of atmospheric CO_2 would be insignificant.

The Carnot cycle model prediction of maximum intensity is quite sensitive to the relative humidity of near-surface air. The predictions of this quantity in large-scale numerical models are very sensitive to physical assumptions about evaporation from the sea, cumulus convection, radiation, etc. In fact, most models do not do very well at predicting relative humidity. Nature does not appear to be as sensitive, however, since the relative humidity over the ocean is remarkably constant over a wide range of sea surface temperatures, cloud cover, and precipitation. It varies from near 75% over colder ocean waters in middle latitudes to about 80% in the main tropical cyclone–producing regions. In many of the simpler climate models a fixed relative humidity is assumed. In view of these observations and lack of confidence in model predictions, we assume that the relative humidity over the tropical ocean remains fixed at the current values. This assumption also ties the temperature at each level in the troposphere to sea surface temperature.

With these assumptions, changes in maximum tropical cyclone intensity are strictly related to changes in sea surface temperature. The Carnot cycle is particularly sensitive to sea surface temperature as the latent heat content of air at fixed relative humidity is a strongly exponential function of temperature, approximately doubling for each 10°C increment of temperature above 0°C. Table I shows the minimum sustainable central pressures of tropical cyclones as a function of sea surface temperature for a value of ε characteristic of the Carribean and Western Pacific tropical cyclone regions, assuming a surface relative humidity of 78% and an ambient surface

Table I
Minimum Sustainable Central Pressure and Maximum Wind Speed as a Function of
Sea Surface Temperature with $\varepsilon = .33$ and an Ambient Relative Mumidity of 78%.
(Ambient pressure = 1015 mb)

T(°C)	P_c (mb)	V_{max} (m s^{-1})
27	911	72
28	902	75
29	891	79
30	879	83
31	865	88
32	849	93
33	829	99
34	805	106

pressure of 1015 mb. Relatively small changes in sea surface temperature are associated with large intensity changes, with an increase of 3°C leading to a 30–40% increase in the maximum pressure drop. Table I also shows estimates of the maximum wind speed in tropical cyclones based on empirical relations between wind and pressure developed by Holland[5,6]. (The maximum wind is strictly bounded by the maximum pressure drop since the latter serves as the potential energy for inflowing air.) The wind increases as the square root of the pressure drop, yielding a 15–20% increase for an increase of 3°C in sea surface temperature. The destructive potential of the wind, it should be added, varies as the square of the wind speed and thus as the total pressure drop.

To obtain some idea of the global distribution of the changes in maximum cyclone intensity that might result from predicted increases in atmospheric CO_2 we rely on predicted changes in sea surface temperature produced by the Goddard Institute for Space Studies General Circulation Model II described by Hansen et al.[7] with modifications to account for ocean mixed layer heat capacity and variable ice cover as described in ref. 8. Broadly, the model solves the conservation equations for energy, momentum, mass, and water and the equation of state on a coarse grid with a horizontal resolution of 8° latitude and 10° longitude, and with nine layers in the vertical. The radiative scheme accounts for the radiatively significant atmospheric gases, aerosols and cloud particles, and cloud cover and height are predicted. Ground hydrology and surface albedo depend on local vegetation and snow and ice cover, which are predicted. The diurnal and seasonal cycles are included in the model. Ocean temperatures and ice cover are computed based on energy exchange between the atmosphere and an ocean mixed layer with specified heat capacity and advective heat transport. The last two vary with season at each grid point.

Figure 3 shows the minimum sustainable central pressures of tropical cyclones based on sea surface changes predicted by the general circulation model averaged over five Augusts of a simulation with twice the present atmospheric CO_2 content. These changes, which in the tropics range between 2.3°C and 4.8°C, have been added to the

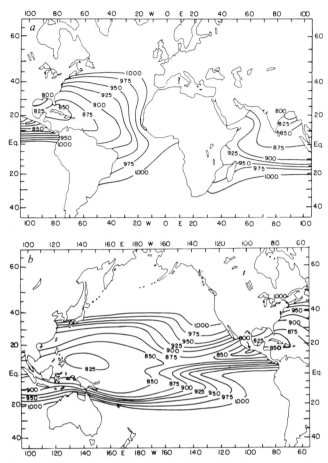

Figure 3. Minimum sustainable central pressures (mb) as in Figure 2, but using sea surface temperature increases predicted by a general circulation model[7,8] averaged over five Augusts in simulations with twice the present CO_2 content.

current August climatological sea surface temperatures, while ε and surface relative humidity remain unchanged from the present climate. Minimum pressures are substantially lower in the tropics, especially in partially enclosed basins such as the Gulf of Mexico and the Bay of Bengal where minimum sustainable pressures are below 800 mb. Were these sea surface temperature increases actually realized, the maximum destructive potential of tropical cyclones would be substantiallly increased, in some places by as much as 60%.

This analysis pertains only to the maximum sustainable pressure drop in tropical cyclones and has no direct implications for either the average intensity of cyclones or their frequency of occurrence. While one intuitively expects that average intensity increases with maximum intensity, the question of frequency is in fact poorly understood. Tropical cyclones do not arise spontaneously even under favorable conditions but appear to originate in disturbances whose origins derive from separate dynamical mechanisms, and will only do so under certain environmental conditions[9]. There is no obvious reason, however, to suppose that frequencies would be substantially diminished in a climate with doubled CO_2.

Several caveats should be mentioned regarding this analysis. In the present climate, only a small fraction of the total number of cyclones which form reach the maximum intensity given by this analysis. Reasons for this include the relatively long spin-up time for storms of this kind, allowing them to move over land or colder water before their maximum potential can be realized, and the tendency for strong cyclonic circulations to induce upwelling of cold water. The latter effect is observed to decrease water temperatures by several degrees near the centers of strong tropical cyclones. The presence of more frequent or intense cyclones in a warmer climate may conceivably induce enough upwelling and mixing to keep sea surface temperatures lower than they might otherwise be. To these caveats we must add the relatively large uncertainties associated with climate simulations. In particular, the coupling of ocean dynamics to atmospheric circulation is presently quite crude; in the simulation discussed here, for example, the horizontal advection of heat is specified. While better estimates of tropical cyclone intensities must await more sophisticated modelling efforts, the present analysis suggests that predicted climate changes associated with increased atmospheric CO_2 will lead to substantially enhanced tropical cyclone intensity.

ACKNOWLEDGMENTS

The author is grateful for the generous assistance provided by James Hansen, Gary Russell and Reto Ruedy of the NASA Goddard Space Flight Center Institute for Space Studies, and by Peter Stone of MIT. The work was completed with the assistance of National Science Foundation Grant ATM-8513871.

REFERENCES AND NOTES

1. *Changing climate*. 1983. Report of the Carbon Dioxide Assessment Committee of the National Academy of Sciences. National Academy Press, Washington.
2. R. A. Anthes. 1982. *Tropical cyclones: Their evolution, structure, and effects.* American Meteorological Society.
3. K. A. Emanuel. 1986. *J. Atmos. Sci.* **43**:585.

4. A certain amount of work must be done to restore the angular momentum of the outflow to its ambient value, but this is generally small. See ref. 3 for details.
5. G. J. Holland. 1980. *Mon. Wea. Rev.* **108**:1212.
6. We take the parameter B defined in (5) to be 1.5.
7. J. Hansen et al. 1983. *Mon. Wea. Rev.* **111**:609.
8. J. Hansen et al. 1984. In *Climate processes and climate sensitivity*. American Geophysical Union, Washington.
9. W. M. Gray. 1982. In *Intense atmospheric vortices*. Springer-Verlag, New York, pp. 3–20.

TROPICAL CYCLONE FREQUENCY AND GLOBAL WARMING

Richard E. Peterson and Thomas E. Warner
Department of Geosciences
Texas Tech University
Lubbock, Texas 79409

ABSTRACT

Under current global weather conditions, tropical cyclones occur across specific stretches of the world's tropical oceans. With climate change, all the factors controlling tropical cyclogenesis may be altered. The twentieth century has already witnessed global warming, punctuated by a warm spell in the 1930s. The response of the atmosphere to these past events in terms of production of tropical cyclones is only imperfectly known. The area second only to the western North Pacific in frequency of tropical cyclones is the eastern North Pacific. The climatology of tropical storm activity for this region for the period since 1900 is being assembled in hopes that it may provide insight into the role of global warming in tropical cyclogenesis, not only earlier this century but into the next.

INTRODUCTION

One of the hypothesized consequences of global warming has been an increased frequency of hurricanes.[1,2] The occurrence of the most intense Atlantic hurricane ever recorded, Gilbert, in the fall of 1988[3] following the extreme summer warmth across the central United States enhances this concept in the media and in the public consciousness. Extremes are common in weather however and must be considered within the context of a longer period of observations.

While much of the attention given to hurricanes focuses on their destructiveness, in most parts of the world affected by such storms the benefits of hurricane-associated rainfall are extensive. One of the ocean regions most frequented by tropical cyclones is the Northeast Tropical Pacific (NETROPAC), off the west coast of Mexico and Central America. Since most of the storms which form there move westward and dissipate before reaching Hawaii, their presence has been underrated. Even more importantly for the south central United States (Texas and its four neighbors, New Mexico, Oklahoma, Arkansas and Louisiana) the role of these storms in providing rainfall has been largely overlooked.

BACKGROUND

Across much of Texas the annual average rainfall regime includes a secondary maximum during the fall. This is sometimes attributed to landfalling tropical storms from the Gulf of Mexico. It may well be that this autumn rain just as often is due to the influence of NETROPAC storms making landfall on the west coast of Mexico.

The significance of Pacific tropical cyclones in understanding the origin of rainfall in the south central United States is far-reaching. This enormously important agricultural region depends on abundant rainfall, especially since its relatively low latitude results in high evaporation rates. On the western, semi-arid margin of the area the years with above-average rainfall associated perhaps with unusual Pacific storm incursions may be the years in which harvests are sufficiently abundant to sustain farmers through the dry years. If global warming is occurring inexorably, the resultant storm pattern may greatly alter the rainfall regime across the region.

Until about 1920 the U.S. Weather Bureau virtually ignored the available evidence of tropical cyclones in the northeastern Pacific Ocean, despite the fact that several cities in western Mexico (e.g., Mazatlán and Loreto) had been devastated by hurricanes in the previous century.[4]

Since the mid-1960s, satellite coverage of the northeastern Pacific has given good statistics on tropical storm development. The period 1966–1980 has been studied in detail regarding the characteristics of the storms, and in particular those making landfall.[5] During that period, 214 tropical storms formed, with 108 achieving hurricane status. The overall average motion was toward the west-northwest; however, during 1966–1980, there were 38 storms which took on an eastward motion. Of these, 40% curved to the east during September alone and 26% in October (a month with much more storm activity than in the western Atlantic).

The impact of NETROPAC storms was made strongly apparent during the first half of the 1980s. Several hurricane events occurred in the south central Plains. The remnants of Norma (1981) spawned devastating floods with damage estimates of $175 million and causing ten deaths in Texas and Oklahoma; four-day rain totals ranged up to 26". As the remnants of Hurricane Paul (1982) advanced across far West Texas into New Mexico, up to 5" of rain was recorded, leading to flash floods, and damages to homes, businesses and crops. Ahead of Hurricane Tico (1983), intense rainfall up to 17" extended across Texas and Oklahoma into Missouri. One life was lost and property damage was over $93 million. In 1985 Hurricane Waldo led to heavy rains (up to 7") and flooding in southeastern New Mexico and West Texas into Kansas; at least one flood-related death occurred in Kansas. Three storms contributed to flooding in 1986: Newton, Paine and Roslyn. The first two yielded precipitation patterns from West Texas into the mid-Missouri Valley; the last promoted flooding rains across the southern half of Texas into the lower Mississippi Valley.

Were the early 1980s entirely anomalous? Or have NETROPAC storms contributed in the past to the autumn rains across the south central states? There are difficulties in responding to this question due primarily to the sparsity of observations in earlier decades. While the role of Pacific tropical storms in the climate of the southwestern states has been considered,[6] the impact on the south central states has gone uninvestigated. The Rocky Mountains have generally been considered as a block for such storms and only the frequency of eastward penetrations in the 1980s has provided a stimulus to our current research. Preliminary studies have uncovered the existence of earlier Pacific tropical storms which have been strong rain-producers in the south central states.

For example during the first week of October, 1930, a very intense hurricane was encountered by several ships in the Gulf of California,[7] a barometric reading as low as 27.72" Hg was taken. The storm crossed the Mexican coast north of Mazatlán headed northeast. Very heavy rains resulted across most parts of Texas; for the state as a whole the monthly rainfall was about 6" above normal. Del Rio, which was within the direct path, was 10" above normal for the month. The tropical storms of the 1930s have provided convincing evidence that such an event has not been limited to the 1980s.

CURRENT INVESTIGATIONS

Because of the impact of the tropical storm remnants on the climatology of the south central U.S.—particularly rainfall and flooding events—studies have been initiated within the Atmospheric Science Group at Texas Tech University to improve the understanding of this phenomenon. The 1966–1980 climatology of tropical cyclones based on satellite imagery has been followed by a detailed case study of Hurricane Tico (1983).[8] Currently we are gathering all available records for this century in order to provide an assessment of the impact of eastern North Pacific tropical cyclones on the south central Plains. In addition, the extended data set may allow inferences regarding the importance of past and future El Niño events as well as global warming on hurricane frequency and strength.

The objective of our research is to derive from the historical record the paths of landfalling storms which affected the south central states. A variety of data sources are being consulted. Maps for each year are being constructed showing the inferred paths. A summary of the weather events attending the storms of each year is being assembled.

For notable storms, rainfall distribution maps are a primary product; the evolution of the rain events is detailed insofar as possible for each case. Radar depictions are also available for the most significant events of later decades. Upper-level maps and surface maps can be included, along with a selection of satellite images, when available, to indicate the diversity of synoptic conditions for which the storms enter the south central states.

For the earliest portion of the period after the turn of the century, the tropical cyclone information may be taken from the *Historical Weather Map* series (which has been obtained on microfilm from the National Climatic Data Center). *Monthly Weather Review* provided tropical storm summaries beginning in the 1920s and continuing into 1980s, which can be culled for details. Throughout the period, the *Climatological Data* publications for each state record rain events and occasionally have written descriptions of extreme events. After World War II, *Mariner's Weather Log* began to publish annual summaries of tropical cyclone occurrence for the northeastern Pacific. The *Daily Weather Map* series depicts a surface and upper-level weather map for each day up to the present day. In addition, newspapers on microfilm (particularly for Texas) are available for checking on rain episodes possibly spawned by tropical storms.

Contact has been established with the Mexican National Weather Service for obtaining unpublished data and/or studies for the storms as they crossed that territory. Various publications of the Servicio Meteorológico Nacional of México are sources of tropical storm data; i.e., *Boletines Climatológicos* and *Previsión del Tiempo*.

EXPECTED RESULTS AND SIGNIFICANCE

The occurrence of tropical cyclones is not entirely predictable; however enough is known regarding the conditions for formation that forecasts of seasonal activity are produced. Warm ocean temperatures and specific low-level and upper-level wind patterns have been identified which favor development and intensification.

With global warming (or cooling) the resulting sea surface temperature changes and shifts in atmospheric pressure patterns would be expected to alter the frequency and location of tropical cyclone occurrence. Tropical cyclones activity may be taken then as one indicator of global climate change. (It may be noted that for the first time ever tropical cyclones have been observed in the tropical *South* Atlantic, the first during April 1991.[9])

The current research should provide a longer perspective (including the warm decade of the 1930s) against which to judge the relevance of the variation of tropical cyclone activity over the upcoming decades.

In addition, a widely cited source of information on the occurrence of tropical storms in the Atlantic is the monograph by Neumann et al.[10] Along with the study by Smith[6], the current investigation could lead to collaboration on a historical climatology volume for the northeastern Tropical Pacific.

REFERENCES

1. K. Emanuel. *Nat.* **326**:483. 1987.
2. K. Emanuel. *J. Atmos. Sci.* **45**:1143. 1988.
3. R. E. Peterson, D. Perry, and J. M. McDonald. *J. Wind Engineer. Indust. Aerodynam.* **36**:87. 1990.
4. A. Court. Tropical Cyclone Effects on California. NWS WR-159. 41 pp. NOAA, Washington. 1980.
5. R. A. Allard and R. E. Peterson. *Rev. Geofiscia* **26**:33. 1987.
6. W. Smith. The Effects of Eastern North Pacific Tropical Cyclones on the Southwestern United States. NWS WR-198. 229 pp. NOAA, Washington. 1986.
7. W. E. Hurd. *Mon. Wea. Rev.* **58**:432. 1930.
8. J. L. Tumbiolo. Analysis of Northeast Pacific Hurricane Tico and Its Associated Heavy Rainfall Event over the Southern Plains. 182 pp. M.S. thesis, Texas Tech Univ., Lubbock. 1989.
9. C. J. McAdie and E. N. Rappaport. Diagnostic Report of the National Hurricane Center. **4**(1):10. 1991.
10. C. J. Neumann, B. R. Jarvinen, and A. C. Pike. Tropical Cyclones of the North Atlantic Ocean 1871–1986. 186 pp. NOAA-NESDIS, Washington. 1987.

THE RESPONSE OF SEA LEVEL TO GLOBAL WARMING

Andrew R. Solow
Woods Hole Oceanographic Institution
Woods Hole, MA 02543

ABSTRACT

One potential effect of global warming is a change in sea level. This paper outlines the current state of scientific knowledge concerning the response of sea level to global warming.

INTRODUCTION

One of the key challenges in predicting the impacts of global warming on human society is understanding the relationship between temperature change and changes in other climate processes like precipitation, tropical cyclones, and sea level. This paper focuses on the response of sea level to global warming. Like many aspects of the global warming, considerable scientific uncertainty surrounds this response. At the same time, over the past ten years or so, considerable progress has been made in understanding the processes involved. The goal of this paper is to survey the current state of knowledge regarding the response of sea level to global warming.

In discussing changes in sea level, it is critical to distinguish between relative sea level, which refers to the position of the sea surface in relation to land, and eustatic sea level, which refers to the absolute level of the sea surface. Relative sea level can change as a result of either eustatic changes or localized changes (e.g., subsidence or uplift of the coast). The main effect of global warming will be on eustatic sea level, and this will be the primary focus of this paper.

FACTORS AFFECTING EUSTATIC SEA LEVEL

The response of eustatic sea level to global warming will depend primarily on two factors: thermal or steric expansion of seawater and melting of land ice. Archimedes' Principle ensures that the melting of sea ice (e.g., in the Arctic region) will have no direct effect on sea level.

Thermal Expansion of Seawater

As seawater warms, density decreases, the oceans expand, and sea level rises. Despite the apparent simplicity of the underlying process, predicting the magnitude and regional distribution of this response is difficult. The pattern of thermal expansion will depend to a large extent on the way in which heat is mixed in the ocean.

Broadly speaking, there are two ways to study thermal expansion. The first is by observing the steric response of seawater to historic changes in temperature. Unfortunately, existing data on steric anomalies and the temperature structure of the ocean are limited (both in time and space) and exhibit high variability[1]. This limits their usefulness in establishing empirical relationships between temperature and steric expansion. The second approach to understanding thermal expansion is through modelling the key processes directly[2]. Ocean mixing processes are complicated and, consequently, the available numerical models are relatively crude. Moreover, despite their crudeness, these models are expensive to run.

A hybrid approach can be based on combining historic data on sea level (discussed below) with a simplified numerical model in which the thermal response to a given warming is controlled by a single parameter[3]. Although this approach is crude, under a number of assumptions, it can give a general idea of the response of mean global sea level to warming. The Intergovernmental Panel on Climate Change (IPCC) used this approach to estimate the response of sea level under a "business as usual" scenario for emissions of greenhouse gases[4]. The conclusion was that steric expansion would contribute 7 to 15 cm of eustatic sea level rise by 2030, with a "best estimate" of 10 cm.

Melting of Land Ice

The second major component of the response of eustatic sea level to global warming is the melting of land ice. In considering this response, it is important to distinguish between glaciers and the two main ice sheets—Antarctica and Greenland. Because they exist under different climatic conditions, the response of these ice forms to warming will be different. To give an idea of their relative sizes, the volume of water measured in equivalent sea level rise is 65 m for Antarctica, 7 m for Greenland, and around 0.35 m for glaciers.

The mass of an ice form can change through ablation (run-off of melt water and evaporation) and accumulation. The balance between these two processes determines the net mass balance of the ice form. Both ablation and accumulation depend on surface air temperature. A stylized view of this dependence is shown in Figure 1. As this figure shows, net accumulation occurs with warming when temperature is low, while net ablation occurs with warming when temperature is high. A key factor omitted

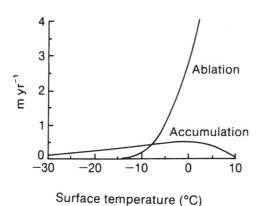

Figure 1. Temperature dependence of ablation and accumulation.

from this figure is precipitation. Increased precipitation tends to increase the temperature at which net ablation occurs.

Most of the world's glaciers occur in climates where warming will lead to net ablation. A combination of empirical and modelling work suggests that the the global sensitivity of glaciers is around 0.5 to 2.0 mm of sea level rise per year per °C warming[5]. However, the great variability in the behavior of glaciers over the past 100 years or so suggests that regional climate is very important.

Estimates of the sensitivity of the Greenland ice sheet to warming range from around 0.3 to 0.5 mm of sea level rise per year per °C. This is less than the estimated global sensitivity of glaciers and is explained by the relatively cold temperatures in Greenland. Any increase in precipitation with warming would tend to reduce the sensitivity of the Greenland ice sheet. For example, if precipitation increases by 5% with each 1°C increase in temperature, sensitivity is reduced by about 30%[6].

Much of the original concern about the response of sea level to global warming stemmed from fears that the West Antarctic ice sheet is unstable and would essentially disintegrate in response to even modest warming and sea level rise. Subsequent studies[7] have put these fears to rest. Current estimates of the sensitivity of the Antarctic ice sheet to warming range from around -0.2 to -0.4 mm per year for a 1°C warming.

Based on a variety of empirical and modelling results, the IPCC estimated the response of sea level to global warming for the period 1985–2030 assuming a "business as usual" scenario for emissions of greenhouse gases. These estimates are shown in Table I.

HISTORIC DATA

Tide gauge records of *relative* sea level are available for a large number of locations over the past century or so. As with other types of climate data, the Northern Hemisphere—Europe and North America, in particular —are over-represented in this sample. A number of attempts have been made to estimate historic changes in eustatic sea level from these data[8,9]. Not surprisingly, these estimates depend critically on the way in which the records are corrected for non-eustatic changes. In addition, questions have been raised about methodology[10]. Most estimates of eustatic sea level rise over the past century are around 1.0 mm per year, although the associated uncertainty is large.

The relatively recent development of technologies such as satellite altimetry and Very Long Baseline Inferometry (VLBI) may allow the measurement of eustatic sea level directly. However, given the magnitude and persistence of natural variability in sea level, many years of data will be needed before useful inferences can be drawn.

Table I. Estimates of Sea Level Rise (cm), 1985–2030[4]

	thermal	glaciers	Greenland	Antarctica	Total
HIGH	14.9	10.3	3.7	0.0	28.9
BEST	10.1	7.0	1.8	-0.6	18.3
LOW	6.8	2.3	0.5	-0.8	8.8

A small number of long tide gauge records have been analyzed in an effort to identify a common increase in the rate of change of sea level[11]. Such an increase could be taken to represent the response of eustatic sea level to a regional or global forcing such as greenhouse warming. This approach, which avoids the problem of decomposing relative sea level into its eustatic and localized components, did indicate some evidence for such a change occurring around 1890. If this change is real, then it is most likely related to the end of the terminal phase of the Little Ice Age.

As noted above, predictions of the response of eustatic sea level to future changes in climate have been based on some combination of empirical and modelling results. The data on different components of sea level are of mixed quality and coverage. For example, observations of the historic behavior of certain European glaciers are fairly good. On the other hand, it is extremely difficult to estimate the contribution of the Antarctic ice sheet to eustatic sea level over the past century: even the sign is unclear.

CONCLUSION

This paper has provided a brief outline of the current state of scientific knowledge regarding the response of eustatic sea level to global warming. It is difficult to argue that a slow increase in mean sea level occurring over many decades constitutes a natural hazard. On the other hand, the impacts of storms on the coastal zone do depend on sea level. In this sense, sea level constitutes an initial condition on which the effects of other natural hazards depend.

Not surprisingly, there has been some economic work on estimating the effects of future sea level rise on the coastal zone[12]. This work, which is necessarily sketchy, suggests that most of the impacts of sea level rise on human society can be mitigated by adaptive responses. The impacts on unmanaged ecosystems, such as wetlands, are less easily avoided.

One of the main points of this paper has been the great uncertainty surrounding projections of future sea level. This is illustrated by the substantial change in thinking about sea level rise over the past ten years. This change, which resulted from solid scientific research, has led to major reductions in the estimate of future sea level rise from catastrophic to manageable. The remaining uncertainties are large enough to ensure that further revisions—up or down—are likely.

REFERENCES

1. D. Roemmich. In *Glaciers, ice sheets, and sea level: Effects of a CO_2-induced climatic change*. National Academy Press, Washington. 1985.
2. T. P. Barnett. In *Detecting the climatic effects of increasing carbon dioxide*. Department of Energy, Washington. 1985.
3. T. M. L. Wigley and S. C. B. Raper. *Nature* **330**:127. 1987.
4. J. T. Houghton, G. J. Jenkins, and J. J. Ephraums. *Climate change: The IPCC scientific assessment*. Cambridge University Press, Cambridge. 1990.
5. J. Oerlemans, Z. Gletscherk. *Glazialgeol.* (in press).
6. J. Oerlemans, R. van de Wal, and L. A. Conrads. *Z. Gletscherk. Glazialgeol.* (in press).

7. W. F. Budd, B. J. McInnes, and I. N. Smith. In *Dynamics of the west Antarctic ice sheet*. Reidel, Dordrecht. 1987.
8. T. P. Barnett. In *Climate variations over the past century and the greenhouse effect*. National Climate Program Office, Washington. 1988.
9. W. R. Peltier and A. M. Tushingham. *Science* **244**:806. 1990.
10. A. R. Solow. *Cont. Shelf Res.* 7:629. 1987.
11. V. Gornitz and A. R. Solow. In *Greenhouse-gas-induced climatic change*. Elsevier, Amsterdam. 1991.
12. J. M. Broadus. In *Global change and our common future*. National Academy Press, Washington. 1989.

THE EFFECT OF RISING SEA LEVEL ON THE HYDROLOGY OF COASTAL WATERSHEDS

William K. Nuttle
Rawson Academy of Aquatic Science
Suite 404, 1 Nicholas Street
Ottawa, Ontario, Canada K1N 7B7

ABSTRACT

Rising sea level will increase surface runoff and decrease groundwater discharge from low-lying coastal areas by raising the water table and thus saturating the soil. As a result, there will be increased upland flooding by freshwater and changes in the productive intertidal and near-shore ecosystems that depend on the lowered salinities and nutrients provided by groundwater discharge. This paper compares these effects of future increases in sea level with the present variation in runoff and groundwater discharge caused by interannual variations in precipitation and evaporation. The analysis relies on a water balance model calibrated using data from a small watershed on Cape Cod (USA). Interannual variation of runoff and groundwater discharge is estimated by driving the model with 40 years of historical weather data. The results indicate that a 20 cm rise in sea level will double the mean surface runoff and reduce groundwater discharge by half. These conditions are significantly different than the present mean hydrologic conditions. The variability of surface runoff also increases with rising sea level.

INTRODUCTION

Global warming of the climate threatens to accelerate the rate of eustatic sea level rise. This fact has harrowing implications for the large number of people worldwide who live in, or otherwise depend on, low-lying coastal areas. Although recent estimates of the rate of sea level rise due to global warming are much lower than earlier estimates,[1] interest remains high in impacts of rising sea level, mitigation options and their costs. For the most part, attention has focused on the direct effects of higher sea level and increased inundation of coastal areas. These are loss of land area (including valuable coastal wetlands), loss of freshwater resources due to the intrusion of salt water up estuaries and into coastal aquifers, and increased exposure to the effects of storm surge.

Rising sea level also will affect low coastal areas by raising the water table. This will increase the prevalence of saturated soil conditions and will increase the amount of surface runoff at the expense of the discharge of fresh groundwater along the coast. Although difficult to quantify, the consequences of these changes to the freshwater hydrology of coastal areas can be serious. Increased soil saturation threatens the stability of roads and other structures and increased freshwater surface runoff may overwhelm the capacity of natural and man-made drainage, leading to flooding. While levees can be used to protect areas against inundation by the sea, the effects of rising sea level on the water table will propagate inland through coastal aquifers and will be impossible to mitigate in many instances. Miami, Florida, which is underlain by a highly porous limestone aquifer, is an example of an area that will be affected by a rising water table[2].

The impact that changes in runoff caused by rising sea level will have depends on the magnitude of these changes relative to the short-term variation in runoff caused by climatic factors. Year-to-year changes in the amounts and timing of rainfall and evaporation also affect the water balance, causing changes in the water table, surface runoff and groundwater discharge. The range of conditions determined by the variability of precipitation and evaporation are considered normal and are anticipated, for example when designing drains. Changes related to increases in sea level will have significant impact only if they exceed the anticipated range of variation. Here, these factors are accounted for by estimating the sensitivity of runoff and groundwater discharge to both changes in sea level and climate variability by using a water balance model. The model is calibrated to water balance data from a small coastal watershed. Information on the variability of precipitation and evaporation is obtained from 40 years of historical weather data.

WATER BALANCE IN A COASTAL WATERSHED

The rate of freshwater discharge to the coast from a watershed is determined by rainfall, evaporation and changes in water stored within the watershed. The phreatic aquifer in low-lying coastal areas probably provides for the majority of the water storage, and fluctuations in the water table elevation reflect changes in this storage. Freshwater discharge includes both surface runoff and groundwater discharge. Nuttle and Portnoy,[3] working in a small watershed on Cape Cod, Massachusetts, found that surface runoff depends on recent rainfall and the depth of the water table below the soil surface. A contributing area runoff model describes the rainfall-runoff response of the watershed. Surface runoff occurs only in response to rainfall; no surface runoff occurs during periods of several days with no rain.

Rising sea level affects the distribution of total runoff between surface runoff and groundwater discharge. Sea level and the rate of groundwater discharge determine the elevation of the water table, and to a first approximation, the rate of groundwater discharge is proportional to the height of the water table above mean sea level. Therefore, as sea level rises, the depth of the water table below the soil surface decreases. The soil is closer to saturation, and surface runoff increases. An increase in surface runoff means that groundwater discharge decreases, assuming that rainfall and evaporation on the watershed are unchanged. Decreased groundwater discharge implies a decrease in the height of the water table above sea level; therefore the elevation of the water table will rise more slowly than mean sea level. The degree to which sea level affects the water balance depends on the sensitivity of surface runoff to changes in the elevation of the water table.

SIMULATION MODEL

The effects of rising sea level and interannual variations in precipitation and evaporation on surface runoff and groundwater discharge were investigated through the use of a simulation model of the water balance on a coastal watershed. The model accepts sea level, precipitation and temperature as inputs and calculates daily runoff and changes in storage (changes in water table elevation), Figure 1. This calculation is based on the water balance equation;

$$S_y A\, dh/dt = P - E - R - G, \quad [1]$$

which equates the rate of change in groundwater storage (LHS) to the net water flux due to precipitation P, evaporation E, surface runoff R, and groundwater discharge to

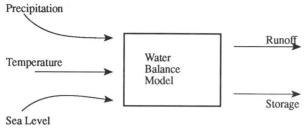

Figure 1. Schematic of inputs and outputs of the simulation model.

the coast G, Figure 2. The rate of change in storage is the product of the area of the watershed A, the specific yield of the soil S_y and the rate of change of the elevation of the water table h. The model uses a linear relation between daily average temperature and daily evaporation in lieu of evaporation data. Surface runoff is calculated using the contributing area, saturated runoff model described by Nuttle and Portnoy.[3] The rate of groundwater discharge out of the watershed is proportional to the difference between the elevation of the water table and mean sea level.

Calibration of the model with the water balance data for a coastal watershed on Cape Cod, Massachusetts,[3] determined the values of the empirical constants in the relations for evaporation and groundwater discharge. Other parameters, such as watershed area, are physically based and could be estimated directly. The calibration procedure optimized the ability of the model to reproduce observed water table elevation and weekly surface runoff, Figure 3.

Forty years of daily precipitation and temperature data for Plymouth, Massachusetts,[4] provided input for model simulations used to characterize the

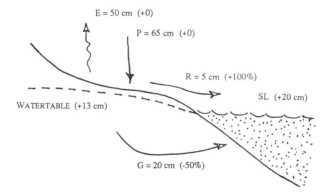

Figure 2. Summary of mean annual water balance fluxes and changes caused by a 20-cm rise in sea level (SL). Fluxes are for the period April to October. A net loss in storage occurs for this period.

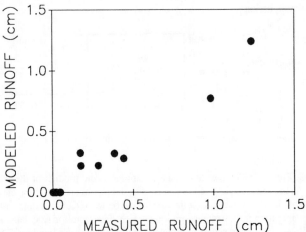

Figure 3. The match between measured weekly surface runoff and modeled runoff obtained with the calibrated water balance model.

variations in the water balance related to climate alone. The sensitivity of runoff to sea level was investigated by repeating the simulations for higher and lower elevations for mean sea level. Sea level was held constant for each 40-year model run. Changes in mean sea level directly affects only the calculation of the rate of groundwater discharge. The effects of increased inundation are believed to be small for the range of mean sea levels investigated and were not included in the simulations.

RESULTS

Simulations show that a significant change occurs in the distribution of runoff in response to a relatively small rise in mean sea level. Mean surface runoff is doubled, and groundwater discharge is decreased by half. The fluxes summarized in Figure 2 are the mean annual fluxes normalized by watershed area for the 40-year simulation period and the changes caused by a 20 cm rise in sea level. Simulations show that the variability of surface runoff will also increase with rising sea level, Figure 4. This suggests that the magnitude of extreme surface flows will increase more quickly than mean surface runoff. The model is calibrated, and simulations cover, the period of April through October. The computed water budget for this period is balanced by a net loss of water from groundwater storage.

DISCUSSION

The following facts will help to illustrate the possible implications of the link between sea level and coastal hydrology that has been established above. Over half of the U.S. fishery depends on the highly productive coastal and estuarine marshes, and there is a growing appreciation for the role played by freshwater discharge along the coast in near-shore ecosystems.[3] Recently, Morris et al.[5] have demonstrated a relation between interannual fluctuations in sea level, productivity of coastal marshes in the

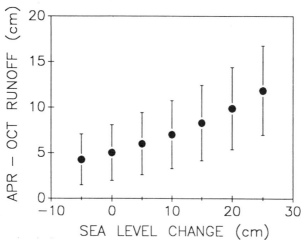

Figure 4. Mean annual surface runoff and the variability in annual runoff increase with rising sea level. Error bars are one standard deviation.

coast in near-shore ecosystems.[3] Recently, Morris et al.[5] have demonstrated a relation between interannual fluctuations in sea level, productivity of coastal marshes in the southern U.S. and total annual landings of shrimp and menhaden. One can speculate on a role for sea-level driven changes in coastal hydrology in this, but as yet there is no established link between freshwater runoff and marsh productivity.

In conclusion, a mechanism exists through which changes in sea level can cause changes in surface runoff and groundwater discharge from coastal watersheds. Model simulations reported here suggest that these changes can be significant compared to the normal interannual variation in runoff. The possible impacts of these changes include increased flooding and erosion, caused by increases in surface water runoff, and disruption of productive near-shore ecosystems. The link to near-shore ecosystems is speculative, but it merits further attention because potential impacts on the human population worldwide are large.

REFERENCES

1. K. O. Emery and D. G. Aubrey. *Sea Level, Land Levels, and Tide Gages* (Springer-Verlag, Berlin, 1991).
2. J. B. Smith and D. Tirpack. *The Potential Effects of Global Climate Change on the United States* (Hemisphere Publishing Corp., New York, 1989).
3. W. K. Nuttle and J. W. Portnoy. *Estuar. Coast. Shelf Sci.* **32**, 203 (1992).
4. J. R. Wallis, D. P. Lettenmaier, and E. F. Wood. *Water Resour. Res.* **27**, 1657 (1991).
5. J. T. Morris, B. Kjerfve, and J. M. Dean. *Limnol. Oceanogr.* **35**, 926 (1990).

GLACIER-RELATED HAZARDS AND CLIMATIC CHANGE

Stephen G. Evans
Geological Survey of Canada
Ottawa, Ontario K1A 0E8

John J. Clague
Geological Survey of Canada
Vancouver, British Columbia V6B 1R8

ABSTRACT

Climatic warming during the last 100–150 years has resulted in a significant glacier ice loss from mountainous regions of the world. Most glaciers have undergone thinning and their margins have retreated significantly since the Little Ice Age. Natural processes associated with this loss of glacier ice pose hazards to people and the economic infrastructure in mountain areas. These processes include glacier avalanches, landslides and slope instability caused by debuttressing, catastrophic outburst floods from moraine-dammed lakes, and outburst floods from glacier-dammed lakes (jökulhlaups). The total loss of life from glacier-related catastrophic events in the Andes, Himalayas, Alps, and other major mountain systems is in excess of 30,000; damage to the economic infrastructure of the affected regions is probably in excess of one billion dollars. In 1941, for example, a single outburst from a moraine-dammed lake in the Cordillera Blanca of Peru killed over 6000 people.

INTRODUCTION

Glaciers in all mountain regions of the world have undergone both dramatic thinning and retreat during the last hundred or so years[1,2]. Little Ice Age glacier limits, which generally represent the maximum Holocene extent of mountain glaciers, are evident both in clearly visible trimlines above present-day glacier surfaces and morainal complexes beyond present-day glacier termini. The rapid glacier ice loss since the Little Ice Age has destabilised adjacent slopes and glacier ice masses themselves, and has also created unstable natural impoundments of significant volumes of water adjacent to glaciers. Catastrophic landslides and outburst floods (Figure 1) have resulted from these changes, and their incidence is directly related to late nineteenth and twentieth century climatic warming.

The objective of this paper is to review the range of catastrophic processes generated by rapid ice loss since the Little Ice Age. Examples are reviewed from mountainous regions of the world with particular emphasis on the Cordillera of western Canada.

GLACIER AVALANCHES

A glacier avalanche[3,4] is a sudden, rapid, downslope movement of ice following its detachment from the terminus of a glacier. Conditions favourable for ice avalanching are created when the terminus of a glacier retreats up a steep slope. Glacier avalanches

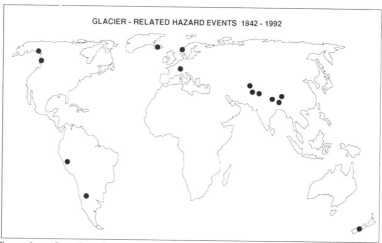

Figure 1.   Location of glacier-related hazard events 1842–1992.

are common in mountainous areas and are hazardous in their own right; for example, at least 124 people were killed by glacier avalanches between 1901 and 1983 in the Swiss Alps alone[5]. In addition, as detailed below, when the avalanching ice plunges into a moraine-dammed lake, it can generate waves that overtop the dam, which may trigger a catastrophic outburst. Glacier avalanches usually occur during the summer months and result from a destruction of tensile strength in the ice mass through progressive fragmentation associated with crevasse development, the melting of parts of the glacier that may be frozen to the substrate, and the reduction of frictional resistance at the ice/rock interface due to increased water pressures.

Three examples are illustrative of the scale and effects of the process:

Altels, Switzerland; On September 11, 1895, approximately 4.5×10^6 m^3 of ice broke away from a glacier just above the 3000-m level on Altels (3629 m)[6,7]. The detachment took place on a smooth bedrock surface dipping towards the valley at 35°. The ice descended the steep slope and became airborne when launched off a ledge at about 2200 m. This generated an air blast, created by the escape of trapped air beneath the avalanche, which killed 4 people in a hut 250 m beyond the margin of the avalanche and also destroyed parts of a mature forest. A total of 6 people and 158 head of cattle were killed and a large part of the summer's produce of cheese, butter, and whey was destroyed. When it hit the valley bottom, 1440 m below the source, the avalanche was travelling at about 120 m/s[7]. The glacier had undergone considerable retreat since 1841 when the first topographic survey was made.

Fallen Glacier, Disenchantment Bay, Alaska; A glacier avalanche involving 29×10^6 m^3 of ice, occurred on the steep western slope bordering Disenchantment Bay, Alaska, in July 1905[8,9]. The ice fell from an altitude of about 300 m, on an average slope of 28°, and plunged into the waters of Disenchantment Bay along a front 800 m wide. The rapid entry (est. velocity 60 m/s[9]) of the ice mass generated a massive displacement wave in the Bay. Eight hundred meters south of the entry zone the wave rose 33 m breaking off alder bushes at that height. About 4.8 km north of the zone vegetation was killed by the wave up to 20 m a.s.l. and 3.5 km across the Bay from the entry zone, it reached a point 35 m a.s.l. at the northwest end of Haenke Island. A similar event in about 1850 reportedly killed about one hundred Indians who were at a summer seal camp in Disenchantment Bay. It appears that Fallen Glacier was destabilized by rapid glacier retreat in the mid to late nineteenth century which occurred in response to the marked regional warming noted by Hansen and Lebedeff[10].

Allalin Glacier, Switzerland; On August 30, 1965, a major disaster took place at the Mattmark dam construction site when about 10^6 m^3 of ice detached from the terminus of the Allalin Glacier[3,4]. The avalanche killed 88 construction workers (Figure 2).

Photographs taken hours before the disaster indicate that the ice mass was heavily fractured. The Allalin Glacier underwent significant retreat after 1923 (Figure 3) up a steep rock slope above the Saas valley and it failed when the terminus reached a steeper portion of the slope.

Figure 2. Oblique aerial view of the 1965 Allalin Glacier avalanche which killed 88 construction workers at the Mattmark Dam construction site, Switzerland (Photo by COMET, Zurich).

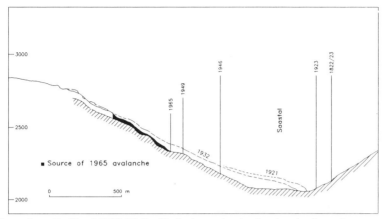

Figure 3. Profile of 1965 Allalin Glacier ice avalanche showing the dramatic retreat of the glacier between 1923 and 1965 (modified from Rothlisberger[4]).

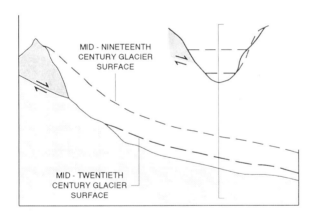

Figure 4. Schematic diagram showing relation between glacier thinning/retreat and landslides.

LANDSLIDES AND DEBRIS FLOWS

Slopes adjacent to glaciers which have undergone retreat and thinning are particularly prone to landslides (Figure 4). Erosion by glaciers during the Little Ice Age and subsequent debutressing during retreat have destabilised slopes.

Two highly destructive landslides from the north peak of Nevados Huascaran[11,12], in the Cordillera Blanca of Peru in 1962 and 1970 were probably due in part to the effects of rapid glacier thinning and retreat from Little Ice Age limits. During the period 1886–1942, the firn limit on Huascaran rose in elevation from 4320 m to 5100 m, including a 500-m rise between 1932 and 1942[13]. In addition, the canopy of ice on the summit ridge of Huascaran thinned dramatically[14]. In 1962 approximately 13×10^6 m^3 of rock and glacier ice detached from the north peak of Huascaran and travelled 16 km down the Rio Shacsa, overwhelming several towns and villages. The average velocity of the landslide was in the order of 47 m/s. The landslide killed about 4000 people, mostly in the town of Ranrahirca.

In 1970, $50-100 \times 10^6$ m^3 of rock and ice, undermined by the 1962 event, detached from a similar position on Huascaran's north peak. The landslide was triggered by an earthquake (M = 7.7), the epicentre of which was located 130 km to the west. The landslide exhibited spectacular mobility: in the upper part of the path, boulders were hurled into ballistic trajectory and impacted 4 km from their launch positions. The debris travelled a vertical distance of about 4200 m in a horizontal distance of 16 km with a mean velocity of 75 m/s. The landslide caused an estimated 18000 deaths, mainly in the town of Yungay.

Mass movements caused by glacier downwasting and retreat are common on steep slopes adjacent to glaciers in the Canadian Cordillera[15]. Fifty-three percent of 30 known historical rock avalanches in the Canadian Cordillera have occurred on slopes adjacent to glaciers. Studies have shown that detachment surfaces of some of these rock avalanches intersect rock slope surfaces below well marked Little Ice Age trimlines and were thus exposed during recent glacier thinning. Examples from the Coast Mountains of British Columbia are the Tim Williams Glacier[16] rock avalanche (estimated volume 3×10^6 m^3, Figure 5) and the 1975 Devastation Glacier debris avalanche (estimated volume ca. 12×10^6 m^3)[15].

A particularly good example of a landslide caused by Little Ice Age steepening and subsequent debutressing is the 1992 rock avalanche (estimated volume $5-10 \times 10^6$ m^3) from the flanks of Mount Fletcher in the Southern Alps of New Zealand[17]. The adjacent glacier had undergone approximately 250 m of thinning since the Little Ice Age maximum. This thinning, which accelerated after the middle of the present century, exposed a steeper toe slope through which rupture took place.

Glacier downwasting and retreat may also give rise to non-catastrophic slope deformation. This type of movement is manifested in slope cracking, subsidence at the top of the slope and bulging at its base. Examples have been reported from slopes adjacent to glaciers in the St. Elias Mountains of British Columbia and Alaska[18], a region that has experienced substantial glacier ice losses in the twentieth century. At Melbern Glacier, for example, thinning of 400–600 m since the Little Ice Age

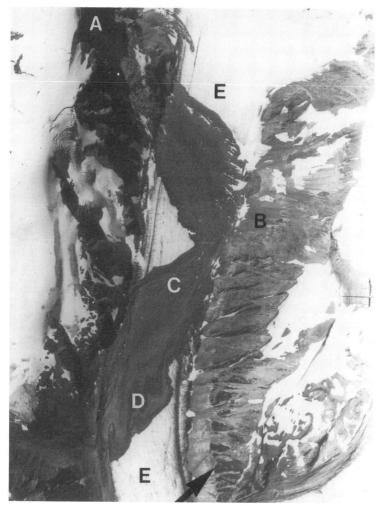

Figure 5. Vertical aerial photograph (BC 2182-52) of Tim Williams Glacier rock avalanche, British Columbia, taken in 1956. A = detachment zone; B = run-up on west valley wall; C = flow lines in debris; D = transverse banding resulting from compositional variation in the debris; E = East Tim Williams Glacier. Little Ice Age trimline is also apparent (arrowed).

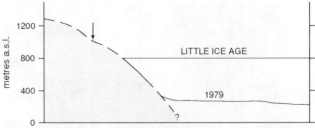

Figure 6. Topographic profile of Melbern Glacier, St. Elias Mountains, British Columbia, and adjacent slope showing glacier thinning between Little Ice Age maximum and 1979. Thinning has destabilised slope which is manifested in sagging (arrowed) and bulging.

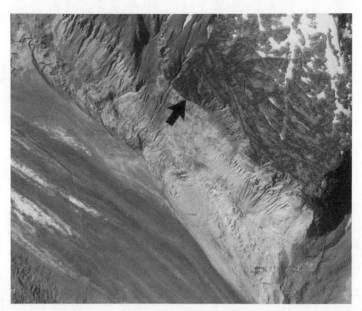

Figure 7. Aerial photograph (Canada A12856-385) of slope deformation adjacent to Melbern Glacier, St. Elias Mountains, British Columbia. Little Ice Age trimline is arrowed and cracking associated with slope deformation is visible on the right hand side of the photograph.

maximum has debuttressed adjacent mountain slopes, causing extensive, non-catastrophic slope deformation (Figures 6, 7).

Debris flows are also associated with recent glacial retreat. In the Swiss Alps during the summer of 1987, numerous debris flows were triggered by intense rainfall of unusually long duration[19]. In a large number of cases, the source of the debris were Little Ice Age terminal moraine complexes exposed during recent retreat. Debris production may also have been related to the decay of ice cores within the moraines. Similar factors have been suggested by Jordan to explain the occurrence of some recent debris flows in the southern Coast Mountains of British Columbia[20].

OUTBURSTS FROM MORAINE-DAMMED LAKES

Moraine-dammed lakes are found in high mountains close to existing glaciers. They formed when glaciers retreated from moraines built during the Little Ice Age and where the moraines dam glacial streams (Figure 8). Moraine dams are susceptible to failure because they are steep-sided, have relatively low width-to-height ratios, and consist of poorly sorted, loose sediment. In addition, these dams and the lakes behind them commonly occur immediately downslope from steep slopes that are prone to glacier avalanches and rockfalls. Moraine dams generally fail by overtopping and incision. The triggering event is most frequently a glacier avalanche[21] from the toe of the retreating glacier which generates waves that overtop the dam (Figure 8). Melting of ice cores and piping are other reported failure mechanisms[21].

Four major moraine dam failures[22,23] and related outburst floods occurred in the Cordillera Blanca, Peru, between 1938 and 1950 following dramatic glacier ice losses initiated in the 1920s and '30s. The most devastating was the 1941 catastrophic emptying of moraine-dammed Lake Cohup. The discharged water entered another lake downstream, causing it to drain. These events triggered a debris flow (estimated volume

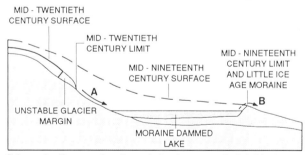

Figure 8. Schematic diagram showing relation between glacial thinning and retreat, and the formation and destruction of moraine-dammed lakes. Avalanching of unstable glacier ice (A) into the lake generates waves that overtop moraine dam (B) initiating catastrophic breaching.

8×10^6 m^3) which destroyed 1/3 of the city of Huaraz, killing more than 6000 people[22,23].

Several moraine dam failures have produced large floods and debris flows in the Canadian Cordillera in recent years[21,24,25]. In the early 1970s, for example, the sudden failure of the moraine impounding Klattasine Lake in an unpopulated part of the British Columbia Coast Mountains released approximately 1.7×10^6 m^3 of water and triggered a massive debris flow (estimated volume $2-4\times10^6$ m^3) that travelled 8 km to block the Homathko River.

In the same region, ca. 6×10^6 m^3 of water was released from Nostetuko Lake when the moraine impounding the lake failed in 1983 (Figure 9). The breach was initiated by waves generated by a glacier avalanche into the lake which overtopped the moraine. The resulting flood wave devastated the valley below the moraine and travelled more than 100 km to the sea.

JÖKULHLAUPS

Glacier-dammed lakes are found mainly at the margins of valley glaciers, although some occur within or beneath cirque and valley glaciers and mountain ice caps. Some of the largest lakes are situated in main valleys dammed by tributary glaciers and at the mouths of tributary valleys blocked by trunk glaciers. These lakes may drain suddenly and rapidly by the formation and enlargement of subglacial and englacial tunnels, and occasionally by overtopping; the resulting flood is termed a jökulhlaup.

Some of the world's largest documented historical jökulhlaups occurred in the Karakoram Himalayas in the first half of this century[26,27]. The damming of the Upper Shyok River by the Chong Kumdan Glacier formed a lake with an estimated volume of 1.4×10^9 m^3. A sudden outburst from this lake occurred in 1929, and the flood wave travelled down the Shyok River into the Indus River. A rise in the level of the Indus of 8 m was measured at Attock, 740 km downstream from the ice dam.

Some formerly stable, glacier-dammed lakes have gone through a cycle of jökulhlaup activity during this century as glaciers have retreated from maximum positions achieved during the Little Ice Age. An example is Summit Lake, dammed by the Salmon Glacier in the northern Coast Mountains of British Columbia[28,29] (Figure 10). This lake first drained catastrophically in 1961 after a lengthy period of stability, and has drained annually since 1970, with peak discharges of the largest floods in excess of 3000 m^3/sec[28,29].

In contrast, many lakes that formerly produced jökulhlaups have disappeared since the Little Ice Age due to glacier retreat. Lake Alsek, one of the largest Holocene glacier-dammed lakes in the world, formed behind Lowell Glacier in the Saint Elias Mountains, Yukon Territory, and periodically produced jökulhlaups with peak discharges larger than the mean flow of the Amazon River[30,31]. Lake Alsek formed and emptied many times during the nineteenth and perhaps early twentieth centuries, but has not existed in recent years due to retreat of Lowell Glacier.

Jökulhlaup frequency is related to climate warming through glacier retreat, as illustrated in Figure 11. The inititiation of a jökulhlaup cycle occurs when a threshold of retreat and thinning is reached. Jökulhlaups then take place with decreasing magnitude and frequency until the glacier dam ceases to exist.

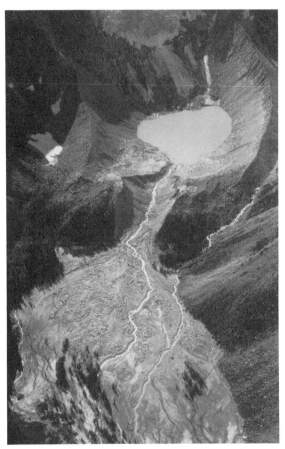

Figure 9. Oblique view of Nostetuko Lake, a moraine-dammed lake in the Coast Mountains of British Columbia which drained catastrophically in 1983. The bulky, sharp-crested Little Ice Age moraine complex formed during the nineteenth century. Approximately 6×10^6 m^3 of water was released from the lake. A glacier avalanche from the Cumberland Glacier, visible above the remnant lake, probably initiated the breaching event. About 1.5×10^6 m^3 of material was eroded from the moraine dam during the outburst, much of which was deposited in a debris fan immediately downstream.

58   Glacier-Related Hazards and Climatic Change

Figure 10. Oblique photo of Summit Lake, British Columbia, looking down the Salmon Valley. Salmon Glacier forms the ice dam which impounds the lake. Summit Lake has drained annually since 1970.

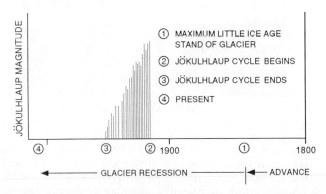

Figure 11. Conceptual model of a jökulhlaup cycle, initiated due to glacier retreat.

CONCLUSIONS

Global warming has caused significant glacier ice loss since the Little Ice Age resulting in both glacier retreat and thinning. Catastrophic natural processes triggered by these glacier changes have been responsible for considerable death and destruction in glaciated mountain areas of the world. These processes include glacier ice avalanches, landslides and debris flows, outbursts from moraine-dammed lakes and jökulhlaups (outbursts from glacier-dammed lakes). Glacier avalanches have occurred where glaciers have retreated up steep rock slopes. Landslides caused by debutressing due to glacier thinning include rapid, mobile rock avalanches and non-catastrophic slope deformation. Sources of debris flows are frequently moraine complexes exposed during glacier retreat, which also may be ice-cored. Outbursts from moraine-dammmed lakes result from the catastrophic breaching of the moraine dam, a process which is commonly initiated by glacier avalanche-generated waves that overtop the moraine. Jökulhlaups occur once a threshold is reached during glacier retreat. A jökulhlaup cycle is thus initiated, during which outburst floods occur with decreasing magnitude and frequency to a point where, because of continued retreat, the glacier dam ceases to exist. Glacier-related hazards are an important, underdocumented response to global warming.

REFERENCES

1. J. M. Grove. *Little ice age* (Methuen, London, 1988).
2. S. C. Porter. *Quat. Res.* **26**:27 (1986).
3. H. Hanke. *Die Bergsteiger* **33**:433 (1966).
4. H. Rothlisberger. *Sonder. Jahrb. der Schweiz. Naturs. Gesell.* **1978**:170 (1978).
5. J. Alean. *J. of Glaciol* **31**:324 (1985).
6. A. Heim. 98 *Neujahr. der Naturf. Gesell. in Zurich* (1895).
7. L. Du Pasquier. *Bull. Soc. des Sci. Nat. de Neuchatel* **24**:149 (1896).
8. R. S. Tarr. United States Geol. Surv. Prof. Pap. 64. p. 183 (1909).
9. R. L. Slingerland and B. Voight. *Rockslides and avalanches*, V. 2 (Elsevier, Amsterdam, 1979). p. 17.
10. J. Hansen and S. Lebedeff. *J. Geophys. Res.* **92**(D11):13345 (1987).
11. B. Morales. *Int. Ass. Hydrol. Sci. Publ.* **69**:304 (1966).
12. G. Plafker and G. E. Ericksen. *Rockslides and avalanches*, V. 1 (Elsevier, Amsterdam, 1978). p. 277.
13. J. A. Broggi. *Bol. de la Socdad. Geol. del Peru* 14–15, 59 (1943).
14. C. M. Clapperton. *Quat. Sci. Rev.* **2**:83 (1983).
15. S. G. Evans and J. J. Clague. Proc. 5th Int. Symp. on Landslides, V. 2:1153 (1988).
16. S. G. Evans and J. J. Clague. Geol. Surv. Can. Paper 90-1E, 351 (1990).
17. M. J. McSaveney. Imm. Rpt. Mount Fletcher rock avalanche, DSIR, N.Z. (1992).
18. D. H. Radbruch Hall. *Rockslides and avalanches*, V. 1 (Elsevier, Amsterdam, 1978) p. 607.
19. W. Haeberli and F. Naef. *Die Alpen* **64**:331 (1988).

20. P. Jordan. Inl. Wat. Dir., Env. Canada, Report IWD-HQ-WRB-SS-87-3, 62 p. (1987).
21. S. G. Evans. Proc. Int. Symp. Eng. Geol. Env. in Mount. Areas (Beijing), V. 2:141 (1987).
22. L. Lliboutry et al. *J. Glaciol.* **18**:239 (1977).
23. A. Heim. *Wunderland Peru* (Verlag Hans Huber, Bern, 1948).
24. J. J. Clague et al. *Can. J. Earth Sci.* **22**:1492 (1985).
25. I. G. Blown and M. Church. *Can. Geotech. J.* **22**:551 (1985).
26. K. J. Hewitt. Int. Ass. Sci. Hydr. Publ. No. 138:259 (1982).
27. K. Mason. *Himal. J.* **2**:40 (1930).
28. W. H. Mathews. *Geog. Rev.* **55**:46 (1965).
29. W. H. Mathews. Int. Assoc. Sci. Hydrol. Pub. **95**:99 (1973).
30. J. J. Clague and V. N. Rampton. *Can. J. Earth Sci.* **19**:94 (1982).
31. G. K. C. Clarke. *Ann. of Glaciol.* **13**:295 (1989).

CHANGES IN WATER SUPPLY IN ALPINE REGIONS DUE TO GLACIER RETREAT

Mauri S. Pelto
Department of Environmental Sciences
Nichols College
Dudley, MA 01571

ABSTRACT

In the late 1970s global temperature rose abruptly, and between 1977 and 1990 has averaged 0.4°C above the 1940–76 mean. In 1980, 50% of the alpine glaciers observed in the Swiss Alps, Peruvian Andes, Norwegian Coast Range, Northern Caucasus and Washington's North Cascades were advancing. By 1990 in response to the warming only 15% were still advancing. During the peak non-glacier snow melt period glaciers are unsaturated aquifers soaking up and holding meltwater for the first two-six weeks of the melt season. This storage acts as a buffer for spring snow melt flooding, and spreads the peak spring flow over a longer period. In the late summer glaciers buffer low flow periods by providing large volumes of meltwater. As glaciers retreat the amount of water they can store decreases raising spring flood danger and the areal extent exposed for late summer meltwater generation decreases, thus reducing late summer flow.

INTRODUCTION

The North Cascade Glacier Climate Project (NCGCP) has observed the terminus behavior, mass balance and runoff of North Cascade glaciers to demonstrate quantitatively the changes in glacier runoff that occur as glaciers retreat.

North Cascade glaciers provide 800 million m^3 of runoff each summer, 25–30% of the regions' total summer supply. This water is utilized for hydropower generation, irrigation and municipal water supply. In the North Cascade region since 1977, winter precipitation has been 15% below, and annual temperature 0.9°C above, the long term mean. The result has been glacier retreat. In 1975, 8 of the 10 North Cascade glaciers observed were advancing, while in 1991, 41 of 47 glaciers observed were retreating. The result has been decreased summer glacier runoff, which combined with rapid development in the Puget Sound Region has led to water shortages for the first time. Because of these changes intelligent water resource management requires monitoring changes in glacier runoff due to climate change.

MASS BALANCE GLACIER RUNOFF RELATIONSHIPS

How can we utilize glacier measurements to determine the magnitude and timing of glacier runoff? The magnitude of glacier runoff is a function of the temporal changes in mass balance and can be accurately estimated from mass balance measurements. The timing of glacier runoff release depends on daily weather conditions and mass balance. To utilize mass balance data to estimate glacier runoff we first must understand how and why glacier runoff varies.

The firn and snowpack on a glacier in the spring is an unsaturated aquifer. During the early melt season much of the meltwater generated is stored within this glacier aquifer, until the aquifer is saturated.[1] During the infiltration period when the aquifer is being filled, 54% of the total meltwater generated in May on the South Cascade Glacier was temporarily stored by the glacier, to be released during the following months.[2]

The amount of meltwater that a glacier can store is its storage capacity. In terms of water storage, the infiltration period which roughly corresponds to the first six weeks of the ablation season, is the only time when North Cascade glaciers effectively store meltwater for a significant amount of time. After the aquifer is saturated, glacier runoff equals meltwater production. Annual changes in a glacier's storage capacity during the infiltration period are determined by the density, depth and areal extent of snow and firn pack.[3] The thicker and more extensive the snow and firnpack, the greater the aquifer size and the greater the storage capacity. Firn more than four years old has too low a porosity and permeability to act as an aquifer.[1] Storage capacity is then dependent on the mass balance of the previous several years not just the current year. Considerable meltwater is also stored in englacial and subglacial areas. This storage capacity is difficult to determine and is less variable from year to year.

In the North Cascades glacier runoff peaks in July, and runoff from non-glacierized basins peaks in May and early June[4]. This delay is due to a glacier's ability to store meltwater and the fact that ablation is highest during July and August.

To calculate the Storage capacity (R) of the glacier snow-firnpack aquifer equation 1 is used:

$$R = (Ds)(As)(Ps) + (Df)(Af)(Pf) \qquad (1)$$

where Ds and Df are the mean snowpack and firnpack depth respectively, As and Af are the areal extent of snowpack and firnpack respectively, and Ps and Pf are the mean snowpack and firnpack water content. Only Ps and Pf are not determined directly from mass balance measurements. Due to the present impossibility of precisely determining Ps and Pf[7], neither is measured. Ps and Pf are dependent on the density of the snow or firnpack, and on the liquid water content of the pore spaces at saturation, and both are consistent from year to year. The best estimate for Ps and Pf, at this time, is 0.41 multiplied by the snow-firn density[5,6].

The lag between peak glacier runoff and non-glacier runoff is determined by glacier storage capacity and weather conditions during this period. A simple equation has been developed to relate storage capacity to the lag in glacier runoff:

$$\text{lag (days)} = [(8SC/10^2 m^3) - (DD/20)] \qquad (2)$$

where SC is the storage capacity in m^3, and DD is the number of degree days at Stevens Pass during the lag period. Before the melt season begins SC is known from mass balance measurements, DD is an adjustment during the melt season. The lag is the time from the onset of the ablation season, the first three consecutive days with the mean temperature above 5°C at Stevens Pass, to the time when measured runoff approximately equals or exceeds measured meltwater generation. Table I presents the data for Columbia Glacier near Monte Cristo in the North Cascades (Figure 1). The success of equation 2 in predicting lag indicates that the delay in glacier runoff versus non-glacier runoff can be accurately estimated, even though subglacial storage change is not considered. This suggests that subglacial storage is influenced by fluctuations in

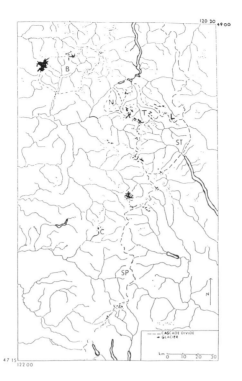

Figure 1. Map indicating the location of glaciers and drainage systems in the North Cascades. The four principal basins of study are the Newhalem (N), Thunder Creek (T), Stehekin River (ST), and Baker Lake (B). The two primary glaciers are Columbia Glacier (C) and Lewis Glacier (L). Climate data for equation 2 is from Stevens Pass (SP).

degree days and winter snowpack thickness also. A 52% drop in storage capacity results in a three-week decrease in the lag of glacier runoff, indicating the potential for decreased storage capacity to increase spring flood danger.

CHANGING GLACIER RUNOFF

The timing of glacier runoff is dependent on glacier mass balance and spring weather conditions. The change in timing of peak runoff between adjacent glaciated and unglaciated basins is illustrated in Table II. In this example, the peak spring

Table I
The measured storage capacity of Columbia Glacier, degree days at Stevens Pass, and measured delay in runoff from Columbia Glacier, compared to the calculated delay in runoff using equation 2

Year	Storage capacity ($10^2 m^3$)	Degree days	Calculated lag (days)	Observed lag (days)
1984	4.13	21	32	31
1985	2.74	37	26	21
1986	2.57	14	20	23
1987	2.01	105	11	8
1988	2.66	27	20	20
1989	2.51	33	19	16
1990	2.16	12	16	14

discharge was delayed an average of four weeks in the heavily glaciated Thunder Creek Basin, versus Newhalem Creek Basin which has one small glacier.

The impact of reduced glacier mass balance on spring glacier runoff is also illustrated by examination of changing runoff into Baker Lake, North Cascades. Baker Lake is fed by 60 glaciers with an area of 26.5 km^2. Mass balance measurements completed by the NCGCP on six glaciers in this drainage basin, Rainbow, Lower Curtis, Easton, Watson, and Hidden Creek glaciers, indicate that 3.21m^3/m^2 of runoff is yielded on average from these glaciers. This amounts to over 85 million m^3 of glacier runoff for Baker Lake between May 15 and October 1. This is 20–25% of the total summer runoff into Baker Lake. Since 1980 the area of glacier cover has been reduced 3% and the length of the infiltration period was less than three weeks during four years between 1984 and 1990. Thus, peak spring glacier and non-glacier snow melt runoff overlapped, increasing peak spring stream flows. A primary objective of Upper Baker Dam, which impounds Baker Lake, is to provide flood control. Each year the Army

Table II
The date of peak spring discharge as marked by the highest four-day total discharge from three North Cascade Basins with varying glacier cover.

	Newhalem	Stehekin	Thunder
% glaciated	0.4%	3.1%	14.1%
1984	May 29	May 29	June 20
1985	May 17	May 14	May 18
1986	May 18	May 19	May 26
1987	Apr 28	Apr 29	May 7
1988	Apr 13	Apr 14	May 11
1989	Apr 5	Apr 15	June 1
1990	Apr 11	Apr 14	June 16
Mean	Apr 30	May 2	May 28

Corps of Engineers buys a certain amount of reservoir volume, which is left empty, to prevent flooding. Since 1985, the average date of peak glacier runoff from Rainbow Glacier has been May 28, compared to the long term average of mid-June. If the recent trend of increased early glacier runoff and reduced late summer runoff continues, a larger storage volume will have to be bought to protect against flooding. If the glaciers are monitored this practice need be adopted only when a warm summer reducing glacier snowpack is followed by a moderate to dry winter and a wet-warm spring rain event. Thus, forewarning can be provided.

As glaciers shrink the meltwater they generate in late summer declines, this combined with earlier runoff due to reduced aquifer size results in lower late summer stream flow. Glaciers provide natural buffers against late summer low flow periods, since they release more meltwater during warm dry periods. However, as they shrink their buffering capacity during droughts declines. The following are examples of the effect of shrinking glaciers on late summer stream flow. Table III compares the drop in stream flow during drought conditions in adjacent drainages, glaciated Thunder Creek Basin and unglaciated Newhalem Creek Basin.

In Thunder Creek during drought periods stream flow dropped 18% versus a 34% drop in Newhalem Creek, indicating the substantial buffering ability of glaciers. The buffering capacity of glaciers is also noted in comparing runoff measurements. First during non-drought August's non-glaciated alpine areas in the North Cascades released an average of 0.10 m^3/m^2 month, and glaciers released 1.25 m^3/m^2 month, a 1250% increase for glacier covered areas. During August droughts mean non-glacier flow is 0.066 m^3/m^2 month and glacier runoff is 1.44 m^3/m^2 month, a 2200% increase for glacier areas. A small area of glacier cover is then important to total basin runoff.

The loss of a glacier will not result in a 1250–2200% decline in late summer runoff, because the formerly glaciated area is still a high alpine basin that will retain much more snow pack than most non-glaciers areas, so how much is the decline? In August 1985, Lewis Glacier had an area of 0.09 km^2 and released 0.15 million m^3 of runoff. By August 1990, Lewis Glacier had disappeared, runoff from the former glacier basin was 0.04 million m^3, only 27% of the glaciated flow. This is despite the fact that

Table III
Stream flow during drought conditions in unglaciated Newhalem Creek Basin and glaciated Thunder Creek Basin.

Date	Newhalem m^3	Thunder m^3
Aug. 1969	2.0	20.1
Aug. 1970	1.7	23.4
Aug. 1979	1.8	27.3
Aug. 1985	1.8	22.0
Aug. 1987	1.9	22.6
Avg. Flow	2.8	28.1
% Depletion	34%	18%

some relict glacier ice still existed and precipitation was the same for the two months. Since glaciers contributed 35–42% of the stream flow to Baker Lake during late summer droughts (July 1985, August 1986, August and September 1987) glacier retreat will cause significant drops in late summer stream flow. This will be critical only during late summer droughts, and forewarning can be provided, by monitoring the changing area of glaciers, which will provide a good estimate of glacier runoff.

THE EFFECTS OF DECLINING GLACIER RUNOFF

During the initial stages of retreat a decline in runoff is masked by the release of meltwater from ice that has been stored for an extended period. In the Wind River Range the shrinkage of Dinwoody Glacier and Gannet Glacier since 1950 has resulted in a water volume loss of $123 \times 10^6 m^3$.[8] These two glaciers provide 13% of the total runoff to Dinwoody Creek. The volume of ice lost from the glaciers since 1950 has comprised one third of their annual contribution.[8] Thus, the warming causing additional melting from each m^2 of the glacier has in part offset the decrease due to reduced glacier area.

In 1989 the U.S. Supreme Court awarded the Wind River Indian Reservation Indians $617 \times 10^6 m^3$ of additional runoff. This is slightly less than the total water runoff from Wind River Range glaciers. A 10% reduction in glacier runoff has already occurred due to glacier retreat. Irrigation headgates have been closed to non-Indian irrigators as late summer water shortages occurred in 1988, 1989 and 1990. In addition Indians are increasing their use of water for fishery maintenance, a wise move. In order to meet the needs of all involved it is clear that we will have to understand what changes in runoff will occur in the near future.

In recent years North Cascades summer glacier runoff has declined, while water usage has climbed dramatically due to population and industrial growth in the Puget Lowland. Peak spring runoff from the Cle Elum Basin has been fours week earlier than normal. Cle Elum Reservoir and several others have been perpetually low during the last 10 years because of this change in the timing and magnitude of the alpine glacier. For this reason it is no longer possible to intelligently manage the area's water resources without considering changing glacier contributions.

This same problem will be much more acute in the Himalayan and Andes areas where agriculture is much more reliant upon glacier meltwater. In the Andes all 10 glaciers monitored in 1980–1985 were retreating.[9] Table IV lists some basins in which the glacier runoff is heavily used for irrigation and the contribution of glaciers to each basin. In each of these basins late summer runoff is greater than 30% of the total runoff. Again long term planning will be required to adapt the local economies to the reduced water supply.

REFERENCES

1. A. B. Bazhev. Academy of Sciences of the U.S.S.R. *Data of Glaciological Studies*, **57**, 50–56 (1985).
2. R. M. Krimmel, W. V. Tangborn, M. F. Meier. *IAHS* **104**, 410–416 (1973).
3. M. S. Pelto. *J. Glaciol.* **34**, 194–200 (1988).
4. A. G. Fountain, W. V. Tangborn. *Wat. Res. Res.* **21**, 579–586 (1985).

5. S. C. Colbeck. *J. Glaciol.* **20**, 189–201 (1978).
6. A. Denoth. *J. Glaciol.* **28**, 357–364 (1982).
7. A. G. Fountain. *Ann. Glaciol.* **13**, 69–75 (1989).
8. R. M. Marston. *Phys. Geogr.* **12**, 115–123 (1991).
9. World Glacier Monitoring Service. Fluctuations of Glaciers 1980–1985 (IAHS-UNESCO, 1988)
10. World Glacier Monitoring Service. World Glacier Inventory, Status 1988 (IAHS-UNESCO, 1989).

Table IV
Large glaciated basins where the water supply is heavily used for agriculture, indicating the volume of glacier runoff and the percentage of the total runoff provided by glaciers[10]

	Glaciated area km^2	Annual glacier runoff $\times 10^8$m^3	Percentage of total summer runoff
Chile			
Cachopoal	317	9.5	45%
Maipo	426	13.0	40%
Argentina			
Baker	1221	30.0	55%
Mendoza	865	21.5	50%
Santa Cruz	2765	65.0	65%
Peru			
Maronon	272	5.5	60%
Santa	503	10.0	70%
Afghanistan			
Ab-i-Panja	342	8.8	40%
Pakistan			
Chitral	1353	32.5	35%
Nepal			
Ganges	1640	50.0	25%
Bhutan			
Gaddhar	1340	40.0	20%
China			
Hengduanshan	1455	38.0	30%

LIGHTNING AND FOREST FIRES IN A CHANGING CLIMATE

Colin Price and David Rind
NASA Goddard Institute for Space Studies and
Columbia University, 2880 Broadway,
New York, NY 10025

ABSTRACT

Future climate change could have significant repercussions on two related natural hazards: lightning and forest fires. The Goddard Institute for Space Studies (GISS) general circulation model (GCM), has been used to study possible changes in lightning and forest fires as a result of climate change. Initial model results show that for an atmosphere containing twice today's CO_2 concentration, the global lightning activity increases by approximately 32%, while the likelihood of severe drought conditions increases from 1% in today's climate to nearly 50% by the year 2060. Conditions favorable for forest fires are strongly linked to climate, and particularly to drought frequencies. Therefore, increases in both lightning activity and the frequency of droughts could result in dramatic changes in forest fire frequencies and intensities in the future.

INTRODUCTION

Lightning and forest fires are two related natural hazards that could be significantly influenced by future climate change.

Lightning results in electrical power interruptions, aircraft damage, property damage, deaths by direct strikes, as well as forest fires. In the United States alone, more people are killed every year by lightning than by tornadoes, hurricanes or floods. Thousands are killed around the world every year as a result of direct lightning strikes. In addition, lightning influences the NO_x concentrations in the atmosphere[1] and the earth's electric circuit [2].

Lightning-caused fires result in considerable damage to wilderness areas, as well as contributing to the trace gas and aerosol concentrations in the atmosphere[3]. In the United States lightning starts 10–15 thousand forest fires each year. Even when fire does not occur, lightning can cause damage to forests, thereby enhancing the spread of insects and disease[4].

One of the methods of studying the effects of future climate change on the above two natural hazards, is by using global climate models. The model utilized for this study is the Goddard Institute for Space Studies (GISS) general circulation model (GCM)[5]. The model has a horizontal resolution of 8°x10°, with nine layers in the vertical.

LIGHTNING

Lightning Parameterization

In order to model lightning activity in a GCM it is necessary to find a relationship between large scale meteorological parameters and lightning frequencies. Strong vigorous updrafts in the mixed phase region of convective clouds help the electrification process in clouds by increasing the rate of charge buildup in thunderstorms[6]. Since convective cloud top heights are positively correlated with updraft intensity, it can be shown that lightning frequencies in thunderstorms are strongly related to the height of the convective cloud tops[7]. With this knowledge we developed a parameterization that could use calculated convective cloud top height from the GCM to simulate the global lightning distributions for a specific climate.

The parameterization involves two separate formulations, one for continental clouds and another for maritime clouds[8]:

$$F_c = 3.44 \times 10^5 \, H^{4.9} \tag{1}$$

$$F_m = 6.40 \times 10^{-4} \, H^{1.73} \tag{2}$$

where F = lightning flash frequency (flashes/min), H = convective cloud top height (km)

The necessity for two separate formulations is due to the fact that the updrafts in marine thunderstorms are much weaker than those in continental thunderstorms[9]. Since the terminal velocities of large cloud drops in marine thunderstorms are often greater than the weak updrafts, large supercooled drops cannot be carried to high altitudes within these thunderstorms. Recently it has been shown that the existence of both ice and liquid particles together above the freezing level in thunderstorms is a requirement for the generation of lightning[10]. This difference in updraft velocity could explain the minimal lightning activity in ocean thunderstorms observed both from satellite measurements[11] and in-situ measurements[12].

Climate Change experiments
Control Run

Before any future climate experiments can be done, it is first necessary to determine how well the model can simulate current lightning distributions.

Figure 1a shows observed lightning data from the DMSP satellite for January–March 1977[13]. Due to the poor detection efficiency of the satellite sensor, the observations had to be calibrated before comparing them with the model. In Figure 1b the model's control run is shown for the months of January–March. This distribution represents a lightning climatology for these months. Considering that the observations were made only at dusk and only for one year, and that the detection efficiency of the satellite sensor was so low (2%), the agreement between the model and observations (after calibration) is quite remarkable.

Both the spatial distribution and the absolute intensities are very similar. The observations give a global lightning frequency of approximately 40 flashes/sec (Jan–March) whereas the model gives a value of 50 flashes/sec.

2×CO_2 Climate

In our model doubling the CO_2 concentration to 630 ppm results in an equilibrium global warming of 4.2°C. The net effect of this warming on the global lightning frequencies is shown in Figure 2. The vast majority of regions show increases in lightning activity, especially over continental regions. In fact lightning over continental

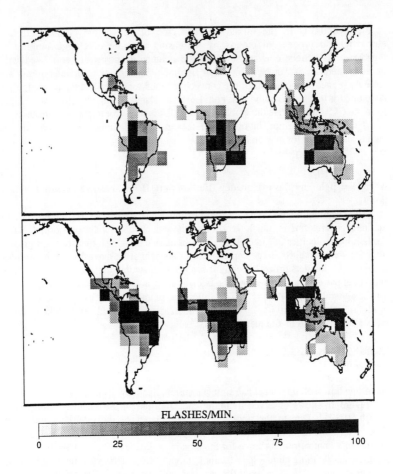

Figure 1. a) January-March 1977 observed dusk lightning[13];
b) GISS GCM January–March lightning climatology (daily mean).

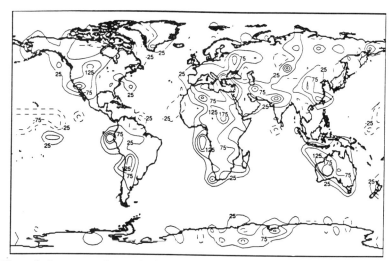

Figure 2. Percentage changes in lightning frequencies for a modeled 2×CO_2 climate. Solid/dashed contours represent increases/decreases in lightning activity.

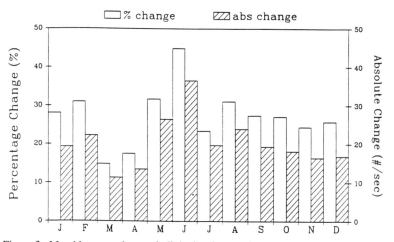

Figure 3. Monthly mean changes in lightning frequencies as a result of a doubled CO_2 atmosphere. Both percentage change and absolute change are given.

regions (grid boxes with more than 50% land) increases by more than 40%, whereas over maritime regions the lightning increases by only 12%. This is two orders of magnitude larger than the interannual standard deviation calculated from the control run.

Figure 3 shows the monthly mean changes in lightning frequencies as a result of a doubled CO_2 atmosphere. The largest increases tend to be during June, although large increases in lightning activity are found throughout the year.

FOREST FIRES

The occurrence of natural forest fires depends on a combination of three factors: climatic conditions, fuel loading and ignition sources. In the tropical rainforests where fuel is abundant and lightning storms are frequent, few fires occur due to the moist climate[14]. On the other hand, in drier midlatitude regions where lightning and fuel are less abundant, lightning-fires dominate in some regions[15]. In addition, in the southeast United States, which is normally moist, the fire frequency increases when dry conditions prevail, whereas in California, which is dry, fire frequency increases when the frequency of thunderstorms increases.

<u>Climatic condition</u>

The environmental factors most likely to effect fire regimes are related to water balance, since water balance is correlated with fuel moisture[16]. It has been shown that the historic fire records are best correlated with fluctuations in effective precipitation, or the difference between precipitation and potential evapotranspiration $(P-E_p)$[17].

A drought index using the effective precipitation has been developed for GCM studies[18], where a supply-demand drought index (SDDI) is given by:

$$SDDI = P - E_p - (P-E_p)\text{clim} \qquad (3)$$

This formulation implies that for today's climate the value of SDDI will be zero for all model grid boxes, since the present day climatological value is always subtracted from the computed value. In this way climate experiments can be conducted to observe changes in SDDI from the present climate.

Most GCMs actually show increases in precipitation as one goes to a warmer climate[19]. However, the increase in potential evapotranspiration due to a warmer atmosphere is far greater, resulting in negative values for the SDDI, implying moisture deficiency.

Using climate change projections for continued exponential growth of atmospheric trace gases (scenario A)[20], Figure 4 shows the modeled drought occurrence as a function of time, averaged over all land points (June-August). Extreme drought conditions that occur less than 1% of the time in the control run increase in frequency to nearly 50% by 2060, the time the global mean temperature has increased by 4.2°C in the model.

Drought conditions initially appear at low latitudes in the tropics. The reason for this is that warm air has a larger moisture-holding capacity than cooler air. Thus whereas an increase in potential evapotranspiration in cool regions initially results in an

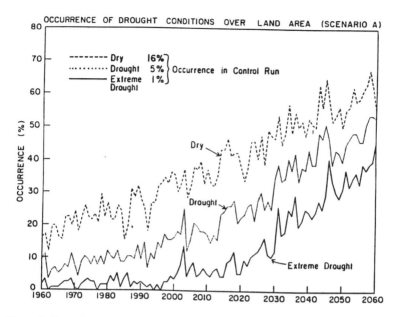

Figure 4. Drought occurrence as a function of time during June–August generated from the SDDI.

increase in precipitation, in warmer regions the increase in potential evapotranspiration results in a moister atmosphere, but a drier surface.

Given the historical data relating moisture balance to fire frequency[17], the above results would imply dramatic increases in the frequency of fires.

Fuel Loading

Vegetation can respond in two possible ways to climatic change. Either the species can migrate to more climatically favorable regions, or succession occurs resulting in new more adaptable species replacing the existing ones.

As a result of climate change vegetation zones can be expected to migrate poleward at a maximum rate of 100 km per 1°C of warming[21]. Future global warming is expected to occur at a rate of 0.5–1°C per decade. This implies that vegetation zones would have to migrate poleward 50–100 km per decade. This rate of migration is far too rapid for the majority of species[21], and therefore one would expect the onset of wilting and desiccation of vegetation, with the degradation of species from forest to shrubland to grassland.

In addition, the increase in the frequency of drought could significantly decrease photosynthesis while increasing respiration. A decline in the photosynthesis to respiration ratio will reduce productivity, and resistance to insects and disease, resulting in increased stress-induced mortality[22]. This would result in an acceleration in the fuel buildup.

Ignition Sources

We have already discussed the effect of climate change on lightning frequencies. However, it is also likely that anthropogenic fires will increase in the future due to increases in land use, as well as increases in population around the globe.

The changes in the above three factors that influence forest fires, as a result of climate change, all appear to heighten the probability of fires in the future. Increased drought, moisture stressed fuels and higher frequencies of lightning storms all imply that future climate change could result in increased fire frequencies, increased fire intensities and an increase in the length of the fire season.

SUMMARY AND DISCUSSION

In this paper the effect of future climate change on two related natural hazards, lightning and forest fires, has been studied.

Lightning frequencies have been shown to increase by more than 40% over continental regions for a 2×CO_2 climate (4.2°C global warming in the GISS model). In some locations the increase is greater than 100%.

In addition to lightning activity, the frequency of fires also depends on climate and fuel loading. The effect of future global warming on these two parameters was also examined. The GISS GCM shows that the frequency of severe drought increases from 1% in today's climate to nearly 50% by the year 2060, for a "business as usual" scenario.

Vegetation cannot be expected to migrate at the same rate as predicted climate change, which will result in the wilting and desiccation of vegetation, leading to an increase in fuel buildup.

These factors all point toward an increase in fire frequency, fire severity and the length of the fire season, although these predictions will obviously depend on the rate and magnitude of future climate change.

Unlike the present climate where natural fires are fairly rare in the tropics, the model calculations imply that the largest increases in both lightning activity and drought conditions will occur in the tropics. This would dramatically affect the frequency of lightning-fires in those regions.

It has been shown from paleoclimatic data that during past warm dry periods fire frequencies were higher than during cool moist periods[17]. The fire records also show that the intensities of fires were greater during warmer periods.

A change in fire frequency can have dramatic effects on the composition of ecosystems, since certain species are more adaptable to short fire intervals than others. By changing the frequency of fires in a specific forest stand, the complete composition of the forest could change[23,21]. In fact, the direct effects of climate change on forest composition may be swamped by the more drastic effects of climate change on forests due to changes in fire frequency.

An aspect not considered in this paper is that of climate feedbacks induced by lightning and forest fires. Lightning is a major source of NO_x in the atmosphere, which can affect the concentration of O_3 in the troposphere[1]. Increases in lightning frequencies could therefore potentially affect the concentration of greenhouse gases in the atmosphere. Forest fires result in the emissions of both trace gases and particulate matter that can influence regional and perhaps global climate[3]. In addition, changes in albedo due to fires, and succession to different vegetation types, can also provide feedbacks on the climate system[24].

It therefore appears that future climate change as a result of increasing concentrations of greenhouse gases could potentially have major implications for both global lightning activity and forest fires. Furthermore, these possible changes in lightning and forest fires could act as positive feedbacks on the climate system.

REFERENCES

1. E. Franzblau, C. J. Popp. *J. Geophys. Res.* **94**:11089 (1989).
2. F. J. W. Whipple. *Quart. J. Roy. Met. Soc.* **55**:1 (1929).
3. D. E. Ward, W. M. Hao. In *Proceedings of Air and Waste Management Association*, Vancouver, Canada (1991).
4. D. G. DeCoursey, W. L. Chameides, J. McQuigg, M. H. Frere, A. D. Nicks. In *Thunderstorms in human affairs*. E. Kessler, ed. University of Oklahoma Press, Norman (1983), p 67.
5. J. Hansen, G. Russell, D. Rind, P. Stone, A. Lacis, S. Lebedeff, R. Ruedy. *Mon. Wea. Rev.* **111**:609 (1983).
6. R. Lhermitte, E. R. Williams. *Rev. Geophys. Space Phys.* **21**:984 (1983)
7. E. R. Williams. *J. Geophys. Res.* **90**:6013 (1985).
8. C. Price, D. Rind. *J. Geophys. Res.* (1992) in press.
9. D. P. Jorgensen, M. A. LeMone. *J. Atmos. Sci.* **46**:621 (1989).
10. C. P. R. Saunders, W. D. Keith, R. P. Mitzeva. *J. Geophys. Res.* **96**:11007 (1991).
11. R. E. Orville, R. W. Henderson. *Mon. Wea. Rev.* **114**:2640 (1986)
12. T. Takahashi. *Geophys. Res. Lett.* **17**:2381 (1990).
13. B. N. Turman, B. C. Edgar. *J. Geophys. Res.* **87**:1191 (1982).
14. P. N. Fearnside. In *Fire in the tropical biota*. J. G. Goldammer, ed. Springer Verlag, Berlin (1990), p. 106.
15. J. S. Barrows. In *Lightning fires in southwestern forests*. USDA Forest Service, Ogden, Utah (1978).
16. L. S. Bradshaw, J. E. Deeming, R. E. Burgan, J. D. Cohen. In *The 1978 national fire-danger rating system*. Technical Report, General Technical Report INT-169 (1983).
17. J. S. Clark. *J. Ecolo.* **77**:989 (1989).
18. D. Rind, R. Goldberg, J. Hansen, C. Rosenzweig, R. Ruedy. *J. Geophys. Res.* **95**:9983, (1990).
19. S. L. Grotch. In Proceedings of DOE Workshop on Greenhouse-Gas-Induced Climate Change, Amherst, Mass. (1989).
20. J. Hansen, I. Fung, A. Lacis, D. Rind, S. Lebedeff, R. Ruedy, G. Russell. *J. Geophys. Res.* **93**:9341 (1991).

21. K. C. Ryan. *Environ. Int.* **17**:169 (1991).
22. R. H. Waring. *Bioscience* **37**:569 (1987).
23. J. T. Overpeck, D. Rind, R. Goldberg. *Nature* **343**:51 (1990).
24. T. W. Jurik, D. M. Gates. *J. Clim. Appl. Met.* **22**:1733 (1983).

THE GLOBAL ELECTRICAL CIRCUIT AS GLOBAL THERMOMETER

Earle Williams and Stan Heckman
Department of Earth, Atmospheric, and Planetary Sciences
Massachusetts Institute of Technology
Cambridge, MA 02139

ABSTRACT

Local observations in the tropics show that lightning activity increases strongly with the wet bulb potential temperature, θ_w, of boundary layer air. The surface wet bulb temperature controls the vigor of convection and the accumulation of ice phase condensate in the mixed phase region of tropical convection, the apparent seat of lightning activity. The Schumann resonance is a global electromagnetic phenomenon driven by global lightning activity. The Schumann resonance amplitude is shown to follow the fluctuations in mean temperature for the entire tropical belt with a sensitivity consistent with the local observations. The manifestation of the ENSO (El Niño–Southern Oscillation) signal in the Schumann amplitude at a single location is consistent with a synchronous warming and cooling over the entire tropical belt.

INTRODUCTION

In recent years there has been considerable interest in temperature fluctuations in the earth's atmosphere. The temperature variability on a global scale, based on a century's worth of surface (dry bulb) temperature records, amounts to several tenths of $1°C$[1,2,3]. On an absolute scale of temperature, the reported global warming amounts to only a few tenths of one percent.

A worthwhile goal in global change research is the identification of measurable physical parameters which are non-linearly dependent on the fluctuations in atmospheric temperature so that some "gain" in the detection of these subtle temperature changes can be achieved. This paper is concerned with the idea that the Schumann resonance, a global electromagnetic phenomenon driven by worldwide lightning activity, is one such measurable parameter. Lightning is linked with cloud electrification and the accumulation of ice particles in the upper troposphere. This non-linear process is controlled by buoyancy, the modest departures from hydrostatic equilibrium caused by temperature differences of the order of $1°C$. Buoyancy in turn is controlled primarily by surface air temperature, the principal datum in current studies of global change.

TROPICAL PREDOMINANCE OF GLOBAL LIGHTNING ACTIVITY

Convection is systematically deeper and more frequent in the tropics than at higher latitudes. This behavior is essentially the result of the pole-to-equator temperature increase and the Clausius-Clapeyron relationship. Observations disclose that lightning activity increases dramatically with the depth and vigor of convection[4]. Figure 1 shows the latitudinal distribution of lightning at midnight from a polar-orbiting satellite, taken from Orville and Henderson[5], and illustrates the dominant role of the tropics in

© 1993 American Institute of Physics

78   The Global Electrical Circuit as Global Thermometer

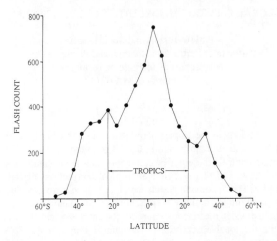

Figure 1. Latitudinal distribution of lightning from space (from Orville and Henderson, 1986)[5] showing a dominant contribution form the tropics (±23°).

terrestrial lightning. Approximately two of every three lightning flashes occur in the latitude interval ±23°. A similar statistic is found in the global lightning data set of Kotaki[6] which includes daytime observations.

DEPENDENCE OF LIGHTNING ACTIVITY ON SURFACE TEMPERATURE IN THE TROPICS

Observations at a large number of land stations in the tropics show that lightning activity increases nonlinearly with surface air temperature. In the following comparisons, wet bulb temperature rather than dry bulb temperature will be used, because the former records simultaneously the effect of temperature and moisture, both of which are important to the thermodynamics of moist convection. Needless to say, many parameters in addition to surface temperature influence the development of deep convection and lightning on any given day and place (e.g., temperature inversions, dry layers, wind shear and lateral gradients in surface temperature). Consequently, single parameter (i.e., wet bulb temperature) correlations with lightning flash counts are quite variable on a day-to-day basis. When monthly mean values are considered, however, consistent lightning-temperature dependencies emerge and show reasonably well defined seasonal variations. Examples of these dependencies for stations in Darwin, Australia, Kourou, French Guyana, and Orlando, Florida are shown in Figures 2–4 respectively.

The interpretation of the sensitive relationship between lightning and wet bulb temperature is based on observations of the thermodynamic structure of the tropical

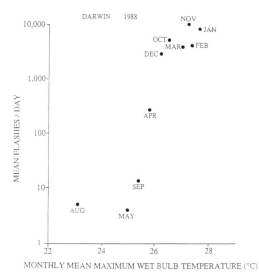

Figure 2. Monthly lightning counts for Darwin, Australia (12°S) versus monthly mean maximum wet bulb temperature for 1988.

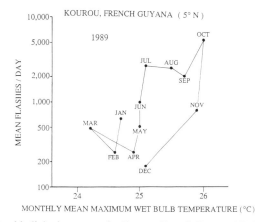

Figure 3. Monthly lightning counts for Kourou, French Guyana (5°N) versus monthly mean maximum wet bulb temperature for 1989.

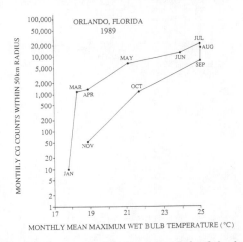

Figure 4. Monthly lightning counts (cloud-to-ground) for Orlando, Florida (28°N) versus monthly mean maximum wet bulb temperature for 1989.

atmosphere[7] and the assumption that ice particle collisions and charge separation by differential particle motions are causing lightning.

As far as thermodynamic structure is concerned, observations at many locations in the tropics show that the energy stored in the atmosphere and made available for convection, lightning and the excitation of Schumann resonance (Convective Available Potential Energy (CAPE)) is determined largely by the wet bulb potential temperature θ_w of boundary layer air. Changes in shape of the temperature profile play only a secondary role in determining CAPE. In an atmosphere with a fixed temperature profile but with a variable θ_w of boundary layer air, the positive area on a tephigram increases monotonically (and approximately linearly) with increases in the wet bulb adiabat. For calculations of CAPE in the tropical atmosphere which incorporate the ice phase (consistent with the assumption linking ice and lightning production), a 1°C increase in wet bulb potential temperature in a tropical wet season is equivalent to about 1000 joule/kg of CAPE[7]. Observations also disclose that CAPE vanishes in the tropical atmosphere when θ_w drops below about 23°C for tropical land stations.

According to parcel theory, the maximum achievable updraft velocity in deep convection is $\sqrt{2 \cdot CAPE}$. This dependence is not sufficient to explain the strongly nonlinear relationships evident in Figures 2–4. Radar observations of deep tropical convection[8,9] provide another clue, and suggest that the real nonlinearity in the lightning-temperature relationship involves the growth of ice particles in the mixed phase region (where 0°C ≥ T ≥ –40°C) of the updraft. Comparisons between monsoon (when $\theta_w \approx$ 24–25°C) and "break-period" continental-type (when $\theta_w \approx$ 26–28°C) convection show hundred-fold (20 dB) differences in radar reflectivity in the mixed phase region and order of magnitude differences in lightning activity. Williams et al.[9]

suggest that modest changes in updraft velocity can affect large changes in the mass of ice-phase condensate and the gravitational energy available for an ice-based charge separation process. The mechanism, which is still poorly understood in its microphysical details, may provide the documented amplification from temperature to lightning.

THE GLOBAL CIRCUIT AS GLOBAL TROPICAL THERMOMETER

The evidence that lightning responds sensitively to temperature at many sites throughout the tropics suggests a global response. The global electrical circuit, which integrates the electrical effects of disturbed weather the world over[10] is expected to provide a natural global thermometer. This global spherical capacitor is a resonant cavity for extremely low-frequency electromagnetic waves excited by global lightning, a phenomenon predicted by Schumann[11] and now named the Schumann resonance (SR). The theory of SR is described in Bliokh et al.[12] and Polk[13]. The fundamental mode of the Schumann resonance is a standing wave in the earth-ionosphere cavity with a wavelength equal to the circumference of the earth. Two complications of SR which work against the presence of a globally representative signal at a single measurement site are 1) the nodal structure inherent in resonant wave phenomena and 2) the changes in cavity shape caused by changes in ionization in the upper atmosphere (largely a local diurnal effect). The most compelling evidence to date that globally representative signals can be extracted from measurements at single stations is found in the recent studies of diurnal variations by Sentman and Fraser[14], who corrected for local ionospheric effects, summed contributions from several simultaneously recorded resonant modes, and thereby produced records at distant sites showing considerable similarity. The corrected diurnal records show distinct peaks which can be associated with the three major tropical zones of convection.

Evidence for a global signal on time scales of months to years

As a test of the idea the SR should behave as a sensitive global tropical thermometer, a five-and-a-half-year time series of SR magnetic field data[15] was examined. This record from Kingston, Rhode Island (71°W, 41°N), is nearly continuous and the magnetic field measurements are well calibrated. Monthly mean values of the magnetic field H (for the fundamental ~8 HZ resonant mode) were extracted from the continuous daily records and plotted in Figure 5 along with the same monthly mean fluctuations in surface (dry bulb) temperature DT for the entire tropics[3] for corresponding months. A correlation plot of H and ΔT using the same data in Figure 5 is shown in Figure 6.

The SR amplitude follows the temperature variation quite closely in Figure 5, particularly for the long period variation. Warmer periods are associated with enhanced magnetic field amplitude; cooler periods with suppressed amplitude. The long period (~40 month) temperature anomalies show a global coherency[16,3] and are associated with the El Niño–Southern Oscillation phenomenon. The annual signal in Figure 5 appears to be significantly smaller, but there is some tendency for minima in January and maxima in July. Part of the noisiness in the comparisons of magnetic field and temperature on seasonal time scales may arise from the latitudinal migration of tropical

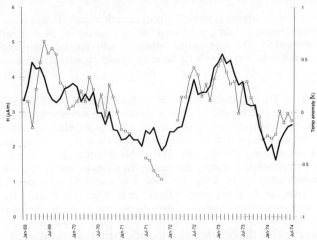

Figure 5. Time series of monthly tropical surface air temperature anomaly, solid line (from Hansen and Lebedeff, 1987)[3] and monthly mean magnetic field for the fundamental mode (8 HZ) of the Schumann resonance in Kingston, Rhode Island, symbols (from Polk, 1969–1975).

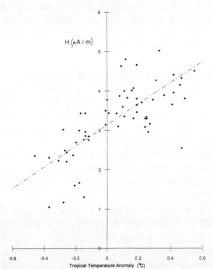

Figure 6. Correlation plot of monthly Schumann resonance amplitude and monthly tropical temperature anomaly (same data as in Figure 5).

lightning sources which significantly affect the amplitude at single stations when only the fundamental resonant mode is measured[13].

The slope of the correlation plot in Figure 6 gives the sensitivity of the global "thermometer." The trend in these results shows an approximate doubling of magnetic field amplitude for a 1°C increase in temperature. According to theory[17], the instantaneous global lightning activity is proportional to the energy density in the resonant cavity, and hence is proportional to amplitude squared. The four-fold increase in lightning per 1°C is consistent with the sensitivity determined on the basis of local measurements in Figures 2 and 3 for tropical stations, and reinforces the connection between a local and a global phenomenon. Here we assume that fluctuations in surface dry bulb temperature (studied by Hansen and Lebedeff[3]) are commensurate with the more physically meaningful wet bulb temperature fluctuations.

DISCUSSION AND CONCLUSION

Evidence has been presented that global lightning activity is dominated by convection in the tropics and that lightning increases strongly with surface wet bulb temperature. The SR provides a natural integration of global lightning activity and the observations show that SR increases with temperature on a global tropical scale in a manner consistent with the observed sensitivity of lightning to temperature in local measurements, thereby providing a sensitive global tropical "thermometer." The good correlation between SR amplitude in Kingston, R.I., and fluctuations in global tropical surface temperature strongly suggests that single station measurements contain globally representative signals. Single station measurements in noise-free areas can therefore serve as a valuable real-time diagnostic of both temperature variability and deep convection in the tropical atmosphere. Improvements in achieving global signals are possible by averaging over higher order modes, as Sentman and Fraser[14] have done in diurnal studies.

The amplification which makes SR a sensitive tropical thermometer is believed to originate in the interaction between ice microphysics and temperature-dependent buoyancy forces during deep convection. Lightning rate is a sensitive indicator of the transport of ice to upper levels of the atmosphere. The ice particles which are believed to participate in the electrification process aloft[18] also become the major players in the radiative feedback effects which sensitively regulate the temperature of the tropical atmosphere. Use of the Schumann resonance may therefore prove to be a useful device for understanding and quantifying these feedback effects.

The mechanism proposed for the SR "thermometer" depends critically on the presence of convective available potential energy (CAPE) in the tropical atmosphere, and the correlated behavior between the global circuit and tropical temperature fluctuations on a wide range of time scales reinforces this result. If the tropical atmosphere continually adjusted to a state of moist neutrality (i.e., zero CAPE), as is assumed in a number of cumulus parameterization schemes in current general circulations models of the atmosphere, it is unlikely that the tropics would provide any contribution to the Schumann resonance. The closest approximation to a moist neutral condition in the real atmosphere is seen in the well-established tropical monsoon when little or no lightning is observed[8,9]).

Finally, one can speculate about trends of lightning activity on longer time scales. The five-year record in Figure 5 is far too scant to make definitive statements about lightning's dependence on climate change. However, the correlated behavior of CAPE, temperature and lightning at the longest time scales available makes it plausible that global lightning activity will increase substantially in a warmer climate.

ACKNOWLEDGMENTS

Long-standing discussions with T. Madden, R. Markson, C. Polk, and D. Sentman on SR and V_I have been very valuable. C. Polk of the University of Rhode Island at Kingston provided vital access to the SR data set. Graduate students S. Heckman, N. Renno, D. Boccippio, and E. Rasmussen performed analyses of electrical data sets and tropical soundings. H. Wilson of NASA/Goddard Institute for Space Studies kindly provided the tropical temperature anomalies. The lightning–wet bulb relationships at low latitudes were established through the generous assistance of A. Aka, A. Bhattacharya, I. Butterworth, J. Core, R. de Araujo, F. de la Rosa, M. Ianoz, R. Jayaratne, A. Liew, D. Mackerras, R. Orville, P. Richard, and H. Torres. Ongoing studies of tropical lightning in DUNDEE have been supported by the Physical Meteorology section of the National Science Foundation on Grant ATM-8818695 with the assistance of Dr. R. Taylor.

REFERENCES

1. J. K. Angell. *Mon. Wea. Rev.* **114**, 1922–1930, 1986.
2. P. D. Jones, T. M. L. Wigley, and P. B. Wright. *Nature* **322**, 430–434, 1986.
3. J. Hansen, and S. Lebedeff. *J. Geophys. Res.* **92**, 13345–13372, 1987.
4. E. R. Williams. *J. Geophys. Res.* **90**, 6013–6025, 1985.
5. R. E. Orville and R. W. Henderson. *Mon. Wea. Rev.* **114**, 2640–2653, 1986.
6. M. Kotaki, I. Kuriki, C. Katoh, and H. Sugiuchi. *J. of Radio Res. Lab.*, Tokyo, Japan, **28**, 49–71, 1981.
7. E. R. Williams and N. Renno. *Mon. Wea. Rev.*, to be published November 1992.
8. S. A. Rutledge, E. R. Williams, T. D. Keenan. *Bulletin of the American Meteorological Society*, **73**, 3–16, 1992.
9. E. R. Williams, S. A. Rutledge, S. G. Geotis, N. Renno, E. Rasmussen, and T. Rickenback. *J. Atmos. Sci.*, **49**, 1386–1395, 1992.
10. F. J. W. Whipple. *Quart. J. Roy. Met. Soc.* **55**, 1–17, 1929.
11. W. O. Schumann. *Z. Naturforsch*, Teil A7, 149–154, 1952.
12. P. V. Bliokh, A. P. Nikolaenko, and Yu. F. Filippov. *Schumann resonances in the earth-ionosphere cavity*, Peter Perigrinus, London, 1980.
13. C. Polk. Schumann resonances, in *CRC handbook of atmospherics*, Hans Volland ed., Vol. 1, CRC Press, Boca Raton, Fla., 1982.
14. D. D. Sentman, and B. J. Fraser. *J. Geophys. Res. (Space Physics)* 15973–19584, 1991.
15. C. Polk. *Geophysics and Space Data Bulletin*, Space Physics Laboratory, AFCRL, Hanscom AFB, Mass., 1967–1974.
16. A. H. Oort. Global atmosphere circulation statistics, 1958–1973, NOAA Prof. Pap. 14, U.S. Dept. of Commerce, April, 1983.

17. M. Clayton and C. Polk. In *Electrical processes in atmospheres*, 440–449, H. Dolezalek and R. Reiter, Eds., Steinkopff, Darmstadt, 1977.
18. E. R. Williams. *Geophys. Res.* **94**, 13151–13167, 1989.

THE RELATIONSHIP BETWEEN ENSO EVENTS AND CALIFORNIA STREAMFLOWS

Ercan Kahya and John A. Dracup
4532 Boelter Hall
Civil Engineering Department, UCLA
Los Angeles, CA 90024

ABSTRACT

This study examines the relationships between types of extreme negative index phases of the Southern Oscillation and unimpaired stream volumes over California. Of particular interest in this investigation is the identification of subregions that appear to have consistent and strong El Niño/Southern Oscillation (ENSO) related streamflow signals. The analysis of 40 stations in the California by an empirical approach indicates that apparent Type 1 ENSO-related streamflow responses exist explicitly within the southern California and implicitly within the central and the northern California. The overall impact of the Type 1 ENSO events on the state is to produce a wet spell in streamflow commencing at the onset time of warm events continuing to the summer or early fall season of the year that follows the episode year. The relevant dynamical mechanisms concerning the observed relationships are discussed. Once an El Niño event sets in and the type (profile) of the event is identified by observing the zonal distribution of sea surface temperature in the equatorial Pacific during the summer, a long-term prediction potential one or two seasons in advance may be available for California. The results of this analysis, confirming previous precipitation studies, also exhibit regionally specific midlatitude hydrologic responses to the tropical ENSO forcing.

INTRODUCTION

The tropical El Niño/Southern Oscillation (ENSO) phenomenon is known to be a major disturbance on the large-scale atmospheric circulations through a marked eastward shift of the active area of convection in the Pacific ocean. Both the oceanic and atmospheric aspects of ENSO events associated with many climatic and geophysical parameters have been extensively analyzed over low latitudes and high latitudes on regional and global domain[1,2,3]. Of particular interest in this study are the relationships between ENSO and streamflow discharges over the midlatitudes of the Northern Hemisphere. The extratropical teleconnections in conjunction with warm events are well reviewed in the recent studies[4,5].

The spatially consistent streamflow fluctuations during the lifetime of warm events in the Southeastern and in the Pacific Northwest (PNW) regions have been recently demonstrated[6]. The existence of ENSO-related responses previously found in the Gulf region for precipitation[7] has been confirmed by the results of another hydrologic parameter. The PNW was not designated as a region where precipitation is sensitive to ENSO events, however it does appear to be a core area for streamflow. A complete analysis[8] using 1009 stream gauging stations throughout the U.S. revealed four core areas that are assumed to have strong consistent ENSO-streamflow relationships, namely South Atlantic Coast (SAC), Gulf of Mexico (GM), Central North (CN), and PNW. The typical seasonal ENSO signal for each region may be expected as a dry period of January–December for GM, a wet period of April–September for CN, and a wet period of January–December for PNW during the ENSO year. For the SAC region, the signal seems to be a wet period from January to May during the subsequent year.

The interpretations of ENSO influences on precipitation on the West Coast are disputed in the literature. For example, various indices of ENSO events and California rainfall are not well correlated over a long period, although the 1982–83 episode had great impacts on California[8]. Ropelewski and Halpert[7] did not find any indication of linkages in California due, possibly, to inadequate data coverage for the region and to the nature of one of their objectives, which was the identification of a large candidate geographical region. In contrast, some studies show ENSO linkages over California[10]. These mixed conclusions about the ENSO linkages to California rainfall can be attributed to the discrepancy of the various types of development in the morphology of individual warm events and the climatic complexity of the state. Therefore, precipitation responses to ENSO phenomena are regionally specific[11]. The dispute is due to the fact that correlations between conditions over the western United States and El Niño are not usually very obvious, since severe and mild winter weather have occurred during warm events with more or less equal frequency[5].

A few studies have been concerned with streamflow's relation to ENSO occurrences, but none of them was specifically focused on the associations between types of ENSO (or even ENSO occurrences in general) and California streamflows. When the April–September seasonal streamflow and the winter precipitation records are correlated to the July–September averaged Southern Oscillation Index (SOI) time series for the eleven western United States, the weakest correlations occurred in northern California[12]. This is consistent with the results of another study[13] that suggests significant negative correlations between December–August streamflow anomalies and the SOI in the southern part, but not in the northern part of California. In part this is due to the fact that rainfall in Northern California is directly associated with midlatitude circulations and is less influenced by ENSO events[11].

Our objective in this study is to analyze the California streams on the monthly scale in relation to the types of ENSO events based on the definition of Fu et al.[14]. The results presented here offer not only a rough picture of streamflow anomaly behaviors during a typical Type 1 ENSO, but also potential long-term forecasting tools for water resources management.

DATA SET AND METHODOLOGY

The monthly unimpaired streamflows used in this study were compiled by Wallis, Lettenmaier and Wood[15]. The data set contains 39 high-quality stream gauge records in California and one in Nevada (very close to the central California border), each with 41 years of observation (1948–1988). Each record contains a total of nine moderate/strong low-index phase of the SO years (El Niño), namely 1951, 1953, 1957, 1965, 1969, 1972, 1976, 1982, and 1986, based on the earlier studies[16,1]. Another total of nine La Niña years in our data are 1950, 1955, 1956, 1964, 1970, 1971, 1973, 1975, and 1988 in which the SOI (Tahiti-Darwin) remained in the upper 25% of the distribution for five months or longer[17]. According to the categorization[14], five of nine ENSO episodes (1951, 1957, 1965, 1972, 1982) are Type 1, two of nine episodes (1953, 1969) are Type 2, and only the 1976 episode is Type 3.

Fu et al.[14] described three types (profiles) of ENSO patterns based on zonal sea surface temperature distribution within the equatorial Pacific band of 4°N–4°S, 120°E–

80°W. We were more interested in Type 1 ENSO events for California, which can be described during the June-August season as: i) much warmer than normal ocean surface water east of the date line, ii) close to normal condition in the western Pacific, and iii) the warmest water stretching from the eastern Pacific to 150°–160°W longitudes.

An empirical methodology is used, which is described by Ropelewski and Halpert[7,18] and applied by Kahya and Dracup[8] with some changes and extensions, to search ENSO-streamflow relationships over the contiguous United States. The method of analysis involves two main phases. It starts with the transformation of monthly streamflows to percentile ranks and the construction of a 24-month warm event composite, which is based on the number of ENSO episodes under consideration, at each station. In order to account for the spatial variability and the skewness of the data, a percentile rank method was used. This method is better than common standardization procedures as a data presentation when averaging over a spatial domain[19]. The first harmonic extracted from such a composite is assumed to be the ENSO signal appearing in the streamflow anomaly. The amplitude and phase of the harmonic indicate the strength of the relationship and time of maximum anomaly within the ENSO cycle, respectively. The statistical significance of the amplitude is assessed by Schuster's test for autocorrelated series[21]. The vectorial display of the harmonics, calculated from each station's composite, over a map enables us to subjectively identify some regions that are the areal extents of ENSO influence on the streamflow field. The vectorial coherence (or constancy) is calculated for each region: if this is larger than 0.80, then that region is named as a candidate region and is subjected to further analyses.

The second phase of the analysis is carried out within each candidate region. First, the time series of monthly original streamflow volumes for each station, which are included in a candidate region, are transformed to percentiles based on the log-normal frequency distribution. The quantiles of a hydrometeorological variable, based on the proper theoretical frequency distribution, measure the relative importance of the variable amounts for a given month or season[20]. Then, ENSO composites are constructed for each station and averaged within the region. As a result, an *aggregate* ENSO composite is obtained to detect subjectively a single *season* within the ENSO cycle. Finally, the index time series, abbreviated as ITS, that are the spatial averaged percentiles for the region are plotted against the season detected and examined for the consistency of the ENSO-related streamflow signal.

RESULTS

To better demonstrate our motivation in this study, a part of Figure 3 of Kahya and Dracup[8] illustrates the harmonic vectors map of California based on compositing nine ENSO episodes for each stream gauging station. As seen, there is no evidence of a striking coherence in phases and amplitudes, that is a group of vectors pointing the similar direction. However, the subregional ENSO signals in California come to light when compositing based on the types (profiles) of warm episodes, which is the purpose of the present study.

In the first phase of the analysis, three types of ENSO composites have been constructed at each station and have been separately subjected to the harmonic analysis. Thus three sets of the amplitude and phase angle of the first harmonics are computed for each of 40 streams in California. The maps of harmonic dials are plotted in Figure 1 and no spatially coherent subregions are observed for the case of Type 2 and Type 3

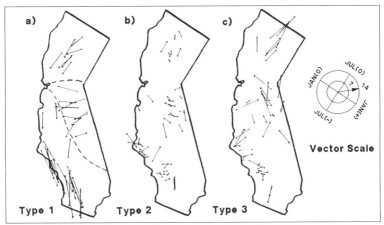

Figure 1 a) Streamflow station vectors based on the 24-month harmonic fitted to Type 1 ENSO composite of the seven events. The outlined areas of coherent response, so-called subregions, were selected for the analyses in Part 2. b) Same as in a), except for the compositing bases on the two Type 2 ENSO events and no identified subregions. c) Same as in b) except for one Type 3 ENSO event. The symbols (–), (0), and (+) in the vector scale indicate the event year, the year before, and the year after, respectively.

episodes. It is self-evident that the map of Type 1 ENSO delineates three distinct subregions where a group of vectors have almost the same direction and magnitude. Even when comparing Figure 1a with the part of California of Figure 3 of Kahya and Dracup, there is a similarity in terms of the phase, but not the magnitude of the dials. It is easy to make relevant mathematical comments in the following manner. In the case of the earlier study, the computation of each individual month within the 24-month ENSO composite at a particular station requires first the summation of nine values corresponding to that month. The structure of streamflow anomalies during the period of an event varies from episode to episode and dramatically from one type of ENSO occurrence to another (e.g., sign reversed). Therefore the summation contains plus and minus values with reference to fiftieth percentile, median, for a single member of the composite. In contrast, in the case of Figure 1, the summation in the composite of Type 1 ENSO consists of mostly one-signed values resulting in a large absolute numerical value, thus resulting in large values in magnitude for each member of the composite. The remaining of the study will be only concerned with Type 1 ENSO events.

The steps of the second phase of the analysis have been performed for three subregions of California identified from Type 1 composites. Table I summarizes the total number of stations included in each subregion and the number of stations with respect to the degree of significance (DOS) based on the Schuster's test (see Kahya and Dracup[8] for details). Aggregate composites shown in Figure 2 indicate that the central and the northern regions seemingly have similar anomaly sequences within the Type 1 ENSO cycle, except for the enhanced flow during late winter/early spring and an

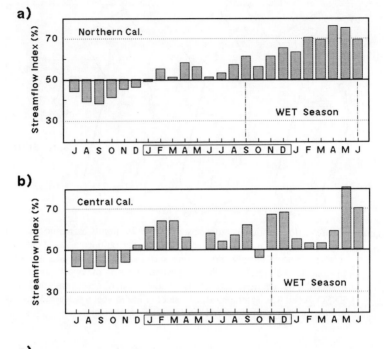
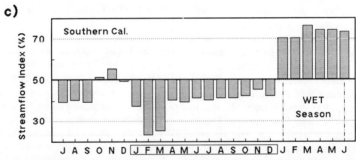

Figure 2. Aggregate composite based on the log-normal distribution for the three subregions, namely Northern, Central, and Southern California. The dashed lines delineate a season of possible Type 1 ENSO-related responses. The months in the box refer to ENSO or (0) year.

Table I
The number of stations in the candidate regions for which the degree of significance (DOS) of the first harmonics is smaller than specified limits, and the other properties of the subregions in California. The symbols (0) and (+) designates the year that a Type 1 ENSO event occurs and the subsequent year, respectively.

Regions	S. Calif.	Cen. Calif.	N. Calif.
Sta. Total No.	20	13	7
DOS < 5	16	–	–
DOS < 15	18	–	1
DOS < 20	19	–	4
DOS > 20	20	13	7
Coherence	0.985	0.904	0.980
Season	Jan(+)–June(–)	Nov(0)–June(+)	Sep(0)–June(+)
Consistency	86	100	86

indication of upward trend in Northern California. The lower diagram in Figure 2 illustrates a different structure from the aforementioned two regions so that year-long negative anomalies during the event year are followed by a period of highly wet conditions, which is coincident with the onset of the mature phase of a warm event. A single season representing the ENSO signal of Type 1 is selected for each region and is indicated by the dashed lines. To make sure that the seasons previously detected are germane to the hypothesized tropical thermal forcing on stream volumes, the subregional ITS are plotted to see whether consistent relationships exist between streamflow and Type 1 ENSO phenomena (Figure 3). In order to do so, the monthly time series of the percentiles of log-normal of the stations in each subregion are aggregated to obtain the relevant ITS with 492 (41×12) data points that is representative for that subregion. Then seasonal averages for every year based on the previous step are calculated and plotted over the yearly study period.

In addition to the five years listed in Section 2, two Type 1 ENSO episode years, 1968 and 1977, listed in Table 4 of Schonher and Nicholson are also marked as crosshatches in Figure 3. Overall, an immediate impression is the prevailing above-normal conditions over California during Type 1 events. For the northern part, the seasonal relation implied by Figure 2a is confirmed by 6 out 7 cases. Two out of 4 extreme wet conditions occurred during Type 1 El Niño years, which are categorically strong events[16]. For Central California, all seven episodes in the ITS (Figure 3b) give positive anomalies ratifying the seasonal signal shown in Figure 2b. In this region, three out of the seven wettest years in the 41-year period also coincided with the occurrences of Type 1 ENSO. Finally, the results in Figure 3c for Southern California are very impressive. Except the moderate 1965 episode, all others are consistent with the implication of apparent seasonal anomaly behaviors of Figure 2c. During the subsequent year of six episodes, the January-June averaged streamflow percentiles are all above the seventieth percentile. Even the three most extreme wet seasons (above the ninetieth percentile) in the 41-year period are observed in conjunction with Type 1 ENSO events. In this region, the number of vectors whose DOS is equal to or greater than 3% is 13 out of 20 and the vectorial coherence (0.985) is the highest among the three subregions.

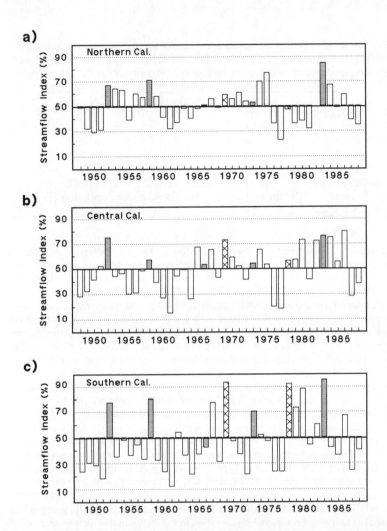

Figure 3. The index time series of streamflow for the three subregions based on the seasons detected in Figure 2. Streamflow is represented by the average of the log-normal percentiles for each of the stations within the area. Type 1 ENSO years are shown by the dark and cross-hatched bars. See the text for the details.

In actuality, the wet seasonal anomalies for all subregions in the following year of Type 1 ENSO actually persist to September of the same year. From the above evaluations, the Type 1 ENSO forcing modulates streamflows in the form of wet conditions over the whole state. This generalization is consistent with the results of precipitation studies that imply wet conditions for rainfall in California during all Type 1 episodes, except the case of 1965, which produced normal conditions statewide[11].

The probability of occurrence of wettest season during Type 1 warm events (Figure 3) by chance was tested using a binomial model. A *success* is defined as the occurrence of a wet season in any year with a positive departure from the median is equal to or greater than the seventieth percentile in an appropriate ITS. Thus, the definitions of the model parameters involved are: p is the probability that a success occurs in any year, n is the sample size (the number of year whose seasonal departures are equal to or greater than the seventieth percentile), m is the number of success out of n trials (the number of Type 1 ENSO years at which seasonal departures are equal or greater than the seventieth percentile), and P_m is the probability that a success happens exactly m times in n trials computed by the binomial distribution function. Table II summarizes the application of this test. The ITS of southern California region reflects a significant insight that highly wet streamflow conditions are expected during the years following Type 1 events.

DISCUSSION

The implied relationships presented in the previous section are possible when the extratropical atmospheric circulation is influenced by dramatic changes in the tropical Pacific ocean, such as the intensity, size, and position of the tropical heat source. This influence is realized in the form of either barotropic Rossby wave propagation or triggered barotropic modes of the atmosphere. Consequently a Pacific North American (PNA) type of circulation pattern, which alters the zonal flow, is set up over North Pacific and American sectors[4]. The observational studies have also shown the interaction between the extreme phases of Southern Oscillation (SO) and the PNA pattern in such a way that the strengthening of the PNA pattern is usual during the warm events as the reverse is true for the cold events[22]. The development of an upper-level anticyclonic couplet over the central equatorial Pacific is pronounced during the mature phase of ENSO. The anomalous westerlies on the poleward sides of the couplet intensify the subtropical jets that are displaced toward the equator by the presence of a deep Aluetian low. The jet on the Northern Hemisphere and associated low level westerlies bring subtropical moist air to the West Coast at the surface[5]. This phenomenon and the strong teleconnections between the low and high latitudes are more likely at the time midlatitude westerlies extend over the tropical heat source

Table II
The probabilities of the highly wet conditions associated with Type 1 ENSO events based on the binomial distribution. See the text for the definition of symbols.

Regions	p	n	m	P_m
S. Calif.	0.219	9	6	0.004
Cen. Calif.	0.171	7	3	0.081
N. Calif.	0.097	4	2	0.046

during the winter half of the year[23]. Moreover the eastward migration of the warmest water pool is more prone to be accompanied by the extension of equatorial westerlies to the east during the developing phase of Type 1 ENSO events[14]. In summary, the physical mechanism regarding to Type 1 ENSO events is based specifically on two main conditions that occur during only Type 1 episodes; the persistent June–August high SST anomalies into winter season and the presence of strong westerlies in the vicinity of the heat source[11].

The outcomes of this investigation, which are consistent with the brief dynamical mechanisms explained above, revealed i) that the strongest seasonal signal of the Type 1 warm events appearing as wet conditions is found in Southern California (this may be related to the deepening of the Aleutian low, the intensification of westerly jet streams, and the increase in frequency of Pacific storms), and ii) that the initiations of wet seasons correspond to the Northern Hemisphere winter season with some lag caused by land-surface processes.

In addition, we constructed an aggregate 24-month La Niña composite (not shown here) centered on the years of 1950, 1955, 1964, 1971, 1973, and 1975 for each subregion. For Southern California, the La Niña signal appears as a strong dry season with the same duration as in El Niño event. This sign-opposition in the streamflow anomalies during the apparent seasonal signal is expected. The weak north Pacific winter atmospheric circulation (which is defined as the pattern associating with the weakened and westwardly displaced Aleutian low, the smaller cutoff low in the Gulf of Alaska and the high pressure zone over the southeast) occurs at the same time with the cold phase of the SO, as the near-normal or strong north Pacific circulations dominate during the winter of Type 1 episodes[24]. These opposite type of circulations inherently bear different surface responses over the same area in the North American continent. However, the opposite sign of the seasons was not noticeable for the La Niña composites in the Central and Northern California.

CONCLUSION

The streamflow anomaly fluctuations during the evolution of Type 1 ENSO events have been considered in the study domain of California. Compositing based upon the nine most recent ENSO episodes (not discriminating by the characteristic features of the events) for California streams did not reveal satisfactory results, but compositing based on the five Type 1 ENSO events showed excellent association. California has three distinctive regions that are assumed to response to the tropical forcing in different manners. Among the identified subregions, Southern California exhibits a very strong and consistent Type 1 ENSO-related signal. A simple statistical model has been applied to test the significance level of the implied relations, and again Southern California has shown the highest level of significance. Also, the individual Type 1 ENSO composite at each station in this region has been represented by the highly significant first harmonic, thus supporting the assumption made that one maximum (or minimum) corresponds to the impact of the forcing within the event cycle. All this information indicates that streams in the southern part of the state are modulated by the Type 1 ENSO occurrences.

Although the central and the northern subregions have a high percentage of agreement (Table 2), their fluctuating magnitude of seasonal anomaly during the episodes suggests that they may be strongly modulated by other types of influence, such as topography and local circulation. The earlier studies also implied that precipitation responses in Central California may vary dramatically, that is, dry and wet conditions are equally likely to be produced by an ENSO event. Northern California has been previously indicated to be the region least influenced by an ENSO and solely related to midlatitude circulation systems. This is essentially consistent with the results of the present study, which gives an indication of regionally specific warm event–streamflow relationships over California. The regional aggregate composites in conjunction with the index time series might be utilized to anticipate the streamflow behavior to be expected by the summer of the year when a Type 1 ENSO phenomenon occurs.

REFERENCES

1. E. M. Rasmusson and T. H. Carpenter. *Mon. Wea. Rev.* **110**:354. 1982.
2. E. M. Rasmusson and T. H. Carpenter. *Mon. Wea. Rev.* **111**:517. 1983.
3. R. S. Bradley, H. F. Diaz, G. N. Kiladis, and J. K. Eischeid. *Nature* **327**:497. 1987.
4. B. Yarnal. *Prog. Phys. Geogr.* **9**:315. 1985.
5. S. G. Philander. In *El Niño, La Niña, and the southern oscillation.* Academic Press, San Diego. 1990. p. 293.
6. J. A. Dracup and E. Kahya. Proc. 16th Annual Climate Diagnostics Workshop. U.S. Govt. Printing Office, Washington, D.C. 1992. p. 359.
7. C. F. Ropelewski and M. S. Halpert. *Mon. Wea. Rev.* **2352**. 1986.
8. E. Kahya and J. A. Dracup. Submitted to *Water Resources Res.* 1992.
9. E. M. Rasmusson and J. M. Wallace. *Science* **222**:1195. 1983.
10. J. Michaelson and J. T. Dally. Proc. 8th Annual Climate Diagnostics Workshop. U.S. Govt. Printing Office, Washington, D.C. 1983. p. 140.
11. T. Schonh er and S. E. Nicholson. *Jour. of Climate* **1258**. 1989.
12. R. W. Koch, C. F. Buzzard, and D.M. Johnson. Presented at the Western Snow Conference, Juneau, Alaska. 1991.
13. D. R. Cayan and D.H. Peterson. AGU Monograph **375**. 1989.
14. C. Fu, H. F. Diaz, and J. O. Fletcher. *Mon. Wea. Rev.* **114**:1716. 1986.
15. J. R. Wallis, D. P. Lettermaier, and E. F. Wood. *Water Resources Res.* **27**:1657. 1991.
16. W. H. Quinn, D. O. Zopf, K. S. Short, and R. T. W. Kuo Yang. *Fish. Bull.* **76**:663. 1978.
17. C. F. Ropelewski and C. F. Jones. *Mon. Wea. Rev.* **115**:2161. 1987.
18. C. F. Ropelewski and M. S. Halpert. *Mon. Wea. Rev.* **115**:1606. 1987.
19. H. F. Diaz, R. S. Bradley, and J. K. Eischeld. *Jour. of Geoph. Res.* **94**:1195. 1989.
20. C. F. Ropelewski and J. B. Jalickee. 8th Conference on Probability and Statistics in Atmosphere Sciences, Hot Springs, Ark. 1983.
21. V. Conrad and L. W. Pollak. *Methods in climatology.* Harvard Univ. Press. M.A. 1950.
22. B. Yarnal and H. F. Diaz. *Jour. of Climatology* **6**:197. 1986.
23. J. D. Horel and J. M. Wallace. *Mon. Wea. Rev.* **109**:813. 1981.
24. W. J. Emery and K. Hamilton. *Jour. of Geoph. Res.* **90**:857. 1985.

DROUGHTS AND CLIMATE CHANGE

Ignacio Rodríguez-Iturbe
Instituto Internacional de Estudios Avanzados
P.O. Box 17606—Parque Central
Caracas, Venezuela
and
Ralph M. Parsons Laboratory
Department of Civil and Environmental Engineering
Massachusetts Institute of Technology
Cambridge, MA 02139, U.S.A.

ABSTRACT

Drought is a complex phenomenon whose central characteristic is the notion of water deficit. Droughts and desertification are intimately linked to specific components of the hydrologic cycle and thus their occurrence as well as their intensity in time and space can be greatly influenced by climatic changes. The primary forcing factors are changes in precipitation and evapotranspiration. They cause alterations not only in the mean conditions of water availability but very importantly also in the variability and seasonality of supply (and demand) of the resource. In addition to these direct effects, climatically induced changes in vegetation may have a considerable impact on the hydrologic cycle. As a feedback effect, the storage of heat and moisture in the soil is a key factor in determining the spatial and temporal character of climate. The strong coupling between the forcing factors and the soil moisture conditions over large spatial scales may undergo shifts and alterations of pronounced character as a result of climatic changes. The consequences of such changes can be of great importance for the persistence of droughts and the occurrence of desertification conditions.

INTRODUCTION: SOME KEY FEATURES OF DROUGHTS

Drought is a complex phenomenon which different disciplines define in different ways. All definitions agree in that a drought occurs when there is a moisture deficiency which results in a water scarcity for human activities and related earth ecosystems. The water deficit involved in a drought usually occurrs over large spatial scales ($> 10^5 \text{km}^2$) and over time scales of months and years. A well known and dramatic example of a major drought is the one in the sub-Saharan regions. Figure 1[18] shows the standardized time series of annual rainfall where it is observed that in the early part and in the end of the record the drought persists for over a decade.

In drought analysis it is not only the total amount of the water deficit integrated over periods of, say, one year that is of major importance. The timing of such deficit inside the annual period is also crucial. Thus the total amount of the resource may not show a major deficit but if all or most of the deficit is concentrated in a period where the moisture is most needed then one usually talks about the existence of a drought. A relatively minor shift in the temporal march of the precipitation or the snowmelt season may have very large and adverse consequences if it persists for several years.

Ignacio Rodríguez-Iturbe 97

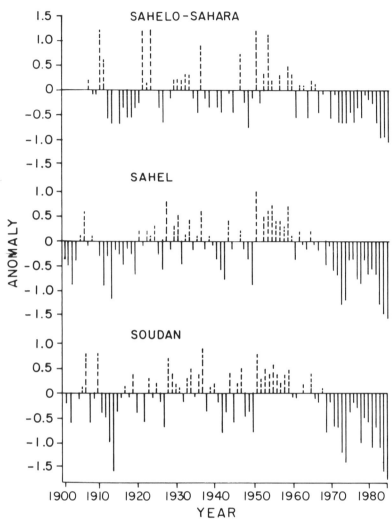

Figure 1. Standardized time-series of annual rainfall in sub-Saharan Africa (from Nicholson and Entekhabi, 1986).

With respect to the occurrence of droughts, Namias[17] writes: "As with many meteorological and climatological phenomena, there is never one, but multiple causes." It is important to distinguish between the *initiation* of the phenomenon and the maintenance of the anomaly, e.g., persistence, which will transform it into a drought. The *initiation* of droughts over large continental regions is generally related to anomalous features in the general circulation of the atmosphere. An example of this is large scale heating anomalies associated with shifts in seasonal sea-surface temperature patterns which may be a forcing factor for the development of general circulation features that initiate droughts over large continental regions (see Voice and Hunt[28] for a review). Many other examples may also be given which relate drought initiation with anomalies in the general circulation of the atmosphere. These causal connections are indeed quite important in the ladder of scientific explanations for the occurrence of droughts. Nevertheless the prediction and causal explanation of such anomalies is what becomes crucially important if one wishes to progress in the understanding of droughts. From the prediction viewpoint it is not at all clear that such a goal can be attained with sufficient anticipation to make it operationally useful. More probable is a causal explanation of the anomalous features in the general circulation of the atmosphere that may lead to the initiation of a drought, this is a goal which will likely be fulfilled. Nevertheless, the extreme complexity of the problem, its highly non-linear character and the uncertainty in the initial conditions which are of global character may lead to a chaotic type of behavior in the evolution of the system. Such behavior involves a system whose paths of evolution implies exponentially in time when starting from arbitrarily close initial conditions.

An anomalous general circulation feature may be a necessary condition but it is not a sufficient one for the development of a severe drought. Feedback mechanisms are what transforms an anomalous perturbation into a persistent and prolonged drought. The identification and understanding of these feedbacks is one of the main scientific challenges of the field. Thus vegetation cover may be reduced in response to an anomalous rainfall deficit; this results in a higher surface albedo; the land surface may then lose more radiative energy to space that it gains from absorbed solar radiation; the local region becomes a net sink of energy with respect to the global energy balance and the overlaying atmosphere develops into a state that favors the further inhibition of precipitation (Charney biogeophysical hypothesis[4]).

The water deficit characteristic of a drought manifests itself firstly in the precipitation signal, either in its overall intensity or in the shift of its monthly characteristics. The anomalies present on the precipitation signal propagate through other phenomena of vital hydrologic importance such as runoff, soil moisture, streamflow, groundwater levels, etc. Figure 2, after Chagnon[3], illustrates the propagation of perturbation in precipitation through the land branch of the hydrologic cycle. It is in this propagation process where a wide range of crucial feedbacks between the different phenomena and processes occur. These feedbacks may, in turn, affect the atmosphere and the precipitation signal and thus cause conditions which lead to a persistent and prolonged drought.

A key process in the above analysis is the soil moisture supply. This variable is crucial in the large scale surface heat and moisture balance between the land and the

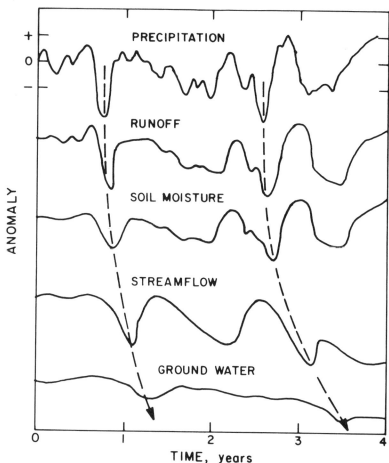

Figure 2. The propagation of a perturbation in precipitation amount through the land branch of the hydrologic cycle (after Chagnon, 1987, in McNab, 1989).

atmosphere. Soil moisture also has a much longer memory than the original input precipitation. This longer memory will be crucial, through large scale spatial feedbacks, to the temporal persistence of droughts.

The perturbations in key forcing parameters that influence the general circulation and initiate droughts may be of natural origin (e.g., shift in position of the intertropical convergence zone) or of anthropogenic character (e.g., increase of carbon dioxide). The structure of these perturbations in space and time may have a predominant stochastic structure or a strong deterministic component in their evolution. A most important anthropogenic perturbation is the atmospheric increase of CO_2. Its possible impact on droughts is the topic of the next section of this paper.

CO_2 INCREASE AND ITS POSSIBLE IMPACT ON DROUGHTS

Carbon dioxide and temperature have been very closely correlated over the past 160,000 years. Whether there is a causal relationship is still a hotly debated topic although it seems fair to say that a majority of atmospheric scientists tend to favor such a hypothesis..

The study of the impact of CO_2 increase in the land branch of the hydrologic cycle involves a very broad range of spatial and temporal scales among a large number of interrelated atmospheric and land surface processes. Global Circulation Models (GCMs) are the most commonly used tool for the study of the effects of different scenarios of CO_2 increase on variables related to drought problems. Although quite crude in many of their assumptions and parameterization GCMs are considered to provide an indication of the impact of CO_2 in hydrology when focusing on large spatial scales

An indication of the complexity which exists among the processes occurring in different places at global scale is given in Figure 3 from Koster[12]. It shows the region of influence of Sudd January evaporation and it was obtained by means of GCMs simulations "marking" the water evaporated in the region colored in black and following it in the model in space and time.

The change in soil wetness in response to an increase of atmospheric concentration of carbon dioxide is of major interest for the estimation of anthropogenic effects on droughts. As mentioned before, soil moisture plays a crucial role both in the feedbacks from the land processes to the atmosphere as well as in the direct links of these processes to agriculture and water supply. CO_2 induced changes of climate and hydrology are commonly evaluated by a comparison between two quasi-equilibrium climates of a GCM model corresponding to "normal" and above normal concentrations of atmospheric carbon dioxide. An example of the type of results obtained is shown in Figure 4 from Manabe and Wetherald[13] which gives the CO_2 induced change in soil moisture in summer, expressed as a percentage of soil moisture obtained when doubling CO_2 over a situation with the present amount of CO_2. The changes over extensive mid-continental regions of both North America and Eurasia are indeed dramatic.

The impact of climate fluctuations and changes like the ones described above extends over most of human activities. Schneider[23] discusses many of these impacts. As an example, just in rain-fed agriculture it has been estimated that to maintain crop

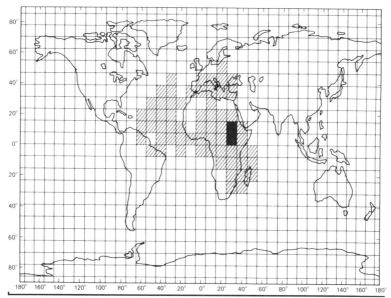

Figure 3. Region of influence of Sudd January evaporation (from Koster, 1985).

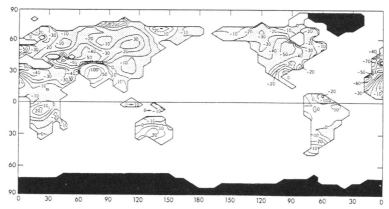

Figure 4. CO_2-induced changes in soil moisture in summer, expressed as percentage of soil moisture obtained when doubling CO_2 amount (from Manabe and Wetherald, 1986).

production at current local levels in the face of a 3°C warming, irrigation would have to increase by about 15% at 39 states distributed all over the United States. If this warming were combined with a 10% decrease in precipitation, then the increase in irrigation would go up to 26%. If the warming were combined with a 10% increase in precipitation, then the increase in irrigation would be only 7%. Those values may be much more extreme at a local scale rather than averaging over 39 states[20].

It is important to remark that results of the kind described in this section obtained from GCM simulations are clouded by a considerable degree of uncertainty. There is considerable disagreement between model predictions, even from the point of view of just the sign of the expected change at regional scale[11].

DEFORESTATION AND CLIMATE CHANGE

A clear example of the impact on regional climate change by the land branch of the hydrologic cycle comes from the large conversion of forests into pastures or annual crops. Extensive studies of the effects of large scale deforestation have been carried out using GCMs coupled with models of the biosphere. Shukla et al.[26] investigated the effects of Amazonian deforestation in the regional and global climate. They found dramatic changes when the Amazonian tropical forests were replaced by degraded grass in the model. These changes were associated with a significant increase in mean surface temperature and sizable decreases of annual evapotranspiration, precipitation, and runoff. The differences were greatest during the dry season and the deforested case was associated with larger diurnal fluctuations of surface temperature and vapor pressure deficit. Such effects have been observed in existing deforested areas in Amazonia. Figure 5 shows an example of the previously described results. Particularly important is to notice that the calculated reduction in precipitation (Figure 5c) is larger than the calculated decrease in evapotranspiration (Figure 5d). This implies that the dynamical convergence of regional moisture flux also decreased as a result of deforestation.

Nobre et al.[19] also found for the case of Amazonian deforestation that the length of the dry season increases with respect to the forested case. They point out that this could have very serious implications for the survival of tropical rain forests since they only occur when the dry season is very short or non-existent.

The above type of results quantitatively indicate that deforestation is clearly a process with strong feedback reinforcement and with direct implications to droughts which may ultimately lead to desertification conditions. Desertification is an extreme condition characterized by the diminution or destruction of the biological potential of the land which leads finally to desert-like conditions. The principal desertification processes are: degradation of the vegetative cover, accelerated water and wind erosion, and salinization and water logging. The resulting temporary or permanent loss of soil productivity is only one part of the environmental damage. Downstream flooding and lowering of groundwater tables alter local components of the hydrologic cycle[5].

ON THE PERSISTENCE OF DROUGHTS

A crucial characterization for the existence of a drought is the persistence in time of the water deficit. Thus, for both scientific and operational reasons, it is quite

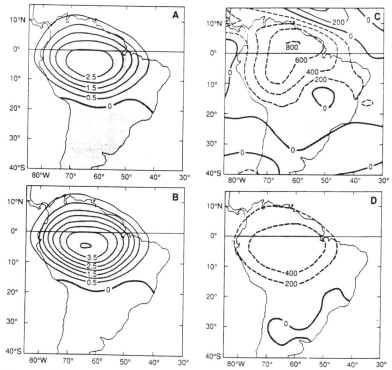

Figure 5. Differences between 12-month means (1 January to 31 December) of deforestation and control cases (deforested-control) for the South American sector: a) surface temperature increase in degrees centigrade; b) deep soil temperature increase in degrees centigrade; c) total precipitation changes (dashed lines indicate a decrease) in millimeters; and d) evapotranspiration decrease in millimeters. Model results were smoothed before plotting. (From Shukla, Nobre, Sellers, 1990.)

important to discern if the persistence of a fluctuation is due to the natural variability of the process itself or if it can be attributed to reasons external to the process and most importantly to anthropogenic factors. This is not an easy task in view of the very large natural variability of the processes involved.

When analyzing the risks associated with droughts and many other climatic fluctuations, it is important to emphasize that such fluctuations are frequently characterized by a strong persistence embedded in the structure of the processes. Thus it becomes necessary to objectively describe this persistence in a quantitative manner.

A useful manner to measure persistence is through the adjusted range.

$$\text{A.R.} = \frac{R^*(n)}{s(n)} \tag{1}$$

where

$$R^*(n) = \text{Sup } S^*(i) - \text{Inf } S^*(j)$$
$$i \in (0,1,2,\ldots,n), \quad j \in (0,1,2,\ldots,n)$$

and

$$S^*(k) = \sum_{i=1}^{k} X_i - \frac{k}{n}\sum_{i=1}^{k} X_i, \quad k = 1,2,\ldots,n$$

The standard deviation of n values is designated by s(n).

It is possible to give an intuitive meaning to the adjusted range. Think of X_i as the water flow into a reservoir in the ith year and $\sum_{i=1}^{n} X_i$ as the total water inflow in n years. If $(1/n)\sum_{i=1}^{n} X_i$ units are released each year, then $(k/n)\sum_{i=1}^{n} X_i$ will be the amount released in the first k years and $S^*(k)$ is the surplus or deficit relative to the amount released during the kth year and to the starting storage of the reservoir. The difference between the largest surplus and the greatest deficit gives the capacity that a reservoir must have to maintain a constant release equal to the mean of the river without overflows or deficits during the n year period. This concept, as a practical idea for the design of a reservoir, is not of much use, since it is necessary to know the flows that will occur in the next n years. However, the statistical behavior of $R^*(n)$ provides insight into the range of volumes that should be maintained. Hurst and colleagues[9,10] investigated the behavior of the statistic $R^*(n)/s(n)$, called the rescaled range, where s(n) is the standard deviation of the sample of n values. His studies included various geophysical phenomena in nature, such as tree rings, varves, precipitation series, and streamflow series. He noted that values of n versus $R^*(n)/s(n)$ plotted as straight lines on log-log graph paper. This indicates that the rescaled range follows an equation of the type

$$\frac{R^*(n)}{s(n)} = Kn^H \tag{2}$$

Hurst found that the average value of H was 0.73 and the standard deviation was 0.08. He also found that for large n a standard normal independent series behaves as

$$E[R^*(n)] = 1.25 n^{0.5} \tag{3}$$

The behavior H = 0.5 is also asymptotically true for any other process with an integrable correlation structure. The tendency of geophysical time series to produce values of H greater than 0.5 has become known as Hurst's phenomenon which has a long and distinguished history of research in hydrology[15,16,1]. A value of H > 0.5 indicates a particularly strong kind of persistence associated with an infinite memory process. Mandelbrot and Wallis[15] give the results of the adjusted range analysis for a series of the minimum annual water levels of the Nile River. They found a Hurst coefficient of 0.91. Such result implies a very strong persistence, which is directly related to the natural occurrence of droughts in a nonperturbed environment. Processes with this kind of persistence "pose perplexing puzzles and are prone to frequent misinterpretation"[25]. They are characterized by a self-similar fractal type of structure with power spectrum proportional to $f^{-\beta}$ for large frequencies f. The relation between H and β is $\beta = 2H + 1$. In the case of the Nile minima the power spectrum decays as f^{-3}. This kind of spectrum diverges when f tends to zero but "if nothing else, finite observation times T would limit observable excesses"[25]. Thus a realistic power spectrum P(f) with asymptotic f^{-3} dependence obtained from data collected over a time period T might look[25]:

$$P(f) = \frac{T^4 f}{1 + T^4 f^4} \quad (f > 0) \tag{4}$$

Extending the observation period, T, in the above spectrum adds a lot of power in the low frequencies of the spectrum. In other words, the more we observe, the more probable the long period fluctuations become and hence droughts and persistence like phenomena appear more common. In summary, before drawing quick conclusions about the origin of seemingly very long duration climatic fluctuations based only on comparisons with past record, it is important to remember the very strong dependence which H > 0.5 type of processes have in the length of observations.

The strong persistence described in this section is rooted in the non-integrable correlation or power spectrum structure of $f^{-\beta}$ type of noises. From the point of view of climatic analyses one may still have strong persistence in processes with H = 0.5 which display exceedingly long excursions above and below their mean values. In this case the probabilistic structure of the process involves fluctuations and transitions between preferred states, e.g., modes, of the climatic variables. Very interesting from the climatologic viewpoint is the fact that the multiplicity of preferred states may arise from feedback effects and nonlinear interactions between the components of the hydrologic cycle in both the land and the atmosphere. This is the topic of the next section of this paper which is taken literally from Rodriguez-Iturbe et al.[21,22] and Entekhabi et al.[8].

NON-LINEAR DYNAMICS OF LARGE-SCALE WATER BALANCE WITH LAND AND ATMOSPHERE INTERACTION

The focus here will be on a specific feedback effect. Although the kind of approach to be presented does not involve the millions of interactions of many variables disaggregated in space and time in the way CGM's do, its analytic tractability and its focus on specific phenomena makes it very attractive for the study of some very basic questions related to droughts.

The surface hydrologic balance for a large-scale continental region is represented by the changes in the areal-mean soil moisture:

$$nZ_r \frac{ds}{dt} = P(s) - Q(s) - E(s) \tag{5}$$

where nZ_r is the product of the soil porosity and the hydrologically active soil depth. The state variable s is the saturated fraction of this total depth that is available for moisture storage; that is, it is the relative soil saturation $0 \leq s \leq 1$. The differential change in storage depth nZ_r is equal to the difference between precipitation rate P(s) and the runoff and evapotranspiration losses Q(s) and E(s).

The runoff and the actual evapotranspiration rates depend on the surface moisture condition as symbolized by the argument s in the functional form of these processes in Eq. (5). As empirical but physically consistent parameterizations,

$$\frac{Q(s)}{P(s)} = \varepsilon s^r \qquad (6)$$

and

$$\frac{E(s)}{E_p} = s^c \qquad (7)$$

are given, where ε, r, and c are nonnegative constants. Equation (6) represents the runoff ratio or the fraction of precipitation that is direct runoff. It is empirically related to the relative soil saturation by a power law; little runoff is generated when the soil is dry ($s \to 0$), and greater runoff is generated for humid conditions ($s \to 1$). The runoff ratio is also confined to the interval [0,1], and it can be nonlinearly related to soil moisture depending on the parameter r.

The evapotranspiration rate is also proportional to the relative soil saturation in a nonlinear manner (linear if $c = 1$). The maximum evaporation rate is confined to be less than or equal to the potential evaporation E_p rate, which is the evaporation rate if the water is nowhere limiting, and the loss rate is restricted by the available energy at the surface. The parameter c depends on soil and vegetation characteristics.

In the balance Eq. (5), the precipitation input itself is conceptualized to be functionally related to the governing soil moisture. The landsurface-atmosphere interaction is intended to be represented through this functional dependence.

The precipitation water over large continental regions is derived from the advected (lateral and vertical) moisture in the overlying atmosphere. By developing a water balance for the overlying atmospheric moisture reservoir and coupling it to the balance for the surface reservoir in Eq. (5), a model is constructed for the surface and atmospheric interaction in the overall moisture supply for large continental regions.

The water vapor balance for the overlying atmosphere is performed by separating the precipitation into that derived from outside of the region (through lateral advection), P_a, and that derived from local evaporative sources, $P_m(S)$. The total precipitation is

$$P(s) = P_a + P_m(s) \qquad (8)$$

The supply of precipitation water by the advection of moisture from outside of the region is independent of the hydrologic conditions over the continent. The local source of precipitation contributing to the amount $P_m(S)$ of the total is, however, related to the surface evapotranspiration over the continent itself. Through this latter moisture supply mechanism, the total precipitation rate is functionally related to the governing areal-average soil moisture s, that is P(s). Rodriguez-Iturbe et al.[21] and Entekhabi et al.[22] based on work by Budyko[2], derived a physically based parameterization for this functional form which allows them to represent the dependence of total precipitation on the continental soil moisture conditions as:

$$P = P_a \left[1 + \frac{P_m(s)}{P_a} \right] = P_a \left[1 + \frac{s^c}{\Omega} \right] \qquad (9)$$

where the nondimensional climatic parameter Ω is defined as $\Omega = LE_p/2wu$, L being the length of the territory, w the vertically integrated water vapor and u a weighted average of the wind velocity vector normal to the boundary of the region.

It is clear that the fraction of the total precipitation that is due to local evaporation and is hence controlled by land surface processes over the region ($P_m(s)/P$) is largest when the value of L is also large. This means that if the airmass streamtube traverses a long distance and gradually reduces its content of advectively derived moisture along the streamline and instead gains moisture that is locally evaporated along this path, then the precipitation resulting from that moisture admixture is largely locally derived. Thus, the fraction of precipitation that is recycled and the strength of this mechanism of landsurface-atmosphere interaction are more significant over large land areas and inner continental regions.

Of all the surface hydrologic fluxes, precipitation is the one most characterized by randomness; this process is especially rich in the high frequencies and may be considered a stochastic process. In the context of the precipitation expression here, the parameters composited in Ω are considered to be the source of the randomness. The winds, streamtube characteristics, and other general circulation-related features represented by the parameter are taken to be characterized by noise fluctuations around a certain mean value. The variable Ω^{-1} is taken to be an uncorrelated white noise process with the expectation $E[\Omega^{-1}]$ and the variance σ^2.

Upon substituting Eqs. (6), (7), and (9) into the continental hydrologic balance, along with the random process representation for the large-scale forcing Ω^{-1}, the following time-differential equation for the soil-moisture process is obtained:

$$ds_t = \left[\frac{P_a}{nZ_r}\left(1+E[\Omega^{-1}]s_t^c\right)\left(1-\varepsilon s_t^r\right) - \frac{E_p}{nZ_r}s_t^c\right]dt + \left[\frac{P_a}{nZ_r}s_t^c\left(1-\varepsilon s_t^r\right)\right]\sigma dw_t \quad (10)$$

Here dw_t is a $N(0,1)$ white noise and it is the derivative of the unit Wiener process. The governing Ito stochastic differential equation for the continental water balance with atmospheric interaction is rewritten as

$$ds_t = G(s_t)dt + g(s_t)\sigma dw_t \quad (11a)$$

where

$$G(s_t) = \frac{P_a}{nZ_r}\left(1+E[\Omega^{-1}]s_t^c\right)\left(1-\varepsilon s_t^r\right) - \frac{E_p}{nZ_r}s_t^c \quad (11b)$$

and

$$g(s_t) = \frac{P_a}{nZ_r}s_t^c\left(1-\varepsilon s_t^r\right)\sigma \quad (11c)$$

It may be shown that the evolution of the probability density function for the state s_t of stochastic differential Eq. (11) is governed by the Fokker-Planck equation

$$\frac{\partial f(s_t,t)}{\partial t} = \frac{\partial}{\partial s}G(s_t)f(s_t,t) + \frac{1}{2}\sigma^2\frac{\partial^2}{\partial s^2}g^2(s_t)f(s_t) \quad (12)$$

(Schuss, 1980).

After a long period, the stationary annual soil moisture stochastic process attains the steady-state probability density function

$$f_{ss}(s) = f(s, t \to \infty)$$

when
$$\frac{\partial f(s,t)}{\partial t} = 0$$
Equation (12) then becomes an ordinary differential equation whose solution is

$$f_{ss}(s) = C \exp\left\{-\frac{2}{\sigma^2} U(s)\right\} \tag{13}$$

where C is a normalization constant to insure that Eq. (13) conserves probability mass and integrates to one. The function U(s) is referred to as the potential

$$U(s) = -\int^s \frac{G(u)}{g^2(u)} du + \sigma^2 \ln g(s) \tag{14}$$

Rodriguez-Iturbe et al.[21] give the expression for $f_{ss}(s)$ in the case of the continental water balance model described here.

The useful interpretation of the potential function U(s) associated with the steady-state probability density function $f_{ss}(s)$ is that it represents the landscape over which a particle, namely the current state of the soil moisture, moves in response to random perturbations. If the probability density function $f_{ss}(s)$ is Gaussian or unimodal, then the corresponding potential landscape would be a depression with its well located at the statistical mode of $f_{ss}(s)$. A particle on this landscape rolls up and down the sides of the well in response to random impulses, but it is ultimately confined to this single well and resides in the well depression most of the time.

As evident in the governing evolution Eq. (11), the surface soil moisture balance contains a nonlinear atmospheric feedback term in the form of g(s) that multiplies the simple white noise random term. Such a multiplicative noise situation may lead to the bifurcation of the potential landscape into a multiplicity of depressions and wells. There will be several new and stable depressions formed that will attract the moving particle. Noise-induced fluctuations will, if strong enough, jolt the particle to escape one depression and fall into the next. Ridges between depressions will be unstable residences for the particle, and the particle will thus spend little time at those locations. Such a landscape will have a multimodal probability density function associated with it. The several modes that arise (the one at the dry end to be called the "drought" and the one at the more moist extremity to be called "pluvial" in the case of soil moisture with a bimodal probability density function) are due to the presence of feedbacks in the water balance equation. Once in a drought (near the drought mode or in the neighborhood of the depression of the potential function at the low-soil-moisture end), the evaporation rate and hence the local supply of precipitation water diminishes. The drought is reinforced and it persists until a strong enough anomaly in the form of external random noise forcing (i.e., general or large-scale circulation forcing on precipitation) occurs that will be the fluctuation-induced route for escaping drought. Similar arguments apply to the pluvial mode and the local depression in the potential landscape associated with it. It must be noted that standard time series model used in climate and hydrologic time series modeling (e.g., ARMA) are additive noise models, and they are thus restricted to unimodal statistical distribution.

The moving particle on the landscape is agitated by random noise whose intensity or energy level is measured by the variance σ^2. It follows that the stronger the noise (larger σ^2), the more frequent will be the jumps from one depression (statistical mode) to another. There are, however, a number of other factors involved. The transition time of the particle between modes depends not only on the level of its excitation (i.e., σ^2) but on the shape of landscape as well. The deeper the wells and the higher the ridges between these depressions, the more energy it takes to overcome them. Thus, the random fluctuation that results in a successful escape must have a higher magnitude. In terms of the white noise, this means less probable. Besides the topography of the landscape, the other factor that affects transition time between statistical modes is the multiplicative nature of the noise itself. The position of potential landscape itself is randomly fluctuating in time as the noise multiplies the state variable. The intensity of these landscape fluctuations also depends on the variance of noise σ^2.

An example of the development of the multimodal situation is given in Figure 6 from Entekhabi et al.[8] where details are given regarding the values of the physical variables used in the characterization of the so-called semi-arid and semi-humid climates.

Different scenarios of climate change will greatly affect the feedbacks explicitly incorporated in the above methodology. The nonlinear stochasticity of the system will then compound the effect of such changes and their effect on the maintenance of droughts or pluvial modes.

FINAL COMMENT

The intent of this paper has been to present a quick overview of the relationship between droughts and climate change rather than an in-depth perspective of any of the two. The relationship is fundamentally influenced by the strong feedbacks among the variables and the nonlinearities of the system. The problem is far from being well understood and presents a fertile ground for collaboration among hydrologists, climatologists and atmospheric scientists in one of the most exciting and crucially important research areas of our time.

REFERENCES

1. R. L. Bras, I. Rodriguez-Iturbe. *Random functions and hydrology.* Addison-Wesley, 1985.
2. M. I. Budyko. *Climate and life.* Academic Press, 1974.
3. S. Chagnon. Detecting drought conditions in Illinois, ISWA/CIR-169-87, 1987.
4. J. G. Charney, P. H. Stone, W. J. Quirk. Drought in the Sahara: A biogeophysical feedback mechanism. *Science* 187:434-435, 1975.
5. H. E. Dregne. Desertification. In *Resources and world development*, D. J. McLaren and B. J. Skinner, eds. Wiley-Interscience, 697-725, 1987.
7. P. S. Eagleson. The emergence of global scale hydrology. *Water Resour. Res.* 22(9):65-145, 1986.
8. D. Entekhabi, I. Rodriguez-Iturbe, R. L. Bras. Variability in large-scale water balance with landsurface-atmosphere interaction. *J. of Climate* 5(8):798-813. 1992

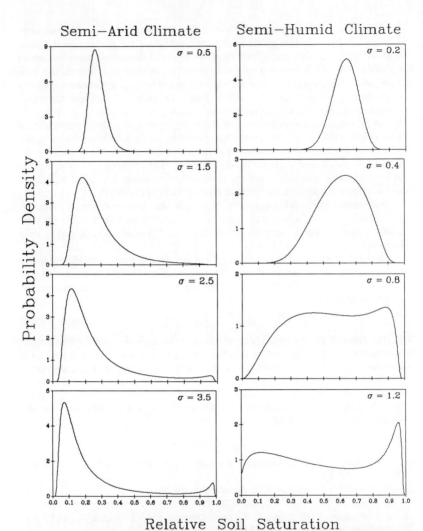

Figure 6. The steady-state probability distribution function for soil moisture in both climate cases (column 1: semiarid; column 2: semihumid). With low amounts of variance σ^2, the distribution is near-Gaussian around the fixed equilibrium value which would result from the deterministic ($\sigma = 0$) situation. With successively larger amounts of variance, the distribution traverses a larger domain and develops multiple modes in both cases. (From Entekhabi et al., 1992.)

9. H. E. Hurst. Long term storage capacities by reservoirs. *Trans. ASCE* **116**:776–808, 1951.
10. H. E. Hurst, R. P. Black, V. M. Simaika. *Long term storage*. London, Constable, 1965.
11. W. W. Kellogg, Z. Zhan. Sensitivity of soil moisture to doubling of carbon dioxide in climate model experiments, Part 1: North America. *J. of Climate* **1**(4):348–366, 1988.
12. R. D. Koster. Unpublished manuscript referred to in Eagleson (1986), 1985.
13. S. Manabe, R. Wetherald. Reduction in summer soil wetness induced by an increase in atmospheric carbon dioxide. *Science* **232**:626–628, 1986.
14. S. Manabe, R. Wetherald. Large-scale changes in soil wetness induced by an increase in atmospheric carbon dioxide. *J. Atmos. Sci.* **44**(8):1211–1235, 1987.
15. B. B. Mandelbrot, J. R. Wallis. Noah, Joseph, and operational hydrology. *Water Resour. Res.* **4**(5):909–918, 1968.
16. B. B. Mandelbrot, J. R. Wallis. Some long run properties of geophysical records. *Water Resour. Res.* **5**(2):321–340, 1969.
17. J. Namias. Some courses of United States drought. *J. Climate Appl. Meteor.* **22**(1):30
18. S. E. Nicholson, D. Entekhabi. The quasi periods behavior of rainfall variability in Africa and its relationship to the Southern Oscillation. *Arch. Meteor. Geophy. Biochem.* **Series A, 34**:311–348, 1986.
19. C. A. Nobre, P. J. Sellers, J. Shukla. Amazonian deforestation and regional climate change. *J. of Climate* **4**:957–988, 1991.
20. D. F. Peterson, A. A. Keller. Irrigation in climate change. Waggoner, referred to in Schneider (1989), 1987.
21. I. Rodriguez-Iturbe, D. Entekhabi, R. L. Bras. Non-linear dynamics of soil moisture at climate scales, 1: Stochastic analysis. *Water Resour. Res.* **27**(8):1899–1906, 1991.
22. I. Rodriguez-Iturbe, D. Entekhabi, J.-S. Lee, R. L. Bras. Non-linear dynamics of soil moisture at climate scales, 2: Chaotic analysis. *Water Resour. Res.* **27**(8):1907–1915, 1991.
23. S. H. Schneider. *Global warming*. Sierra Club Books, 1989.
24. S. H. Schneider. The changing climate. *Scientific American*, 70–79, September 1989.
25. M. Schroeder. *Fractals, chaos, and power laws*. Freeman, 1991.
26. J. Shukla, C. A. Nobre, P. Sellers. Amazon deforestation and climate change. *Science* **247**:1322–1325, 1990.
27. Z. Schuss. *Theory and applications of stochastic differential equations*. Wiley and Sons, 1980.
28. M. E. Voice, B. G., Hunt. A study of the dynamics of drought initiation using a global circulation model. *J. Geophys. Res.* **89**(D6):9504–9520, 1984.

AN APPROACH FOR ASSESSING THE SENSITIVITY OF FLOODS TO REGIONAL CLIMATE CHANGE

James P. Hughes
Department of Statistics
University of Washington
Seattle, WA 98195

Dennis P. Lettenmaier
Department of Civil Engineering
University of Washington
Seattle, WA 98195

Eric F. Wood
Water Resources Program
Princeton University
Princeton, N.J. 08544

ABSTRACT

A high visibility afforded climate change issues in recent years has led to conflicts between and among decision makers and scientists. Decision makers inevitably feel pressure to assess the effect of climate change on the public welfare, while most climate modelers are, to a greater or lesser degree, concerned about the extent to which known inaccuracies in their models limit or preclude the use of modeling results for policy making. The water resources sector affords a good example of the limitations of the use of alternative climate scenarios derived from GCMs for decision making. GCM simulations of precipitation agree poorly between GCMs, and GCM predictions of runoff and evapotranspiration are even more uncertain. Further, water resources managers must be concerned about hydrologic extremes (floods and droughts) which are much more difficult to predict than "average" conditions. Most studies of the sensitivity of water resource systems and operating policies to climate change to date have been based on simple perturbations of historic hydroclimatological time series to reflect the difference between large area GCM simulations for an altered climate (e.g., CO_2 doubling) and a GCM simulation of present climate. Such approaches are especially limited for assessment of the sensitivity of water resources systems under extreme conditions, since the distribution of storm inter-arrival times, for instance, is kept identical to that observed in the historic past. Further, such approaches have generally been based on the difference between the GCM altered and present climates for a single grid cell, primarily because the GCM spatial scale is often much larger than the scale at which climate interpretations are desired. The use of single grid cell GCM results is considered inadvisable by many GCM modelers, who feel that the spatial scale for which interpretation of GCM results is most reasonable is on the order of several grid cells. In this paper, we demonstrate an alternative approach to assessing the implications of altered climates as predicted by GCMs for extreme (flooding) conditions. The approach is based on the characterization of regional atmospheric circulation patterns through a weather typing procedure, from which a stochastic model of the weather class occurrences is formulated. Weather types are identified through a CART (Classification and Regression Tree) approach. Precipitation occurrence/non-occurrence at multiple precipitation stations is then predicted through a second stage stochastic model. Precipitation amounts are predicted conditional on the weather class identified from the large area circulation information.

INTRODUCTION

General Circulation Models (GCMs) have been widely used to make predictions about long-term changes in the earth's climate, such as those associated with greenhouse warming due to increased emissions of CO_2 and other trace gases over the last century. The spatial scale of GCMs, presently on the order of several hundred kilometers, is much larger than the local scale at which most assessments are conducted. Further, for hydrologic studies, precipitation is the variable of greatest interest, and GCM simulations of precipitation for present climate conditions are notoriously poor (see for example, Rind et al.[1], Wood et al.[2], Grotch[3]). One approach that has been explored recently for interpreting GCM predictions at the regional scale is the use of nested, higher resolution (e.g., mesoscale) models, for which the GCM provides the boundary conditions. Giorgi[4] describes such an approach using NCAR's MM4 mesoscale model for the Western U.S. and Europe. While this approach is conceptually consistent with the GCM representation of the climate system, it suffers from high computational costs (hence, restrictions on the length of scenario that can be simulated), and, because it uses GCM results as the boundary conditions, inaccuracies in the GCM may be transmitted to the local scale. Further, computational constraints limit the nested (fine scale) grid mesh to several tens of kilometers, which may still be too coarse for local assessments.

An alternative strategy is to translate directly from the GCM simulations to point surface variables, such as precipitation and temperature. Generally, GCMs represent free atmosphere variables much better than surface variables. Stochastic transfer models offer a means of linking the large-scale circulation conditions as inferred from the GCM to local surface meteorological conditions. Such an approach was followed by Storch[5] who used a canonical correlation model to relate seasonal precipitation over the Iberian Peninsula to GCM simulated surface pressure over the North Atlantic. Wilson et al.[6] simulated daily precipitation at three locations in the northwestern U.S. using an empirical orthogonal function classification of historical observations of surface pressure, 850 mb temperature, and 850 mb wind for a large area of the North Pacific. A semi-Markov model was used to represent the transitions between daily weather states, and a hierarchical Polya urn model was used to simulate multiple site daily precipitation conditional on the climate state in such a way that the statistics of historical wet-dry occurrences conditioned on the climate state were preserved. Such approaches, while not a substitute for more physically based dynamic models, do have the advantage of translating the climate model signal directly to the local level, and are much simpler.

In this paper we describe an initial approach for quantitatively evaluating climate change impacts on flood frequencies. The procedure described in this paper consists of three steps. The first step is the development of a stochastic transfer function model which links large scale atmospheric variables to local precipitation. The second step is to generate local precipitation conditional on GCM derived atmospheric circulation patterns and the historical conditional precipitation distribution, and to generate corresponding temperature sequences. This is done for both $1 \times CO_2$ and $2 \times CO_2$ climate scenarios. The third step is the application of a hydrological model to transform the precipitation and temperature series into stream discharges and a flood frequency distribution.

The procedure was applied to two basins in the Olympia Peninsula area of Washington state as shown in Figure 1. The two basins are the Satsop River (12-0350) which is dominated by heavy rainfall floods, and the Hoh River (12-0412) which is dominated by mixed snowmelt–rainfall floods. The elevation distribution for the two basins is given in Table I.

APPROACH

Weather Classification Model

The weather classification model links large-scale atmospheric variables, observed and assimilated (by the National Meteorological Center) for a 2.5° latitude by 3.75° longitude grid over the Eastern Pacific Ocean and Western United States, to precipitation occurrences (rain/no rain) and amounts at two precipitation stations (Forks and Aberdeen). In the first stage, each day is classified into one of four weather states using principal components of the gridded surface presure field, 850 mb temperature and the one-day lagged values of the same variables at a few selected grid nodes. The four weather states consist of rain/no rain combinations at the two rain gauge stations: rain/rain, rain/no rain, no rain/rain, no rain/no rain.

Nine years of historical data were used to develop the weather classification algorithm. The atmospheric data consisted of surface pressure, 850 mb temperature and 1-day lag values of these measures at 20 NMC grid nodes over the northwestern U.S. and the eastern Pacific Ocean (see Figure 2 for the location of the NMC grid nodes). Surface pressure and 850 mb temperature were used because they were the only variables for which we had daily GCM simulation results. Our preliminary results suggest that these two variables capture the general character of large scale circulation patterns, and should be sufficient for this preliminary analysis.

Precipitation data for the same time period were available for two rain gauge stations in western Washington. The weather classification scheme was developed with the objective of producing weather classes which gave maximal separation of the joint precipitation occurrence distributions at these two stations. The joint precipitation distribution was parameterized by the relative frequencies of the four possible precipitation states corresponding to rain- no rain at each of the two stations. For each season, stepwise linear discriminant analysis was used to select the 8–12 atmospheric measures × grid node combinations (out of 4 atmospheric measures × 20 grid nodes = 80 variables) which were most predictive of the precipitation state. A Classification and Regression Trees (CART) procedure[7] was then used to define three or four (depending

J. P. Hughes *et al.* 115

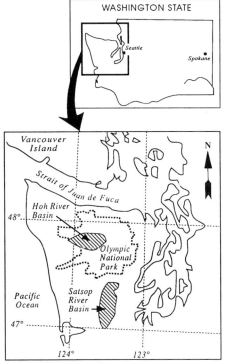

Figure 1. Location of study basins.

Table I: Elevation distribution for study basins

F(z)	Satsop Basin	Hoh Basin
0.001	60.0	
0.050	60.0	
0.100	92.0	
0.125		372.0
0.375		502.0
0.500	146.0	
0.625		879.0
0.875		1435.0
0.900	609.0	
0.950	713.0	
0.990	853.0	
0.995	909.0	
0.999	1042.0	

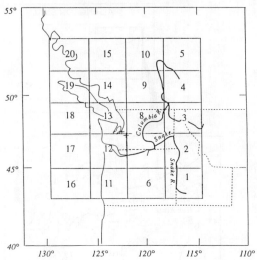

Figure 2. Location of NMC nodes.

on season) weather states based on these 8–12 variables. Table II presents the results for the Winter and Spring seasons. This table gives the relative frequencies of rainfall occurrence conditional on the weather classes. From the table its possible to relate the weather class to meteorological conditions; for example Winter weather class 3 corresponds to cold, high pressure systems which produces little precipitation. A separate classification algorithm was developed for each of four seasons: Winter (January–March), Spring (April–June), Summer (July–September) and Fall (October–December).

The above model was then applied to general circulation model (GCM) alternative climate simulations. The GCM results that were used from the analysis were 5 years of simulation of current ($1 \times CO_2$) and doubled CO_2 from NOAA's Geophysical Fluid Dynamics Laboratory (GFDL) R-30 GCM. The GCM simulation output was available at a daily time step. The CART algorithm was applied to the large scale GCM generated atmospheric variables which resulted in a sequence of weather states. Tables II and III provide the results for the weather classification step of the analysis. Table III gives the relative frequencies for the weather states for the historical (NMC) data, the GCM current climate simulation (designated GFDL in Table III) and the GCM–$2 \times CO_2$ climate simulations. Because of our interest in floods, which occur in the Winter and Spring in these basins, only these two seasons are reported here. From Table III it appears that the weather state frequency (and thus large scale circulation patterns) are reasonably well preserved in the current climate GCM simulations. For the Spring season, weather state 4 occurs more often in the GCM simulations than were found in the historical data.

Table II. Relative Frequencies of Rainfall Occurrence Conditional on Weather Class

Winter

Weather Class	1	2	3
No rain	0.092	0.186	0.538
Aberdeen rain	0.059	0.106	0.073
Forks rain	0.058	0.163	0.132
Both rain	0.791	0.545	0.257

Spring

Weather Class	1	2	3	4
No rain	0.134	0.322	0.470	0.545
Aberdeen rain	0.097	0.104	0.073	0.040
Forks rain	0.080	0.125	0.120	0.137
Both rain	0.689	0.449	0.336	0.278

Table III. Relative Weather State Frequencies

Winter

Weather Class	1	2	3
Historical	.61	.14	.24
GFDL	.55	.18	.27
GFDL, 2×CO_2	.686	.118	.196

Spring

Weather Class	1	2	3	4
Historical	.14	.21	.22	.43
GFDL, current	.044	.073	.101	.782
GFDL, 2xCO_2	.042	.110	.105	.748

Generation of Precipitation and Temperature Sequences

A semi-Markov model was then fitted to the weather states sequences by season. It was assumed that within a weather state, the residence time has a mixed geometric distribution and at the end of the "stay" within that weather state, it transits to a new state via a fitted transition matrix. Conditional on the weather state, the daily precipitation states are assumed temporally independent. Daily precipitation at each station are generated through resampling from the historical data. This is done conditional on the weather state and precipitation state. This allowed for us to generate a 40-year sequence of 2-station rainfall, for both current and doubled CO_2, via resampling as discussed above; these results are given in Table IV along with the historical values. Figures 3 and 4 show the frequency distribution for daily precipitation for the two stations. The general conclusion from these results is that the weather classification and conditional precipitation generation scheme works reasonably well.

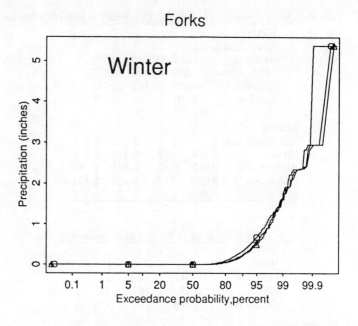
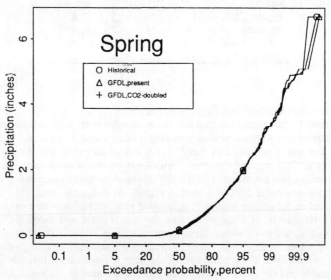

Figure 3. Precipitation frequency curves for Forks gauge.

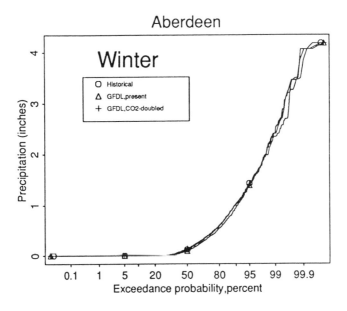
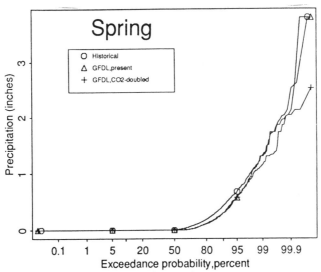
Figure 4. Precipitation frequency curves for Aberdeen gauge.

Table IV. Average Daily Precipitation (inches)

	Winter	Spring	Summer	Fall
Forks				
Historical	.490	.185	.113	.512
GFDL, curr.	.470	.164	.090	.457
GFDL, 2×CO_2	.495	.159	.094	.464
Aberdeen				
Historical	.347	.133	.077	.349
GFDL, curr.	.321	.101	.056	.284
GFDL, 2×CO_2	.348	.097	.057	.307

For some seasons, for example Spring, precipitation estimates for Aberdeen from the current climate GCM is 30% lower than the historical data while for Forks it is about 10% too low. Continued research is needed to better understand the errors in the historical data as well as model-induced errors in the GCM/stochastic transfer model approach.

Daily temperature was also generated because of its need in the hydrological model to simulate snowmelt. The temperature model is conditioned on the climatological mean temperature, T850 (the 850 mb temperature), the present and previous day's rain state at the rain gauge station, and the previous day's residual. To remove temperature biases within the GCM simulation, the mean 850 mb temperature from the GCM was adjusted to the historical temperature. To produce the temperature for a doubling of a carbon dioxide scenario, the historical temperature was increased by the incremental differences in temperature found between the $1 \times CO_2$ and $2 \times CO_2$ GCM runs. The temperatures for 4 different elevation bands were then estimated using a standard lapse rate. The zonal temperatures are needed for the snowmelt model.

Runoff Generation

In the third step two hydrological models were applied. Because runoff in the Hoh River basin (and to a lesser extent in the Satsop River) are snowmelt dominated, the snow accumulation and ablation process must be modeled. The snowmelt model used was developed by Anderson of the U.S. National Weather Service Hydrologic Research Laboratory[8] and has been widely tested in mountainous basins in the western United States and elsewhere. It describes the change in storage of water and heat in the snowpack, based on six-hourly precipitation and temperature data. Because temperature is strongly dependent on elevation, the snowmelt model was implemented using an elevation band approach. The soil moisture accounting model that was used was first developed by Burnash et al.[9], originally for forecating runoff in the Sacramento River basin. It has been extensively evaluated elsewhere[10,11,12]. This model is a deterministic, spatially lumped, conceptual model which accounts for the soil moisture flux between five conceptual storage zones. Transfers of water between the soil moisture zones control the runoff response. The soil moisture zones include an upper free and tension water zone and lower tension, free primary and secondary zones. Rainfall or snowmelt water that does not contribute to direct runoff is split between upper free and tension water. Tension water is removed only through evapotranspiration; free water can be transferred to the lower tension and free water zones. Likewise, the lower tension zone is depleted only through evapotranspiration. The lower free water zones combine to produce a nonlinear base flow recession. The models were calibrated using the observed historical data.

The hydrologic models were applied to the generated 40 year sequences of precipitation and temperature for the following scenarios: observed climate using NMC data, GFDL $1 \times CO_2$ GCM simulation and GFDL $2 \times CO_2$ GCM simulation. The resulting discharge sequence formed the basis for the flood frequency analysis.

Derived Flood Frequency Curves

The series annual maxima were extracted from the derived discharge record and plotted as a frequency curve as shown in Figures 5 and 6. Plotted in these figures are four flood frequency curves based on the historical data, the NMC-observed climate derived curve, a $1 \times CO_2$ GCM derived curve and a $2 \times CO_2$ GCM derived curve. A number of observations can be made from these results. The first is that the historical and NMC-based curves are very close for both basins. This implies that the methodology of transfering large-scale atmospheric features to local variables through the stochastic transfer model is reasonable. The second observation is that the $1 \times CO_2$ GCM derived curve is significantly different than the historical or NMC-based curve for both basins. This implies that the GCM-derived large scale atmospheric variables have errors that, at this time, may preclude using the GCM results to provide absolute values of local climate variables.

The third observation is that in comparing the $1 \times CO_2$ GCM derived curve with the $2 \times CO_2$ GCM derived curve, floods for the Hoh River basin increased for the same return period while floods for the Satsop River basin decreased. We interpreted these results as follows. In the Satsop River basin, the floods are rainfall dominated due to its lower evelation. The precipitation sequence (intensity and duration of the storms) under the $2 \times CO_2$ climate senario resulted in lower floods. For the Hoh River whose floods result from snowmelt (or rain on melting snow), the higher temperatures under the $2 \times CO_2$ climate scenario with the generated precipitation that had slightly higher rates (see Table IV) and similar intensities (see Figure 3) resulted in the higher floods. These results emphasize the complexity of the climate-land hydrology system in which the apparent results from the GCM simulations do not necessarily transfer themselves, even on a relative basis.

Discussion and Conclusions

In this paper we carried out a preliminary analysis for assessing the impact of climate change on the flood frequency distribution for a river basin. The approach was applied to two basins in Washington state, one whose floods are dominated by snowmelt and one whose floods are dominated by rainfall. The approach consisted of three steps: the development of a stochastic transfer model that relates large scale atmospheric variables (pressure and temperature) to precipitation occurrences, the generation of daily rainfall sequences conditional on the weather state, and the

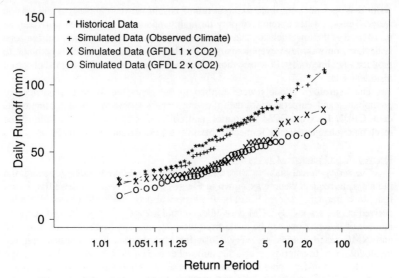

Figure 5. Annual flood frequency curve for Satsop River (12-0350).

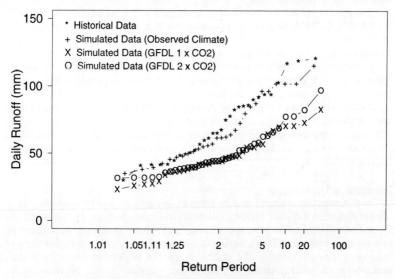

Figure 6. Annual flood frequency curve for Hoh River (12-0412).

estimation of stream discharge using the generated precipitation and temperature sequence and a hydrologic model.

The results show that the stochastic conditional precipitation model described here provides a useful framework for developing daily precipitation sequences at multiple sites corresponding to alternative climate scenarios as simulated by GCMs. When applied to the two stations located in the Olympia Penninsula area of Washington State and a present climate and $2 \times CO_2$ climate scenario from the GFDL R-30 GCM, the model predicts increased precipitation in the winter and less in the Spring. However these results must be viewed with caution. Perhaps the greatest spring concern is that the GCM fails to reproduce the historical climate characteristics, producing less precipitation for the present climate scenario in all seasons. In fact, as seen in Table III, the differences between the historical precipitation and the GCM present climate simulations are larger than the differences predicted between the GCM present and $2 \times CO_2$ climate scenarios. These model biases are transferred through to the derived flood frequency curves presented in Figures 5 and 6.

It could be argued that the effect of the model biases introduced by the GCM deficiencies will be diminished by making relative comparisons between the present climate and $2 \times CO_2$ climate scenarios. To a first approximation this may be true but it is unlikely to be strictly correct since the models are highly nonlinear.

There are many areas where the approach could be improved. These include improvement of the weather state definitions and better modeling of the temporal evolution of the weather states and precipitation process. However, one of the most significant limitations for this study was the constraint on the length and GCM variables made available to us. We are currently in the process of obtaining output from much longer GCM simulations with additional atmospheric variables (wind components, pressure heights and temperature profiles throughout the atmspheric column) which will undoubtedly lead to better weather state definitions.

ACKNOWLEDGMENTS

The research in this paper was supported by the Andrew W. Mellon Foundation for Dr. Wood, by the Pacific Northwest Laboratory under Contract DE-AC06-76RLO-1830 with the Department of Energy for Dr. Lettenmaier, and by a fellowship to Mr. Hughes from the IBM Thomas Watson Research Center.

REFERENCES

1. Rind, D., R. Goldberg, and R. Ruedy. Change in climatic variability in the 21st Century. *Climatic Change* **14**:5–37, 1989.
2. Wood, E. F., D. P. Lettenmaier, and J. R. Wallis. Comparisons of an alternative land surface parameterization with the GFDL high resolution climate model. IBM Research Division RC 16566, Jan. 1991.
3. Grotch, S. L. Regional intercomparisons of general circulation model predictions and historical data. Technical Note DOE/NBB-0084, U.S. Department of Energy, 1988.
4. Giorgi, F., and L. O. Mearns. Approaches to the simulation of regional climate change: A review. *Reviews of Geophysics* **29**(2):191–216, May 1991.

5. Storch, H., E. Zorita, and U. Cubush. Downscaling of global climate change estimates to regional scales: An application to Iberian rainfall in wintertime. Max Planck Institut fur Meteorologie Report No. 64, June, 1991.
6. Wilson, L. L., D. P. Lettenmaier, and E. Skyllingstad. A multisite stochastic daily precipitation model conditioned on large scale atmospheric circulation patterns. *Journal of Geophysical Research*, in press, 1992.
7. Breiman, L., J. H. Friedman, R. A. Olshen, and J. C. Stone. *Classification and regression trees*. Wadsworth and Brooks, Monterey, Calif., 1984.
8. Anderson, E. A. National Weather Service river forecast system: Snow accumulation and ablation model. NOAA Technical Memo NWS HYDRO-17, Natl. Oceanic and Atmos. Admin., Boulder, Colo., November 1973.
9. Burnash, R. J. C., R. L. Ferral, and R. A. McQuire. A generalized streamflow simulation system. In *Conceptual modeling for digital computers*. U.S. National Weather Service, Sacramento, Calif., 1973.
10. Kitanidis, P. K., and R. L. Bras. Real-time forecasting with a conceptual hydrologic model, 1: Analysis of uncertainty. *Water Resources Research* 16(6):1025–1033, 1980.
11. Gupta, V. K., and S. Sorooshian. Uniqueness and observability of conceptual rainfall-runoff model parameters: The percolation process examined. *Water Resources Research* 19(1):269–276, 1983.
12. Lettenmaier, D. P., and T. Y Gan. Hydrologic sensitivities of the Sacramento–San Joaquin River Basin, California, to global warming. *Water Resources Research* 26(1)69–86, January 1990.
13. Efron, B. The jacknife, the bootstrap, and other resampling plans. Monograph 38, Society for Industrial and Applied Mathematics, Philadelphia, 1982.
14. Manabe, S., and R. T. Wetherald. On the distribution of climate change resulting from an increase in CO_2 content of the atmosphere. *Journal of Atmospheric Sciences* 37:99–118, 1980.
15. Raftery, A. E. A model for high-order Markov chains. *Journal of the Royal Statistical Society*, Series B, Vol. 47, Part 3, pp. 528–539, 1985.

BLOWING DUST AND CLIMATE CHANGE

Richard E. Peterson* and James M. Gregory†
*Department of Geosciences and †Department of Civil Engineering
Texas Tech University
Lubbock, TX 79409

ABSTRACT

In order to better understand some of the weather phenomena which may be responsible for blowing dust in diverse regions of the globe, the meteorological record is closely examined for Lubbock, Texas, which records more blowing dust than anywhere else in the United States. The development of a semi-empirical model for wind erosion is also mentioned.

INTRODUCTION

Blowing dust and dust storms are frequent phenomena across broad areas of the earth[1], as well as Mars[2] and possibly Venus[3]. At various times (e.g., the 1930s and 1950s) the occurrence of blowing dust has drawn greater attention[4,5]. At the present time the largest community in the region of maximum blowing dust frequency in the United States is Lubbock, Texas (pop. 200,000); across the world, some very large populations are affected, particularly in North China[6]. Changes in the global atmospheric circulation may lead to shifts in the patterns of precipitation and strong surface winds, resulting in new areas (perhaps quite populous and/or agriculturally important) bearing the burden of blowing dust.

BLOWING DUST ACROSS THE GLOBE

Many regions of the world are affected by problems of blowing dust: for example, the former Soviet Union[7], West Africa[8], Southwest Asia[9], the Middle East[10], Mexico[11], and Australia[12]. Dry periods across the fertile plains of North America each spring lead to widespread and frequent periods of blowing dust; no location however records more hours of visibility reduction due to dust in the air than Lubbock, Texas[13,14].

Duststorms are a recurrent feature of late winter and springtime weather on the High Plains of the United States. Often the dust is only of regional extent (affecting areas several tens of kilometers wide and some hundred kilometers along the direction of the wind); occasionally however the dust carries far beyond the Mississippi Valley to the east into the Great Lakes or to the Atlantic[15]. Such storms were first studied in the 1930s[16] and later during the dry 1950s[17].

The impact of such day-to-day periods of dustiness can be far reaching. The greatest long-term economic impact is the loss of topsoil; however, short-term off-site damages can be 40–60 times that of topsoil loss[18,19]. There are health impacts as well. In the short-term, high concentrations of dust may stimulate or exacerbate respiratory difficulties. Long-term exposure may lead to nasal ailments. Unexpectedly severe blowing dust events have led to curtailment of aircraft operations and vehicular accidents, resulting in deaths. There is even the threat of valley fever (a fungus transported by blowing dust that killed 11 people in 1992 in California) if global warming increases the winter temperatures above the kill temperature for the fungus[20].

WEST TEXAS PHYSICAL SETTING

Historical and geological evidence confirms that dust movement has been a natural and relatively common phenomenon across the High Plains for a very long time[21,22]. The area around Lubbock (33°39'N, 101°49'W), referred to as the Llano Estacado, is almost completely covered by a mantle of aeolian material, a sand sheet grading into loess which is up to 30 m thick. The coarsest (sandy) material is to the west and southwest; the soil is thinner and finer grained (silt clay) to the east and southeast. Averaging about 1000 m in elevation, the terrain slopes only gently (upward to the west) and has few trees, water courses or irregularities. Scattered shallow lakes (playas) tend to become dry during the colder season.

Undoubtedly, variations in land use have contributed to the contemporary occurrence of blowing dust across the southern High Plains[23]; blowing dust is often observed to begin at the north-south border between New Mexico and Texas and intensify eastward. The primary use of the land is for range in New Mexico and for cotton in Texas. The crop is usually planted in the late spring and harvested late fall to early winter. Little plant residue remains; moreover it is common to clear the fields in January. As a consequence, the land in Texas usually is bare from December into May.

The agricultural practices, unless accompanied by periodic intervention, serve to maximize the likelihood of wind erosion. For the Texas South Plains, the average annual precipitation ranges from 400 to 500 mm; for Lubbock, the average is 451 mm. The annual distribution shows sparse average precipitation from November through April; moreover snowcover, which averages 27 cm, is generally short-lived. The southern High Plains are relatively cloudless; Lubbock averages 73% of possible sunshine, with the cold season being sunnier. In most years the polar jet stream meanders southward across Texas during the winter season, with very strong winds not far above the ground. As a result, during the months when the fields are relatively bare, the soil and air are dry, daytime heating and large-scale storm activity are vigorous, and blowing dust becomes most frequent. It is not uncommon to experience rain one day and blowing dust the next; the spring rains followed by low relative humidity reduce soil roughness and thereby increase soil erosion and dust.

LONG-TERM DISTRIBUTION OF BLOWING DUST

Records began to be kept at Lubbock at the U.S. Weather Bureau office in 1947; dust or blowing dust is noted when the horizontal visibility is restricted to less than 10 km. Until recently the airport location has been surrounded by relatively rural conditions. The record shows that Lubbock averages 150 hrs (or about 3%) of blowing dust annually. The monthly frequency is small (just a few hours) from July through October; however there is a rapid increase thereafter, reaching a peak in March-April (more than 30 hrs each). In particular years, very few hours are recorded (e.g., the "El Niño" year 1982–83 with only 40 hrs); however in other years the frequency is much above average (e.g., 1953–54 with 473 hrs, and April 1974, March 1977 with 107 hrs). The dustiest individual month has been March 1954 with 159 hrs.

The occurrence of blowing dust depends on the availability of soil particles and sufficient wind to raise the soil[24]. In turn, the availability of particles is related to the nature of the soil, the type of ground cover (or land use), the relative humidity and soil wetness, which is affected by the precipitation history. During the periods of blowing dust, the annual rainfall was 75–175 mm below the long-term average; during less dusty periods, rainfall totals did not depart greatly from average. The windiness at a particular site depends on the nature and strength of the wind systems affecting the region. (There is some suggestion that the strength of the mid-latitude westerlies was greater in the 1930s and into the early 1950s than in subsequent decades[25].)

GENERATING MECHANISMS

Across the Texas South Plains there are several weather phenomena which may bear strong, dust-raising winds: dust devils, thunderstorm outflows, cold fronts, deep cyclonic storms and daytime mixing-down of high-altitude winds.

Occasionally, particularly during the spring when rapidly warming soil temperatures combined with cold air temperatures aloft produce unstable lapse rates, large vigorous dust devils develop across the South Plains. Although such whirlwinds do inject dust hundreds of meters into the air[26], they are never sufficiently numerous to reduce the general visibility.

Thunderstorms occur across the South Plains on an average of 36 days annually[27]. With the relatively dry subcloud air, evaporation of the precipitation can generate surface winds more than sufficient to raise dust. Spectacular walls of advancing dust (haboobs) may result in visibilities near zero with darkness like night[28]. These relatively limited periods account for most of the summer hours of blowing dust recorded.

The passage of cold fronts leads to similar dusty intervals; the visibilities are usually not as low as for haboobs, but the durations may be several hours rather than minutes.

The most long-lasting episodes of high winds and blowing dust come with deepening low-pressure systems[29]. The Texas South Plains come under the influence of one of the most active cyclogenetic areas of the globe, southeastern Colorado[30]. The dustiness may be severe through the night and day until the storm has moved into the Missouri Valley. With the strong southwest winds locally generated dust is experienced, while the subsequent northerly winds import dust from Colorado and western Kansas.

The most typical blowing dust event across the South Plains occurs without a prominent surface weather pattern. The primary mechanism is mixing down of stronger momentum from aloft. In the winter and early spring, relative humidities are usually low at all levels overhead. As a result, at sunrise the near-ground environment is quite stable under an intense radiation inversion. By late morning under clear skies, heating has eroded the inversion, allowing for deep mixing throughout the remaining day. The surface wind pattern is altered rather quickly as the usually stronger winds aloft are brought to the ground. By noon, dust is being raised, becoming thicker during the afternoon. Towards sunset the mixing mechanism shuts down and the dust settles. Under otherwise quiescent weather patterns, the same sequence may be repeated for many days in succession.

SYNOPTIC CLIMATOLOGY OF BLOWING DUST

A close examination of the surface weather observations, including the upper-level and surface weather maps, allows the evaluation of the relative importance of the different dust-generating mechanisms.

Overall, the mixing-down mechanism has been responsible for over 40% of all the blowing dust episodes; fronts account for 30%, thunderstorm outflows for 20%, and deep cyclones almost 10%. From year to year there is considerable variability in the order of importance; e.g., the mixing down led to 51 out of 98 episodes in 1954, but only 3 out of 21 in 1957.

On average the deep cyclones are most common in the March–April period, while the thunderstorm effect peaks in June (before the cotton crop has developed a good ground cover). Dustiness due to mixing down and fronts tends to follow the annual cycle of total dust hours.

The surface records note the time of onset as well as the end of the reduced visibility; the diurnal variation of the dustiness can thereby be examined. Initiation times rapidly increase toward mid-morning local time, remaining numerous until late afternoon. This coincides with the predominance of the mixing-down mechanism and the low relative humidity necessary for low threshold wind velocity. Afternoon to early evening initiation times are favored for thunderstorm outflows. The frontal passages and deepening of cyclones are not controlled as strongly by daytime heat; their initiation times may come at any time of day or night.

BLOWING DUST MODELING

In order to evaluate the potential for soil erosion due to strong winds, a blowing dust model is being developed at Texas Tech University.

For initialization of the semi-empirical model, numerous input parameters are required: standard surface weather data, friction velocity, threshold friction velocity, erodibility, particle size distribution of the surface soil, bulk density, surface ground cover parameter, length of the erosion segment, incoming mass from upstream soil segments and the duration of the last erosion episode if total soil loss is to be predicted.

The model output includes: soil movement rate, total soil loss, mass concentration with height, kinetic energy of particles with height, particle size distribution with height, surface area per air volume with height and visibility with height.

The variables in the model most affected by climate change are threshold friction velocity, which varies with surface soil moisture (often in equilibrium with air relative humidity), and the ground cover parameter, which varies with rainfall and the land management. Lower absolute humidity and higher daytime temperatures will reduce relative humidity and threshold friction velocity during the day when winds are typically highest. Soil particle sizes near 0.1 mm in diameter are most sensitive to relative humidity effects. Midwest areas north and east from Lubbock with finer-textured soils and currently higher relative humidities are at very high risk of increased blowing dust if climate change produces warmer temperatures, less rainfall, and lower relative humidity. Obviously, less rainfall and higher temperatures with higher evapotranspiration will typically reduce vegetative ground cover. Increased variability of rainfall would increase the occurrence of droughts and also increase the incidence of dust.

With space technology, it may be practical to monitor the two conditions that cause the greatest risk of wind erosion and dust generation. Changing cover conditions can be detected by satellite. Satellites can also observe changes in surface albedo associated with surface soil wetness. If the soil is wet, wind erosion does not occur. If the surface soil appears to be dry, then the relative humidity of the air controls the surface soil moisture and the threshold friction velocity. By coupling surface changes detected by satellites with the wind erosion and blowing dust model, it should be possible to provide some warning (possibly early enough in drought sequences to alter management) to those living and working in the areas affected.

CONCLUSIONS

Blowing dust is more common and the generating mechanisms are more diverse across the High Plains of Texas than anywhere else in the nation; the region is a natural laboratory for diagnosing the factors of wind erosion. The understanding gleaned from these studies should apply to dusty areas elsewhere now and in later decades if climate change shifts the formative areas. Current global circulation models yield conflicting indications about what parts of the globe may be adversely affected. A challenge remains to refine the GCM output to identify areas which may suffer future increases of blowing dust.

REFERENCES

1. A. Goudie. *Prog. Phys. Geog.* **7**:502 (1983).
2. R. W. Zurek. *Icarus* **50**:288 (1982).
3. J. D. Iversen and B. R. White. *Sediment* **29**:111 (1982).
4. H. Koschmieder. *Naturwiss.* **27**:133 (1939).
5. H. H. Finnell. *Sci. Amer.* **191**:25 (1954).
6. C.-K. Ing. *Weather* **27**:136 (1972).
7. L. Klimenko and L. Moskaleva. *Meteor. i Gidro.* **9**:93(1979)
8. M. Kästner, P. Kopke and H. Quenzel. *Adv. Space Res.* **2**:119 (1982).
9. N. J. Middleton. *J. Climatol.* **6**:183 (1986).
10. J. Joseph, A. Manes and D. Ashbel. *J. Appl. Meteor.* **12**:792 (1973).
11. M. A. Blanco. *Soc. Mex. Geog. Estadist. Bol.* **70**:111 (1950).
12. N. J. Middleton. *Search* **15**:46 (1984).
13. M. M. Orgill and G. A Sehmel. *Atmos. Environ.* **10**:813 (1976).
14. M. J. Changery. A Dust Climatology of the Western United States. NUREG/CR-3211 (Nuclear Reg. Comm., Washington 1983).
15. C. H. Vermillion. *Bull. Amer. Meteor. Soc.* **58**:330 (1977).
16. H. R. Byers. *Mon. Wea. Rev.* **64**:86 (1936).
17. G. F. Warn. *Bull. Amer. Meteor. Soc.* **33**:240 (1952).

18. P. C. Huszar. Proc. 1988 Wind Erosion Conf. (Texas Tech Univ., Lubbock, 1988) p. 223.
19. B. Davis and G. Condra. Ibid., p. 239.
20. D. Morain. *Los Angeles Times*, 30 Jan.:A25 (1992).
21. J. C. Malin. *Kansas Hist. Quart.* **14**:129, 265, 391 (1946).
22. V. T. Holliday. *Soils and Quaternary Landscape Evolution*, ed. J. Boardman (Wiley, New York, 1985) p. 325.
23. E. Kessler, D. Y. Alexander and J. F. Rarick. *Proc. Okla. Acad. Sci.* **58**:116 (1978).
24. D. A. Gillette and T. Walker. *Soil Sci.* **123**:97 (1977).
25. R. A. Kalnicky. *Ann. Assoc. Amer. Geog.* **64**:100 (1974).
26. P. C. Sinclair. Atmosphere Surface Exchange of Particulate and Gaseous Pollutants, CONF-740921 (ERDA, Oak Ridge, 1974) p. 497.
27. A. Court and J. F. Griffiths. Thunderstorm Morphology and Dynamics (U.S. Dept. Comm., Washington, 1982) p. 11
28. S. B. Idso. *Smithsonian* **5**:68 (1974).
29. H. W. Brandli, J. P. Ashman and D. L. Reinke. *Mon. Wea. Rev.* **105**:1008 (1977).
30. C. H. Reitan. *Mon. Wea. Rev.* **102**:861 (1974).

BIOMASS BURNING AND GLOBAL CHANGE

Joel S. Levine, Wesley R. Cofer III, and Donald R. Cahoon, Jr.
Atmospheric Sciences Division, NASA Langley Research Center
Hampton, VA 23665-5225

Edward L. Winstead
Hughes/STX Corporation, Hampton, VA 23666

Brian J. Stocks
Forestry Canada, Ontario Region, Sault Ste., Marie, Ontario, Canada P6A 5M7

ABSTRACT

The burning of living and dead biomass, including forests, savanna grasslands and agricultural wastes is much more widespread and extensive than previously believed and may consume as much as 8700 teragrams of dry biomass matter per year. The burning of this much biomass releases about 3940 teragrams of total carbon or about 3550 teragrams of carbon in the form of CO_2, which is about 40% of the total global annual production of CO_2. Biomass burning may also produce about 32% of the world's annual production of CO, 24% of the nonmethane hydrocarbons, 20% of the oxides of nitrogen, and biomass burn combustion products may be responsible for producing about 38% of the ozone in the troposphere. Biomass burning has increased with time and today is overwhelmingly human-initiated.

INTRODUCTION

Our planet is a unique object in the solar system due to the presence of a biosphere with its accompanying biomass and the occurrence of fire.[1] The burning of living and dead biomass is a very significant global source of atmospheric gases and particulates. Crutzen and colleagues were the first to consider biomass burning as a significant source of gases and particulates to the atmosphere.[2,3] However, in a recent paper, Crutzen and Andreae[4] point out that "Studies on the environmental effects of biomass burning have been much neglected until rather recently but are now attracting increased attention." The "increased attention" reference in the Crutzen and Andreae paper was the Chapman Conference on Global Biomass Burning: Atmospheric, Climatic, and Biospheric Implications held in Williamsburg, Virginia in March, 1990.[5,6] Biomass burning and its environmental implications have also become important research elements of the International Geosphere-Biosphere Program (IGBP) and the International Global Atmospheric Chemistry (IGAC) Project.[7]

GASEOUS EMISSIONS DUE TO BIOMASS BURNING

Biomass burning includes the combustion of living and dead material in forests, savannas, and agricultural wastes, and the burning of fuel wood. Under the ideal conditions of complete combustion, the burning of biomass material produces carbon dioxide (CO_2) and water vapor (H_2O), according to the reaction

$$CH_2O + O_2 \rightarrow CO_2 + H_2O \tag{1}$$

where CH_2O represents the average composition of biomass material. Since complete combustion is not achieved under any conditions of biomass burning, other carbon species, including carbon monoxide (CO), methane (CH_4), nonmethane hydrocarbons (NMHCs), and particulate carbon, result through the incomplete combustion of biomass material. In addition, nitrogen and sulfur species are produced from the combustion of nitrogen and sulfur in the biomass material.

While CO_2 is the overwhelming carbon species produced by biomass burning, its emissions into the atmosphere resulting from the burning of savannas and agricultural wastes are largely balanced by its reincorporation back into biomass via photosynthetic activity within weeks to years after burning. However, CO_2 emissions resulting from the burning of forests converted to nonforested areas and other carbon combustion products from all biomass sources including CH_4, CO, NMHCs, and particulate carbon, are largely "net" fluxes into the atmosphere since these products are not reincorporated into the biosphere.

Biomass material contains about 40% carbon by weight, with the remainder hydrogen (6.7%) and oxygen (53.3%).[8] Nitrogen accounts for between 0.3 to 3.8%, and sulfur for between 0.1 to 0.9% depending on the nature of the biomass material.[8] The nature and amount of the combustion products depend on the characteristics of both the fire and the biomass material burned. Hot, dry, fires with a good supply of oxygen produce mostly carbon dioxide with little CO, CH_4, and NMHCs. The flaming phase of the fire approximates complete combustion, while the smoldering phase approximates incomplete combustion, resulting in greater production of CO, CH_4, and NMHCs. The percentage production of CO_2, CO, CH_4, NMHCs, and carbon ash during the flaming and smoldering phases of burning based on laboratory studies is summarized in Table I.[9] Typically for forest fires, the flaming phase lasts on the order of an hour or less, while with the smoldering phase may last up to a day or more, depending on the type of fuel, the fuel moisture content, wind velocity, topography,

Table I
Percentage of production of gases during flaming and smoldering phases of burning based on laboratory experiments[9]

	Percentage in burning stage (%)	
	Flaming	Smoldering
CO_2	63	37
CO	16	84
CH_4	27	73
NMHCs	33	67
NO_x	66	34
NH_3	15	85
HCN	33	67
CH_3Cl	28	72

etc. For savanna grassland and agricultural waste fires, the flaming phase lasts a few minutes and the smoldering phase lasts up to an hour.

EMISSION RATIOS

The total mass of the carbon species (CO_2 + CO + CH_4 + NMHCs + particulate carbon) M(C) is related to the mass of the burned biomass (M) by M(C) = f × M, where f = mass fraction of carbon in the biomass material (40% by weight). To quantify the production of gases other than CO_2, we must determine the emission ratio (ER) for each species. The emission ratio for each species is defined as

$$ER = \frac{\Delta X}{\Delta CO_2}$$

where ΔX is the concentration of the species X produced by biomass burning, ΔX = $X^* - \overline{X}$, where X^* is the measured concentration of X in the biomass burn smoke plume, and $\overline{X}$ is the background (out of plume) atmospheric concentration of the species, and ΔCO_2 is the concentration of CO_2 produced by biomass burning, ΔCO_2 = $CO_2^* - \overline{CO_2}$ where CO_2^* is the measured concentration in the biomass burn plume, and $\overline{CO_2}$ is the background (out of plume) atmospheric concentration of CO_2.

In general, all species emission factors are normalized with respect to CO_2, since the concentration of CO_2 produced by biomass burning may be directly related to the amount of biomass material burned by simple stoichiometric considerations as discussed earlier. Furthermore, the measurement of CO_2 in the background atmosphere and in the smoke plume is a relatively simple and routine measurement.

For the reasons outlined above, it is most convenient to quantify the combustion products of biomass burning in terms of the species emission ratio (ER), i.e., the excess species production (above background) normalized with respect to the excess CO_2 production (above background). Measurements of the emission ratio for CH_4, CO, and NMHCs normalized with respect to CO_2 for diverse ecosystems (i.e., wetlands, chaparral, and boreal) for different phases of burning, i.e., flaming and smoldering phases and combined flaming and smoldering phases, called "mixed" are summarized in Table II.

Some researchers present their biomass burn emission measurements in the ratio of grams of carbon in the gaseous and particle combustion products to the mass of the carbon in the biomass fuel in kilograms. Average emission factors for CO_2, CO, and CH_4 in these units for diverse ecosystems are summarized in Table III and emission factors for CO_2, CO, CH_4, NMHCs and carbon ash in terms of percentage of fuel carbon based on laboratory experiments are summarized in Table IV. Inspection of Tables II–IV indicates that there is considerable variability in both the emission ratio and emission factor for carbon species as a function of ecosystem burning and the phase of burning (i.e., flaming or smoldering). A recent compilation of CO_2-normalized emission ratios for carbon species is listed in Table V. This table gives the range for both field measurements and laboratory studies and provides a "best guess."

Table II
Emission ratios for CO, CH_4, and NMHCs for diverse ecosystems
(in units of $\Delta X/\Delta CO_2$, in percent)[23]

	CO	CH_4	NMHCs
Wetlands			
Flaming	4.7 ± 0.8	0.27 ± 0.11	0.39 ± 0.17
Mixed	5.0 ± 1.1	0.28 ± 0.13	0.45 ± 0.16
Smoldering	5.4 ± 1.0	0.34 ± 0.12	0.40 ± 0.15
Chaparral			
Flaming	5.7 ± 1.6	0.52 ± 0.23	0.52 ± 0.21
Mixed	5.8 ± 2.4	0.47 ± 0.12	0.46 ± 0.15
Smoldering	8.2 ± 1.4	0.87 ± 0.23	1.17 ± 0.33
Boreal			
Flaming	6.7 ± 1.2	0.64 ± 0.20	0.66 ± 0.26
Mixed	11.5 ± 2.1	1.12 ± 0.31	1.14 ± 0.27
Smoldering	12.1 ± 1.9	1.21 ± 0.32	1.08 ± 0.18

Table III
Average emission factors for CO_2, CO, and CH_4 for diverse ecosystems
(in units of grams of combustion product carbon to kilograms of fuel carbon)[24]

	CO_2	CO	CH_4
Chaparral-1	1644 ± 44	74 ± 16	2.4 ± 0.15
Chaparral-2	1650 ± 31	75 ± 14	3.6 ± 0.25
Pine, Douglas fir, and brush	1626 ± 39	106 ± 20	3.0 ± 0.8
Douglas fir, true fir, and hemlock	1637 ± 103	89 ± 50	2.6 ± 1.6
Aspen, paper birch, and debris from jack pine	1664 ± 62	82 ± 36	1.9 ± 0.5
Black sage, sumac, and chamise	1748 ± 11	34 ± 6	0.9 ± 0.2
Jack pine, white and black spruce	1508 ± 161	175 ± 91	5.6 ± 1.7
"Chained" and herbicidal paper birch and poplar	1646 ± 50	90 ± 21	4.2 ± 1.3
"Chained" and herbicidal birch, polar and mixed hardwoods	1700 ± 82	55 ± 41	3.8 ± 2.8
Debris from hemlock, deciduous and Douglas fir	1600 ± 70	83 ± 37	3.5 ± 1.9
Overall average	1650 ± 29	83 ± 16	3.2 ± 0.5

Table IV
Emission factors for gases and ash based on laboratory experiments
(in % of fuel carbon and fuel nitrogen)[9]

	Mean	Range
CO_2	82.58	49.17 - 98.95
CO	5.73	2.83 - 11.19
CH_4	0.424	0.14 - 0.94
NMHC (as C)	1.18	0.14 - 3.19
Ash (as C)	5.00	0.66 - 22.28
Total sum C	94.91	—
N_2	21.60	—
NO_x	13.55	5.27 - 21.69
NH_3	4.15	1.04 - 11.74
HCN	2.64	0.31 - 6.75
CH_3CN	1.00	0.079 - 2.323
Nitrates	1.10	—
Ash (as N)	9.94	1.75 - 45.98
Total sum N	53.98	—

Table V
CO_2-normalized emission ratios for combustion species: summary of field measurements and laboratory studies in units of grams of species per kilograms of C in CO_2[18]

	Field measurements	Laboratory studies	"Best guess"
CO	6.5 - 140	59 - 105	100
CH_4	6.2 - 16	11 - 16	11
NMHCs	6.6 - 11.0	3.4 - 6.8	7
Particulate organic carbon (including elemental carbon)	7.9 - 54	—	20
Element carbon (black soot)	2.2 - 16	—	5.4
NO_x	2 - 8	0.7 - 1.6	2.1
NH_3	0.9 - 1.9	0.08 - 2.5	1.3
N_2O	0.18 - 2.2	0.01 - 0.05	0.1
H_2	33	—	33
SO_x	0.1 - 0.34	—	0.3
COS	0.005 - 0.016	—	0.01
CH_3Cl	0.023 - 0.033	0.02 - 0.3	0.05
O_3	4.8 - 40	—	30

EMISSION OF GASES

Once the mass of the burned biomass (M) and the species emission ratios (ER) are known, the gaseous and particulate species produced by biomass burn combustion may be calculated. The mass of the burned biomass (M) is related to the area (A) burned in a particular ecosystem by the following relationship:[3]

$$M = A \times B \times \alpha \times \beta \tag{3}$$

where B is the average biomass material per unit area in the particular ecosystem (g/m^2), α is the fraction of the average above-ground biomass relative to the total average biomass B, and β is the burning efficiency of the above-ground biomass. Parameters B, α, and β vary with the particular ecosystem under study and are determined by assessing the total biomass before and after burning.

The total area burned during a fire may be assessed using satellite data. Recent reviews have considered the extent and geographical distribution of biomass burning from a variety of space platforms: astronaut photography,[10] the NOAA polar orbiting Advanced Very High Resolution Radiometer (AVHRR),[11,12,13,14] the Geostationary Operational Environmental Satellite (GOES) Visible Infrared Spin Scan Radiometer Atmospheric Sounder (VAS)[15] and the Landsat Thematic Mapper (TM).[16]

Hence, the contribution of biomass burning to the total global budget of methane or any other species depends on a variety of ecosystem and fire parameters, including the particular ecosystem that is burning (which determines the parameters B, α, and β), the mass consumed during burning, the nature of combustion (complete vs. incomplete), the phase of combustion (flaming vs. smoldering), and knowledge of how the species emission factors (EF) vary with changing fire conditions in various ecosystems. The contribution of biomass burning to the global budgets of any particular species depends on precise knowledge of all these parameters. While all these parameters are known imprecisely, the largest uncertainty is probably associated with the total mass (M) consumed during biomass burning on an annual basis (and there are large year-to-year variations in this parameter!). The total mass of burned biomass material on an annual basis according to source of burning is summarized in Table VI.[3,4,17,18] The estimate for carbon released of 3940 Tg/yr includes all carbon species produced by biomass combustion (CO_2 + CO + CH_4 + NMHCs + particulate carbon). About 90% of the released carbon is in the form of CO_2 (about 3550 Tg/yr).

Biomass burning is indeed a significant global source of several important radiatively and chemically active species. Biomass burning may supply 40% of the world's annual gross production of CO_2 or 26% of the world's annual net production of CO_2 (due to the burning of the world's forests).[3,4,5,17,18,19] Biomass burning supplies 32% of the world's annual production of CO; 24% of the NMHCs, excluding isoprene and terpenes; 21% of the oxides of nitrogen (nitric oxide and nitrogen dioxide); 25% of the molecular hydrogen (H_2); 22% of the methyl chloride (CH_3Cl); 38% of the precursors that lead to the photochemical production of tropospheric ozone; 39% of the particulate organic carbon (including elemental carbon); and more than 86% of the elemental carbon.[5,18]

Table VI
Global estimates of annual amount of biomass burning and the resulting release of carbon and CO_2 to the atmosphere[3,4,17,18]

Source of burning	Biomass burned (Tg/yr)[1]	Carbon released (Tg(C)/yr)[2]	CO_2 released (Tg(C)/yr)[3]
Savanna	3690	1660	1494
Agricultural waste	2020	910	819
Fuel wood	1430	640	576
Tropical forests	1260	570	513
Temperate and boreal forests	280	130	117
Charcoal	21	30	27
World total	8700	3940	3546

1   Tg (teragram) = 10^6 metric tons = 10^{12} grams.
2   Based on a carbon content of 45% in the biomass material. In the case of charcoal, the rate of burning has been multiplied by 1.4.
3   Assuming that 90% of the carbon released is in the form of CO_2

HISTORIC CHANGES IN BIOMASS BURNING

It is generally accepted that the emissions from biomass burning have increased in recent decades, largely as a result of increasing rates of deforestation in the tropics. Houghton[19] estimates that gaseous and particulate emissions to the atmosphere due to deforestation have increased by a factor of 3 to 6 over the last 135 years. He also believes that the burning of grasslands, savannas, and agricultural lands has increased over the last century because rarely burned ecosystems, such as forests, have been converted to frequently burned ecosystems, such as grasslands, savannas, and agricultural lands. In Latin America, the area of grasslands, pastures, and agricultural lands increased by about 50% between 1850 and 1985. The same trend is true for South and Southeast Asia. In summary, Houghton[19] estimates that total biomass burning may have increased by about 50% since 1850. Most of the increase results from the ever-increasing rates of forest burning, with other contributions of burning (grasslands, savannas, and agricultural lands) having increased by 15% to 40%. The increase in biomass burning is not limited to the tropics. In analyzing 50years of fire data from the boreal forests of Canada, the U.S.S.R., the Scandinavian countries, and Alaska, Stocks[20] has reported a dramatic increase in area burned in the 1980s. The largest fire in the recent past destroyed more than 12 million acres of boreal forest in the People's Republic of China and Russia in a period of less than a month in May, 1987.[12]

The historic data indicate that biomass burning has increased with time and that the production of greenhouse gases from biomass burning has increased with time. Furthermore, the bulk of biomass burning is human-initiated. As greenhouse gases build up in the atmosphere and the Earth becomes warmer, there may be an enhanced frequency of fires. The enhanced frequency of fires may prove to be an important positive feedback on a warming Earth. However, the bulk of biomass burning worldwide may be significantly reduced. Policy options for mitigating biomass burning have been developed.[21] For mitigating burning in the tropical forests, where much of the burning is aimed at land clearing and conversion to agricultural lands, policy options include the marketing of timber as a resource and improved productivity of existing agricultural lands to reduce the need for conversions of forests to agricultural lands. Improved productivity will result from the application of new agricultural technology, i.e., fertilizers, etc. For mitigating burning in tropical savanna grasslands, animal grazing could be replaced by stall feeding since savanna burning results from the need to replace nutrient-poor tall grass with nutrient-rich short grass. For mitigating burning of agricultural lands and croplands, incorporate crop wastes into the soil, instead of burning, as is the present practice throughout the world. The crop wastes could also be used as fuel for household heating and cooking rather than cutting down and destroying forests for fuel as is presently done.

It is appropriate to conclude this chapter with an observation of fire historian, Stephen Pyne:[22]

> We are uniquely fire creatures on a uniquely fire planet, and through fire the destiny of humans has bound itself to the destiny of the planet.

REFERENCES

1. J. S. Levine. *Scientists on Gaia*. S. H. Schneider and P. J. Boston, editors. MIT Press, Cambridge, Mass., 353–361 (1991).
2. P. J. Crutzen, L. E. Heidt, J. P. Krasnec, W. H. Pollock, and W. Seiler. *Nature* **282**:253–256 (1979).
3. W. Seiler and P. J. Crutzen. *Climatic Change* **2**:207–247 (1980).
4. P. J. Crutzen and M. O. Andreae. *Science* **250**:1669–1678 (1990).
5. J. S. Levine. *EOS Transactions, American Geophysical Union* **71**:1075–1077 (1990).
6. J. S. Levine (editor). *Global biomass burning: Atmospheric, climatic, and biospheric implications*. MIT Press, Cambridge, Mass. (1991).
7. R. G. Prinn. In *Global biomass burning: Atmospheric, climatic, and biospheric implications*. J. S. Levine, editor. MIT Press, Cambridge, Mass., 22–28 (1991).
8. H. J. M. Bowen. *Environmental chemistry of the elements*. Academic Press, London (1979).
9. J. M. Lobert, D. H. Scharffe, W.-M. Hao, T. A. Kuhlbusch, R. Seuwen, P. Warneck, and P. J. Crutzen. In *Global biomass burning: Atmospheric, climatic, and biospheric implications*. J. S. Levine, editor. MIT Press, Cambridge, Mass., 289–304 (1991).
10. C. A. Wood and R. Nelson. In *Global biomass burning: Atmospheric, climatic, and biospheric implications*. J. S. Levine, editor. MIT Press, Cambridge, Mass., 29–40 (1991).
11. J. M. Brustet, J. B. Vickos, J. Fontan, K. Manissadjan, A. Podaire, and F. Lavenue. In *Global biomass burning: Atmospheric, climatic, and biospheric implications*. J. S. Levine, editor. MIT Press, Cambridge, Mass., 47–52 (1991).
12. D. R. Cahoon, Jr., J. S. Levine, W. R. Cofer III, J. E. Miller, P. Minnis, G. M. Tennille, T. W. Yip, B. J. Stocks, and P. W. Heck. In *Global biomass burning:*

Atmospheric, climatic, and biospheric implications. J. S. Levine, editor. MIT Press, Cambridge, Mass., 61–66 (1991).
13. J. M. Robinson. *Int'l. Journ. of Remote Sensing* **12**:3–24 (1991).
14. J. M. Robinson. In *Global biomass burning: Atmospheric, climatic, and biospheric implications.* J. S. Levine, editor. MIT Press, Cambridge, Mass., 67–73 (1991).
15. W. P. Menzel, E. C. Cutrim, and E. M. Prins. In *Global biomass burning: Atmospheric, climatic, and biospheric implications.* J. S. Levine, editor. MIT Press, Cambridge, Mass., 41–46 (1991).
16. J. M. Brustet, J. B. Vickos, J. Fontan, A. Podaire, and F. Lavenue. In *Global biomass burning: Atmospheric, climatic, and biospheric implications.* J. S. Levine, editor. MIT Press, Cambridge, Mass., 53–60 (1991).
17. W. M. Hao, M. H. Liu, and P. J. Crutzen. In *Fire in the tropical biota: Ecosystem processes and global challenges.* J. G. Goldammer, editor. Springer-Verlag, Berlin-Heidelberg, 440–462 (1990).
18. M. O. Andreae. In *Global biomass burning: Atmospheric, climatic, and biospheric implications.* J. S. Levine, editor. MIT Press, Cambridge, Mass., 3–21 (1991).
19. R. A. Houghton. In *Global biomass burning: Atmospheric, climatic, and biospheric implications.* J. S. Levine, editor. MIT Press, Cambridge, Mass., 321–325 (1991).
20. B. J. Stocks. In *Global biomass burning: Atmospheric, climatic, and biospheric implications.* J. S. Levine, editor. MIT Press, Cambridge, Mass., 197–202 (1991).
21. K. J. Andrasko, D. R. Ahuja, S. M. Winnett, and D. A. Tirpak. In *Global biomass burning: Atmospheric, climatic, and biospheric implications.* J. S. Levine, editor. MIT Press, Cambridge, Mass., 445–456 (1991).
22. S. J. Pyne. In *Global biomass burning: Atmospheric, climatic, and biospheric implications.* J. S. Levine, editor. MIT Press, Cambridge, Mass., 504–511 (1991).
23. W. R. Cofer III, J. S. Levine, E. L. Winstead, and B. J. Stocks. In *Global biomass burning: Atmospheric, climatic, and biospheric implications.* J. S. Levine, editor. MIT Press, Cambridge, Mass., 203–208 (1991).
24. L. F. Radke, D. A. Hegg, P. V. Hobbs, J. D. Nance, J. H. Lyons, K. K. Laursen, R. E. Weiss, P. J. Riggan, and D. E. Ward. In *Global biomass burning: Atmospheric, climatic, and biospheric implications.* J. S. Levine, editor. MIT Press, Cambridge, Mass., 209–224 (1991).

MODELING AND MEASUREMENTS

FORECAST CLOUDY: THE LIMITS OF GLOBAL WARMING MODELS*

Peter H. Stone
Earth, Atmospheric, and Planetary Sciences
Massachusetts Institute of Technology
Cambridge, MA 02139

ABSTRACT

Predictions of climate change rest on models that are far from complete. But better observations, more powerful computers, and improved understanding can help us fill the gaps.

INTRODUCTION

A report published in late 1990 by the Intergovernmental Panel on Climate Change (IPCC) warned that global warming could soon force temperatures higher than they have been in hundreds of thousands of years. The report, prepared by 170 scientists from all over the world, concluded that if the world's economies follow a "business-as-usual" scenario, increases in carbon dioxide and other trace gases in the atmosphere will cause the earth's average surface temperature to rise by about 5°F before the end of the next century. Such a rise would come on top of the warming of about 15°F that has already occurred since the last major ice age some 15,000 years ago.

The IPCC report was hardly the first attempt to assign a number to the effect of increases in trace gases. That distinction belongs to the Swedish scientist Arrhenius who, almost 100 years ago, calculated that a doubling of CO_2 would cause a rise of 10°F. Since then, CO_2 doubling has become a standard yardstick for gauging global climate sensitivity. It is also a realistic yardstick, because current trends would produce a level of trace gases equivalent to a doubling of CO_2 by the middle of the next century.

The first modern estimate of the effect of CO_2 doubling was made in 1967 by Syukuro Manabe and Richard Wetherald at the National Oceanic and Atmospheric Administration's Geophysical Fluid Dynamics Laboratory in Princeton, N.J. The warming they predicted: 4°F. In 1979 a National Research Council committee chaired by Jule Charney of MIT, recognizing the uncertainties involved, estimated a range of values: 3°F to 8°F. The most recent estimates, including those by the IPCC, still fall within this range. In fact, considering that they benefit from supercomputers and other advances, it is remarkable that the latest predictions are not farther from the figure that Arrhenius arrived at in 1896.

At first glance, this rough consensus might seem to close the book on the issue of global warming: human-produced greenhouse gases such as CO_2 will cause a serious rise in global temperatures, and that's that. Indeed, the apparent robustness of these numbers is why most scientists believe that global warming is a serious threat. But we

---

*A version of this article appeared in *Technology Review*, February/March 1992. Reprinted with permission

have much more to learn. The IPCC report was quick to point out the many question marks in its predictions, especially regarding the timing, magnitude, and regional patterns of climate change. These uncertainties, common to all climate predictions, stem from the complexity of the physics involved and the coarseness of the models that struggle to simulate it.

A climate model consists of mathematical equations based on the fundamental laws of physics. Solving these equations—a task usually done on large computers—can determine how climate variables such as temperature, humidity, winds, and precipitation will respond to changes in factors like the amount of solar radiation reaching the earth, or the concentrations of trace gases in the atmosphere. The climate system is so complex, however, that a model incorporating all the possible variables for all parts of the globe could not be run on even the fastest supercomputers.

As a result, scientists use a wide variety of models to study climate and climate change. At one extreme are simple models that severely limit the number of variables they try to predict (forecasting only temperature, for example), or that severely restrict the physical and chemical processes they include (omitting, say, heat transport by ocean currents). At the other extreme are the large numerical general circulation models (GCMs) that include as many variables and processes as possible.

Because even these models are not truly comprehensive, the simple models play a valuable role in determining what variables and processes are important, thereby allowing scientists to improve the larger models. Also, there are many problems for which the large models are computationally too inefficient to be practical—for example, problems involving climate changes over hundreds of years. For this reason, the IPCC projections for the next century were based on one of the simplest models. Ultimately, however, only GCMs will be able to yield accurate predictions of changes in all the climate variables anywhere on earth.

A COMPLICATED PLANET

Modeling climate change is inherently difficult. To do so, climatologists must try to simulate the behavior of oceanic and atmospheric systems that are not only fantastically complex in themselves but intricately linked.

Just figuring out how fast greenhouse gases will build up is hard enough. The constituents of the atmosphere that absorb the most infrared radiation, and therefore contribute most strongly to the greenhouse effect, are water vapor and clouds. But other gases contribute as well, and their concentration in the atmosphere is growing, mainly because of human activities that are impossible to predict with certainty even in the short term. In the 1980s, for example, chlorofluorocarbons (CFCs) increased by 40 percent, methane by 10 percent, and CO_2 by 4 percent. At these rates, CFCs would replace carbon dioxide as the major contributor to increases in global warming in 25 years. Yet international agreements such as the Montreal Protocol could slow the increase of CFCs, altering the picture considerably.

There are also major scientific uncertainties about the buildup. A portion of the gases added to the atmosphere does not remain there but is absorbed by the biosphere and the oceans or destroyed by chemical reactions. This happens to about half the carbon dioxide now being added to the atmosphere, but the fraction that is removed may vary as climate changes, thus modifying the climate change. The natural processes that remove these gases are not well understood, and no GCM has tried to include them. Until we can predict this kind of change, our models will be incomplete. Because analyses of deep ice cores drilled in the Antarctic and Greenland show that major changes in atmospheric concentrations of carbon dioxide and methane have taken place in the past, this defect in the models represents a major uncertainty.

Forecasting the buildup of gases is only the beginning. The next step, predicting how the buildup will affect temperatures, is a task of extraordinary complexity. Consider the many factors that govern just one key component of the climate system: the planetary albedo, or the fraction of solar radiation that the earth reflects back to space.

If the albedo increases, all else being equal, temperatures will fall. The albedo is affected by clouds, the polar ice caps, glaciers, snow, vegetation, the surface of the ocean, and dust particles in the atmosphere, to name just a few influences. How much solar radiation each component reflects depends on properties that can vary widely— for example, the water content of the clouds, the composition of the dust particles, the age of the snow, the roughness of the ocean surface, and the health of the vegetation. In principle, all these details must be predicted if one is to model climate change accurately. To complicate matters, albedo can be affected by unforeseeable events such as volcanic activity. Indeed, the recent eruption of Mount Pinatubo in the Philippines is likely to cause global cooling over the next few years until the volcanic particles fall out of the atmosphere.

It is especially difficult to predict the way climate will change in a particular region. For some regions, climate models do not even agree on whether temperatures will rise or fall. A fundamental problem is that the atmosphere and the oceans are fluids that move in response to changes in temperature and pressure. The resulting winds and ocean currents transport heat from one locality to another, modifying temperatures. Because of these fluid motions, every point in the earth-atmosphere-ocean system is coupled to every other point in the system. Climate change at one point cannot be predicted accurately without also predicting changes at other points and changes in the fluid motions that couple them. These fluid motions affect predictions of global average temperatures as well.

Another basic problem is that such motions are chaotic. Although they are governed by the classical laws of physics, which in principle yield predictable results, a small uncertainty in our knowledge of the state of the system at any given time leads to a large uncertainty later. Any gap in our information about the state of the atmosphere, no matter how small, makes it impossible to predict the weather more than about three weeks in advance. The resulting unpredictable fluctuations in weather cause unpredictable fluctuations in climate (which can be defined as the average weather). According to calculations carried out at NASA's Goddard Institute for Space Studies (GISS) in New York City, the chaotic behavior of the atmosphere can cause fluctuations of as much as 1°F over periods of about 30 years.

Although a "noise level" of 1°F is small compared with the projected warming of 5°F before the end of the next century, other possible sources of unpredictable behavior have yet to be assessed. For all we know, chaotic fluid motions in the ocean might produce unpredictable climate changes that are larger still. Indeed, unexplained fluctuations much greater than 1°F have occurred in the past, most recently about 10,000 years ago during the so-called Younger-Dryas cold interval.

Thus any effort to predict climate changes assumes that climate is predictable—but this is not guaranteed. Forecasts of the effects of a rise in greenhouse gases are really just predictions of what will happen in the absence of the unpredictable.

PROBLEMS OF SCALE

Calculating just the predictable part of climate change is still a formidable problem. This is not simply because we don't fully understand the physics of the climate system; it is also because the "resolution" of today's general circulation models is extremely low. Not only are they unable to differentiate between the climates of, say, Buffalo and Boston—a limitation that severely restricts our ability to predict regional climate—but they cannot accurately calculate the effects of a number of important physical phenomena that take place on scales smaller than the models' resolution.

One example is clouds, which contribute greatly to the planetary albedo and the greenhouse effect. Another is moist convection, which both cools the surface of the earth and affects the concentration of water vapor. Also not resolvable are hydrological processes that affect the amount of moisture in the soil—an aspect of climate that is important for agriculture and water resources and that is likely to change as a result of global warming. Because of the complex relationships within the climate system, errors in calculating these processes can seriously compromise a model's ability to simulate climate even on the largest scales.

It is because of doubts over whether the models are simulating the small-scale processes accurately that some scientists, such as Richard Lindzen of MIT, are skeptical of the predictions of global warming. Nevertheless, most scientists, myself among them, believe that the range of values climatologists usually quote—as in the Charney committee's 3°F to 8°F—largely accounts for this uncertainty.

A major reason for the models' shortcomings in calculating the effects of small-scale processes is the limited capacity of computers, which restricts the number of locations in the climate system whose state a general circulation model can describe. All climate models must make trade-offs between the number of locations they simulate, the number of climate processes they calculate, and the accuracy with which they calculate those processes.

Today's highest-resolution climate GCMs specify the state of the atmosphere at the intersections of a three-dimensional grid. This grid is divided into sections that are approximately 250 miles on a side in the horizontal direction—an area the size of New England and New York combined—and about a mile thick in the vertical direction. Since these models incorporate five variables at each intersection of the grid (temperature; wind speed in the latitudinal, longitudinal, and vertical directions; and concentration of water vapor), they must predict about 150,000 numbers to describe the state of the atmosphere at a given time.

To keep up with atmospheric changes, these 150,000 numbers have to be recalculated about eight times an hour. Thus, to determine the evolution of the atmosphere—just one part of the climate system—over one year with such a model requires some 20 to 40 hours of calculation on a supercomputer.

The effects of small-scale phenomena cannot be completely left out of GCMs, or the models could not come close to simulating the current climate, much less changes in climate. The makeshift solution the models employ is to "parameterize" such effects. In other words, they simulate the effects by simplified equations based in part on current climate conditions and in part on approximations deemed reasonable in particular circumstances. By design, these simplified equations can be solved far more efficiently than the exact equations, but at the cost of accuracy; indeed, the simplified equations are often quite crude.

What's more, different models use different parameterizations, which lead to contradictory conclusions about the regional effects of global warming. It's for that reason that GCMs, despite yielding similar predictions for how much global mean temperatures will increase, disagree sharply on the patterns and magnitude of changes in soil moisture. For example, two models—those of the National Center for Atmospheric Research and the Geophysical Fluid Dynamics Laboratory—predict that a doubling of CO_2 would make southern California winters drier, while the GISS and United Kingdom Meteorological Office models indicate that the region's winters would become wetter.

It would be nice if we could get by without parameterizations. Unfortunately, calculating small-scale processes accurately requires much higher resolution than computers will be able to deliver in the near future. For example, the important variations in moist convection occur on scales of 100 yards to half a mile. Resolving these processes would require a grid with a horizontal spacing 1,000 times smaller than today's climate GCMs in both latitude and longitude. And to resolve the rapid evolution of the small-scale features of moist convection, the state of the atmosphere would have to be recalculated about 1,000 times more frequently. Some increase in vertical resolution would also be necessary. All in all, computers would have to be about 10 billion times faster than today's to calculate moist convection accurately. So models must depend on parameterizations for a long time to come.

Even if our models could predict with certainty the ultimate effects of increases in trace gases, such as a doubling of carbon dioxide—and even if we knew precisely how fast the trace gases would increase—we would still need to know how quickly the climate would respond to the buildup. After all, people and ecosystems will adapt to climate change much more easily if it happens slowly.

The oceans play the biggest role in determining the rate of warming, because they are the component of the climate system with the greatest capacity to absorb heat. If warming seeps down only slowly into the ocean's deeper layers, the surface layers—and hence the atmosphere—will heat up rapidly. Conversely, if the deeper layers absorb heat quickly from the surface layers, the atmosphere will take longer to warm up. Thus the 5°F warming predicted by the IPCC report might occur as early as 2040 or as late as 2200.

How fast the warming actually spreads to the deeper layers depends on the ocean's circulations. But because of computer limitations, GCMs that try to calculate this process have inadequate resolution. These models do such a poor job of simulating today's climate—sometimes misrepresenting sea surface temperatures by as much as 15°F—that we cannot have much confidence that they are simulating the physics of heat mixing correctly.

REMOVING DOUBT

Models may never be able to reproduce all climate processes with absolute fidelity. However, they may not need to. It would be enough to construct a model incorporating only the processes that have a significant impact on climate and using good parameterizations of the processes that the model cannot resolve. Achieving this would require a better understanding of many of the physical and chemical processes involved. To this end, the bigger and faster computers likely to emerge over the next decade will let us carry out many more "sensitivity studies" to narrow down the processes that need to be included in the models. But the key ingredient necessary for improving our understanding—as well as for validating the models—is more comprehensive observations of the climate system.

NASA's proposed Earth Observing System (EOS) could make a major contribution to gathering some of the necessary data. The agency's original proposal called for launching two series of polar-orbiting satellites packed with instruments to monitor many climate processes simultaneously. The first satellite would have been launched in 1998, and EOS orbiters would have continued making observations for 15 years, enough time to monitor long-term changes. Major goals included expanding our knowledge of small-scale hydrological processes and the biological processes that affect CO_2 concentrations. Meeting these goals would improve our ability to predict changes in regional climate and bolster our confidence in predictions of global warming. Because of recent congressional budget cuts, however, EOS will have to be scaled down.

Addressing another big limitation of climate models—their rudimentary treatment of ocean dynamics—will require bigger and faster computers. William Holland and Frank Bryan at the National Center for Atmospheric Research in Boulder, Colo., have already carried out preliminary experiments with high-resolution ocean models that can accurately characterize important large-scale processes. Computers that are about 10,000 times as powerful as today's machines would enable us to include one of these ocean models in a global climate model. With the necessary resources, massive parallel processing machines could allow this soon, perhaps in three or four years.

But the ocean models, too, will still have to be validated against observations. Without good ocean models, our ability to predict regional climate changes and the rate of global warming will be severely limited. A project that could provide the necessary data is the World Ocean Circulation Experiment, a multinational project started in 1985. Plans call for mapping ocean circulations by taking measurements from ships, moored arrays, and subsurface floats over a period of 10 or more years. Unfortunately, funding constraints have already brought about so many cuts in the original program that WOCE may not yield enough data for testing ocean models adequately.

In view of the funding difficulties of large projects like EOS and WOCE, scientists are scurrying to come up with less costly ways of gathering the most crucial information for improving climate predictions. One project that holds great promise is an experiment devised by Walter Munk of the Scripps Institution of Oceanography in La Jolla, Calif., and Andrew Forbes of the Commonwealth Scientific and Industrial Research Organization in Australia.

The two researchers propose placing acoustic sources deep in the oceans at different locations around the world and then listening to the signals at a distance with hydrophones. Since the speed of sound in the ocean depends on the temperature of the water, measurements of the time delay between generating and receiving the acoustic signals will reveal the mean temperature along the path traveled by the sound waves. By measuring the temperature along many paths, it would be possible to determine how rapidly the deep oceans are warming and thereby improve predictions of how rapidly global temperatures will rise.

The project, scheduled to start in 1993, has received initial funding from a U.S. interagency group, and its feasibility has already been tested. If all goes well, accurate measurements of ocean warming will be available sometime in the first decade of the next century.

Although research efforts like Munk and Forbes's could lead to more reliable climate predictions within 15 or 20 years, some of the more extreme projections raise the possibility that global warming will outrun our ability to forecast it accurately. But even if that happened, we would still have compelling reasons to continue working on the climate modeling problem. Global warming, if it does occur, is unlikely to be the last environmental change we bring upon ourselves. So if we are ever to learn to foresee the consequences of our actions, we must improve our understanding of climate and our ability to model it.

OCEANIC ASPECTS AND GLOBAL CHANGE

Jochem Marotzke
Center for Meteorology and Physical Oceanography
Department of Earth, Atmospheric, and Planetary Sciences
Massachusetts Institute of Technology
Cambridge, Massachusetts 02139

ABSTRACT

The ocean's role in climate and climatic change is discussed, with emphasis on the consequences of changes in the ocean circulation. Observations indicate variability in the North Atlantic on an interdecadal timescale; it is argued that the variability is caused by interaction between the ocean circulation and the atmosphere. Ocean-only models can be constructed that capture the most fundamental aspects of large-scale air-sea interaction. These models exhibit multiple equilibrium states with dramatically different heat transport and hence important impacts on climate. Whether changes in the ocean circulation, including transitions between different steady states, are likely to occur, can only be understood through a synthesis of oceanic *and* atmospheric dynamics.

INTRODUCTION

The atmosphere has a small heat capacity which is matched by the upper 2 or 3 meters of the ocean, and the entire heat content of the atmosphere is recycled within less than a year by ocean-atmosphere interactions. Consequently, the atmosphere responds to changes in external forcing within a couple of months. In contrast, processes exist in the ocean that act on timescales of decades to centuries, and any understanding of climate change on these longer timescales therefore involves the ocean. Only after the ocean's role has been understood in much greater detail than it is today will it be possible to distinguish, in the observed records, between natural, long-term climate variability and anthropogenic climate change.

Because of the ocean's enormous capacity to absorb heat (and carbon dioxide) one might conclude that its main role in the climate system is that of a buffer, which integrates all the fast disturbances from the atmosphere above. This is the basis of Hasselmann's[1] theory of climate variability, essentially considering the interaction between the atmosphere and the oceanic surface mixed layer (typically between 30 and 100 m deep). Notice that this apparently inert block of water by no means merely averages out atmospheric fluctuations. Rather, slight imbalances in the latter will sum up to considerable excursions around a mean state; in a model simulation of an atmosphere coupled to an oceanic mixed layer, sea surface temperature fluctuated by 0.5°C on interannual time scales[2].

The above view does not include changes in the ocean circulation, (i.e., does not invoke internal oceanic dynamics), hence I will call it the *passive* role of the ocean in climate change. There exists, however, an entirely different scenario, as apparently first formulated by Bjerknes[3] (see also Bryan and Stouffer[4]). Bjerknes speculated that changes in North Atlantic sea surface temperatures might be caused by changes in the North Atlantic circulation, which would alter the oceanic heat transport. This scenario clearly invokes ocean dynamics and changes in circulation, and I will therefore call it the *active* role of the ocean in climate change.

From now on, I will focus entirely on the *active* role since its understanding has made rapid progress in the past few years. As Bryan and Stouffer[4] remark, Bjerknes's[3] hypothesis went largely unnoticed because it was not supported by a convincing physical mechanism. Ironically, three years earlier Stommel[5] had found multiple equilibrium states in an extremely simplified model of thermohaline flow (that is, driven by temperature and salinity differences). Stommel's paper went likewise unnoticed, and it was not until the early and middle 1980s that Rooth[6], F. Bryan[7], and Welander[8] again took up the subject of explaining different climate states in terms of different ocean circulation patterns.

This brief review is organized as follows. In the next section, I will discuss some key observations and physical processes relevant to a discussion of ocean climate and its changes. In section 3, modeling results are presented which stress the role of ocean circulation and dynamics in climate change, particularly in the North Atlantic ocean. A short outlook follows in section 4.

KEY PROCESSES AND OBSERVATIONS

The thermohaline circulation is thought to be driven by the sinking of high-latitude water to depths of 1000 m to 4000 m. In winter, surface water is cooled through heat loss to the atmosphere; whether the water can actually sink, however, depends to a large extent on the salinity with which it had been carried poleward. At low temperatures, salinity variations have a much greater impact on density than has temperature. Thus, North Atlantic water with salinity lower than 34.7 per mil would freeze and float as sea ice on the denser, deeper layers, whereas it would convectively overturn and mix with the underlying water if it started off with a salinity larger than 34.7 per mil. The most fundamental difference in water mass properties between North Atlantic and North Pacific is the difference in salinity. The North Pacific is generally too fresh to form deep water (Warren[9]); in contrast, North Atlantic Deep Water (NADW) is formed at a rate of $(15-20) \times 10^6$ m^{3}s^{-1}, and is transported into the South Atlantic and the Indian and Pacific Oceans, where it upwells and eventually returns into the North Atlantic as warm near-surface water. This "global conveyor belt"[10] transports between 0.5 and 1×10^{15}W of heat into the North Atlantic, keeping its surface temperatures 2°–4°C higher than the North Pacific's, and rendering European climate much milder than corresponding regions on the American continent.

There is observational evidence, albeit in parts indirect, that high-latitude North Atlantic temperatures and salinities vary considerably on interdecadal timescales (see, e.g., the review by Mysak and Lin[11]). For example, the duration of coastal ice on the north coast of Iceland varied between almost zero and 25 weeks per winter, during the last five centuries[11]. In the 20th century, sea ice abundance around Iceland was high in the first two decades and between 1965 and 1975[12]. The latter occurrence coincided with the most dramatic excursion of high-latitude ocean climate that has been directly observed to date. From the mid 1960s until 1982, a patch of anomalously fresh water

(0.1 to 0.15 per mil below normal), extending to at least 500 m in depth and on the order of 1000×1000 km² horizontally, migrated from north of Iceland to the Labrador sea, and returned via the Faroe-Shetland Channel to the Greenland sea. This "Great Salinity Anomaly" of the 1970s[12] disrupted convective sinking in the Labrador sea in the years 1968 to 1972[13], thereby shutting off a major contribution to the formation of North Atlantic Deep Water. Tracer studies indicate that simultaneously with the return of the "Great Salinity Anomaly" to the Greenland Sea around 1980, deep water renewal rates there dropped drastically[14]. Consequently, the second main component of North Atlantic Deep Water was formed at a much reduced rate during the 1980s.

There are three major feedbacks between the thermohaline circulation and the high-latitude temperature and salinity (and hence density) fields[15]. A strong thermohaline circulation transports salinity northward, tends to increase high-latitude salinities and hence surface densities, which strengthens deep water formation and thus the thermohaline overturning. This positive feedback is counteracted by northward heat transport (as discussed above), which acts to raise temperatures and thus lower densities, which in turn weakens convective activity and the thermohaline circulation. Another positive feedback is provided by increased evaporation when surface temperatures rise. Since pure freshwater leaves the ocean, the remaining waters become saltier, and higher surface temperatures may actually produce denser waters if the third feedback is stronger than the second one. A complete understanding of all the feedbacks involves interacting oceanic and atmospheric processes. The picture is complicated further by the transport of sea ice through the Arctic ocean into the North Atlantic, which upon melting substantially lowers surface salinities since sea ice typically contains less than 4 per mil salt.

OCEAN MODELING STUDIES:
MULTIPLE STEADY STATES AND DECADAL VARIABILITY

The radiation balance and atmospheric transport processes together exert a strong control on the sea surface temperature and the flux of freshwater into the ocean (precipitation minus evaporation, P–E). Of the three feedbacks listed above, the first one can thus be expected to dominate: There exists no local feedback to quickly remove salinity anomalies if the ocean circulation and the P–E field collaborate to produce major excursions of the surface salinity field. In contrast, the atmosphere strongly damps sea surface temperature anomalies through enhanced or reduced heat exchange, and P–E is then also strongly controlled. This deduction gives a recipe of how to formulate a conceptual atmosphere in conjunction with a purely oceanic model if, because of computational cost or simplicity of approach, one does not want to employ fully coupled ocean-atmosphere models. Sea surface temperature and surface freshwater fluxes are specified as upper boundary conditions in an ocean model, meaning that the most fundamental aspects of the large-scale air-sea interaction are captured. These "mixed boundary conditions" on temperature and salinity represent a conceptual formulation of a coupled ocean-atmosphere system.

All models with this type of boundary conditions exhibit multiple equilibrium states, with the deep water formation sites being radically different between them. For example, in a three-dimensional global ocean model with highly idealized geometry (two identical basins representing the Atlantic and Pacific, and a circumpolar channel in the south), four different stable equilibria are found under the same set of boundary conditions[16]. Two of the equilibria show both oceans in the same state, with high-latitude deep water formation occurring in both northern or in both southern oceans. Two additional equilibria exist in which the thermohaline circulations of the basins differ fundamentally from each other: one ocean forms deep water at northern high latitudes, while the other has a much weaker circulation with sinking in the Southern Hemisphere. One of these equilibria qualitatively corresponds to today's global thermohaline circulation pattern (conveyor belt).

It has been speculated that radical changes in the earth's climate were associated with, or caused by, transitions between different equilibrium states of the global thermohaline circulation[17]. We know now that such a transition is dynamically possible, under moderate perturbations of the surface freshwater fluxes[16]. A recent computation with a comprehensive climate model[18] (comprising ocean-ice-atmosphere-land interactions) shows, as a result of a gradual increase of the atmospheric carbon dioxide concentration, an increased atmospheric water vapor transport from the subtropics to high latitudes. As a consequence, particularly North Atlantic surface salinities decrease, and the Atlantic thermohaline overturning weakens by 35% within 80 years. One may speculate whether eventually a threshold is passed beyond which the global conveyor belt collapses altogether. Clearly, to predict climate change one must understand the ocean circulation and its changes very thoroughly.

As a step into this direction, attention is increasingly focused on oceanic processes involved in inter-decadal variability. An idealized three-dimensional ocean model, representing crudely the North Atlantic, shows spontaneous decadal variability in its deep water formation rate, although it is driven by (mixed) boundary conditions which are constant in time[19,20]. Weaver et al.[21,22] also show that the existence of the internal variability critically depends on the exact P–E fields by which the model is forced. P–E is a quantity notoriously badly known over the oceans, so the question remains unresolved whether the observed climate variability can actually be explained by purely oceanic processes.

OUTLOOK

The global thermohaline circulation can change radically on timescales of a century or less, as demonstrated by three-dimensional ocean models. Therefore, any discussion of global change due to increased concentration of trace gases in the atmosphere must include ocean dynamics. How strong the climate system must be perturbed to induce a flip between different equilibrium states, or a long-term excursion before returning to the old state, is not known reliably. There seems to be a consensus that the ocean models employing mixed boundary conditions in the traditional way[16,22] are too sensitive, in particular to P–E variations. The sensitivity to P–E might be overestimated by the models because only the feedback between circulation and the salinity field is accounted for; the circulation-temperature and temperature-evaporation-salinity feedbacks are very likely to affect the stability of the models. Efforts seem to concentrate on two strategies. At the Geophysical Fluid Dynamics Laboratory in Princeton, for example, the long-term variability in a 1000-year run of the global climate model is investigated, with special attention on high-latitude, large-scale air-sea interactions[23]. Still in its infancy are attempts to include one or both of the additional feedbacks in an ocean-only model, through a more sophisticated formulation of the thermohaline surface boundary conditions. Related to this approach is the coupling of simple ocean-atmosphere models as done by Stocker et al.[24]. To include sea ice formation and transport would also be essential for a deeper understanding of the stability of the thermohaline circulation.

The *active* role of the ocean is a major dynamical factor in climate change, and warrants still more intensive research.

ACKNOWLEDGMENTS

I wish to thank the organizers of the Symposium and the MIT lecture series "Global Change," Rafael Bras, Ronald Prinn, and Paola Malanotte-Rizzoli, for providing a forum to present these ideas. This work was supported by Grant No. OCE-8823043 from the National Science Foundation.

REFERENCES

1. K. Hasselmann. 1974. Stochastic climate models, Part I: Theory. *Tellus* **28**:289–305.
2. J. Hansen and S. Lebedeff. 1987. Global trends of measured surface air temperature. *J. Geophys. Res.* **92**:13345–13372.
3. J. Bjerknes. 1964. Atlantic air-sea interaction. *Adv. Geophys.* **10**:1–82.
4. K. Bryan and R. Stouffer. 1991. A note on Bjerknes' hypothesis for North Atlantic variability. *J. Mar. Systems* **1**:229–241.
5. H. Stommel. 1961. Thermohaline convection with two stable regimes of flow. *Tellus* **13**:224–230.
6. C. Rooth. 1982. Hydrology and ocean circulation. *Progr. Oceanogr.* **11**:131–149.
7. F. Bryan. 1986. High-latitude salinity effects and interhemispheric thermohaline circulations. *Nature* **323**:301–304.
8. P. Welander. 1986. Thermohaline effects in the ocean circulation and related simple models. In *Large-scale transport processes in oceans and atmosphere*. J. Willebrand and D. L. T. Anderson, Eds. D. Reidel, 163–200.
9. B.A. Warren. 1983. Why is no deep water formed in the North Pacific? *J. Mar. Res.* **41**:327–347.
10. A.L. Gordon. 1986. Interocean exchange of thermocline water. *J. Geophys. Res.* **91**:5037–5046.
11. L.A. Mysak and C. A. Lin. 1990. Role of the oceans in climatic variability and climatic change. *The Canadian Geographer* **4**:352–369.
12. R.R. Dickson, J. Meincke, S.-A. Malmberg, and A. J. Lee. 1988. The "Great Salinity Anomaly" in the northern North Atlantic, 1968–1982. *Progr. Oceanogr.* **20**:103–151.
13. J. Lazier. 1980. Oceanic conditions at Ocean Weather Ship BRAVO, 1964–1974. *Atmos.-Ocean* **18**:227–238.

14. P. Schlosser, G. Boenisch, M. Rhein, and R. Bayer. 1991. Reduction of deepwater formation in the Greenland Sea during the 1980s: Evidence from tracer data. *Science* **251**:1054–1056.
15. J. Willebrand. 1992. Forcing the ocean with heat and freshwater fluxes. To appear in: *Energy and water cycles in the climate system*, E. Raschke, Ed. Springer Verlag.
16. J. Marotzke, and J. Willebrand. 1991. Multiple equilibria of the global thermohaline circulation. *J. Phys. Oceanogr.* **21**:1372–1385.
17. W.S. Broecker, D. M. Peteet, and D. Rind. 1985. Does the ocean-atmosphere system have more than one stable mode of operation? *Nature* **315**:21–26.
18. S. Manabe, R. J. Stouffer, M. J., Spelman, and K. Bryan. 1991. Transient responses of a coupled ocean-atmosphere model to gradual changes of atmospheric CO_2, Part 1: Annual mean response. *J. Climate* **4**:785–818.
19. A.J. Weaver, and E. S. Sarachik. 1991. The role of mixed boundary conditions in numerical models of the ocean's climate. *J. Phys. Oceanogr.* **21**:1470–1493.
20. A.J. Weaver, and E. S. Sarachik. 1991. Evidence for decadal variability in an ocean general circulation model: An advective mechanism. *Atmos.-Ocean* **29**:197–231.
21. A.J. Weaver, E. S. Sarachik, and J. Marotzke. 1991. Internal low frequency variability of the ocean's thermohaline circulation. *Nature* **353**:836–838.
22. A.J. Weaver, J. Marotzke, P. F. Cummins, and E. S. Sarachik. 1992. Stability and variability of the thermohaline circulation. *J. Phys. Oceanogr.* In press.
23. T. Delworth. 1992. Personal communication.
24. T.F. Stocker, D. G. Wright, and L. A. Mysak. 1992. A zonally averaged coupled ocean-atmosphere model for paleoclimatic studies. *J. Climate*. In press.

BRIDGING THE GAP BETWEEN MICROSCALE LAND-SURFACE PROCESSES
AND LAND-ATMOSPHERE INTERACTIONS AT THE SCALE OF GCM'S

Roni Avissar
Department of Meteorology and Physical Oceanography
Rutgers University
New Brunswick, New Jersey 08903

ABSTRACT

Assuming that instantaneous wind can be separated into synoptic scale, mesoscale, and turbulent scale within a grid element of a GCM, the mesoscale kinetic energy per unit mass is defined as half the variance of mesoscale perturbations from the synoptic-scale wind. A simulation of the atmospheric boundary layer that develops above a locally deforested region is used to illustrate the relation that exists between mesoscale kinetic energy and mesoscale latent and sensible heat fluxes. A prognostic equation for mesoscale kinetic energy is suggested to relate subgrid-scale landscape heterogeneity to subgrid-scale convective clouds, which could be used to improve the parameterization of clouds and precipitation in GCMs.

INTRODUCTION

It is widely agreed upon that the parameterization of the Earth's surface is one of the more important aspects of climate modeling, since this surface absorbs over 70% of the energy absorbed into the climate system, and many physical, chemical, and biological processes take place there. Of particular importance are the exchanges of mass (notably water), momentum, and energy between the surface and the atmosphere[1].

State-of-the-art parameterizations are based on the concept of "big leaf" which implies that the land is homogeneously covered by a big leaf within a grid element of the numerical atmospheric model. At the scale of resolution of GCMs, however, continental surfaces are very heterogeneous. This can be readily appreciated, for instance, by examining maps of soil, vegetation, topography, or land use patterns. Thus, recently, parameterizations based on the statistical-dynamical approach have been suggested[2,3,4,5,6]. With such an approach, probability density functions (pdf's) are used to represent the variability of the various characteristics of the soil-plant-atmosphere system that affect the input and redistribution of energy and water at the land surface. Collins and Avissar[7] showed that stomatal conductance and surface roughness are probably the most important parameters to be represented by pdf's.

But extended landscape heterogeneities that result, for instance, from the juxtaposition of land and bodies of water or bare-soil and vegetated areas are likely to produce mesoscale atmospheric circulations[4,8,9]. These circulations may have a significant impact on various atmospheric processes, e.g., cloud formation, aerosols and gas transport, etc. For instance, Avissar (1991)[4] discussed the possible effects of developing agriculture in arid regions and in deforested tropical forests on convective cloud formation. He concluded that these processes, which develop at a smaller scale than the horizontal grid resolution of GCMs, are likely to affect significantly the hydrologic cycle, the climate, and the weather. Consequently, he emphasized the need to parameterize these subgrid-scale processes in GCMs. Furthermore, Pielke et al. (1991)[10] demonstrated that mesoscale heat fluxes produced by such circulations may be more significant than the turbulent heat fluxes in the planetary boundary layer (PBL).

The aim of the present paper is to discuss the kinetic energy produced by mesoscale circulations, and to demonstrate the potential of using this variable for the parameterization of subgrid-scale processes in GCMs.

MESOSCALE CIRCULATIONS IN HETEROGENEOUS LANDSCAPES

To demonstrate the impact of landscape heterogeneity on mesoscale atmospheric processes, the Colorado State University (CSU) Regional Atmospheric Modeling System (RAMS), a three-dimensional state-of-the-art mesoscale model described in detail by Pielke et al. (1992)[11] was used to simulate deforestation of a densely forested area. The model was initialized with the parameters summarized in Tables I-VI (at the end of the paper). The simulated region was 300 km wide in the west-east direction and was represented in the model by 60 grid elements with a grid resolution of 5 km. The deforested area (50 km wide) was located in the middle of the domain. Thus, it was flanked by 125 km of forests. The domain was assumed homogeneous and infinite in the south-north direction. The size of the domain was chosen large enough to ensure that the process analyzed in the present study had no interaction with lateral boundaries. The atmosphere was simulated up to a height of 10 km (i.e., roughly the tropopause), and was represented by 19 grid elements with a high grid resolution near the ground surface (see Table VI).

The land-surface parameterization used for this simulation was the Land-Atmospheric Interactive Dynamics (LAID) developed by Avissar and Mahrer[12] as modified by Avissar and Pielke[2]. This scheme, which is based on the "big-leaf" concept, consists of two layers, a vegetation and a soil layer. Surface energy fluxes of latent and sensible heat are calculated from two energy budget equations, one for the vegetation layer, and one for the soil layer. Heat and moisture diffusion are solved simultaneously in the soil layer, which is represented by 13 grid elements with a high grid resolution near the ground surface (see Table II). Short- and long-wave radiation received at the ground surface is redistributed into emitted long-wave radiation, sensible and latent heat fluxes in the atmosphere, and soil heat flux, according to land-surface characteristics and water availability for evapotranspiration in the plant root zone and the soil surface.

The numerical integration for the simulations was started at 0600 a.m. (local standard time), which corresponds to the time when the sensible heat flux becomes effective in the development of the convective PBL on sunny summer days. Figure 1 presents west-east 2-D sections of the meteorological fields (u—the west-east horizontal component of the wind parallel to the domain, v—the south-north horizontal component of the wind perpendicular to the domain, w—the vertical component of the wind, θ—the potential temperature, and q—the specific humidity) obtained at 8 a.m., 2 p.m., 8 p.m., and 2 a.m. next morning. Forests are indicated in these sections by black underbars, and only the lower 3 km of the atmosphere (i.e., roughly the PBL) is displayed.

The regional circulations depicted by the horizontal and vertical components of the wind result from the differential heating produced by the two very distinct surfaces subjected to the same solar radiation input. In the deforested land (assumed bare and relatively dry), most of the radiative energy received from the sun and the atmosphere is used to heat the atmosphere and the ground. However, in the forest area (assumed unstressed), a large part of this radiative energy is used for evapotranspiration. The faster heating rate above the deforested land surface generates a vigorous turbulent mixing and an unstable, stratified PBL, as can be seen from the sections of potential temperature in Figure 1. On the contrary, the slower heating rate above the transpiring forest limits the development of the PBL, which remains shallow above the forest. This creates a pressure gradient between the two areas, which generates the circulations from the relatively cold to the relatively warm areas.

The transpiration from the forest provides a supply of moisture which significantly increases the amount of water in the shallow PBL. This moisture is advected by the generated mesoscale flow which strongly converges toward the deforested area, where it is well mixed within the relatively deep convective boundary layer. This process may eventually generate convective clouds and precipitation under appropriate synoptic-scale atmospheric conditions.

MESOSCALE KINETIC ENERGY

Assuming that within a grid element of a GCM an atmospheric variable (ϕ) can be separated into synoptic scale ($\tilde{\phi}$), mesoscale (ϕ'), and turbulent scale (ϕ''), the mean mesoscale kinetic energy (MKE) per unit of mass (MKE/m = $\overline{E}$) is defined as

$$\overline{E} = 0.5 \overline{u_i'^2} \qquad (1)$$

where u_i' represents the three wind components of the mesoscale perturbation (in a Cartesian coordinate system). The overbar indicates an average at the GCM grid scale.

Similarly, the vertical mesoscale sensible and latent heat fluxes are defined, respectively, as

$$\overline{H} = \rho c_p \overline{w'\theta'} \qquad (2)$$

$$\overline{\lambda E} = \rho L \overline{w'q'} \qquad (3)$$

where ρ is air density, c_p is air specific heat at constant pressure, L is latent heat of evaporation, and θ' and q' are mesoscale perturbation of potential temperature and specific humidity, respectively.

Figure 2 illustrates the vertical profiles of mesoscale kinetic energy and latent and sensible heat fluxes calculated for the deforestation simulation presented in Figure 1. These profiles are shown at 8 a.m., 2 p.m., 8 p.m., and 2 a.m. Therefore, they correspond to the 2-D sections of atmospheric conditions shown in Figure 1.

As expected, there is an obvious relation between the mesoscale kinetic energy and the three wind components of the mesoscale circulation, which develops over the heterogeneous landscape. The profiles of horizontal averages emphasize clearly the strong impact of the mesoscale circulation at the different elevations within the PBL. In

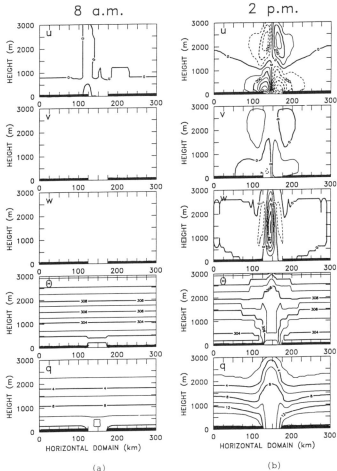

Figure 1. Vertical cross-sections of i) horizontal wind component parallel to the domain (u) in cm s^{-1}, positive from west to east; ii) horizontal wind component perpendicular to the domain (v) in cm s^{-1}, positive from south to north; iii) vertical wind component (w) in cm s^{-1}, positive upward; iv) potential temperature (θ) in K; and v) specific humidity (q) in g kg^{-1}, obtained in the planetary boundary layer that develops over a deforested region located in mid-latitude, during mid-summer time at a) 8 a.m., b) 2 p.m., c) 8 p.m., and d) 2 a.m. Forested land is indicated by a dark underbar. Solid contours indicate positive values, and dashed contours indicate negative values.

160 The Scale of GCM's

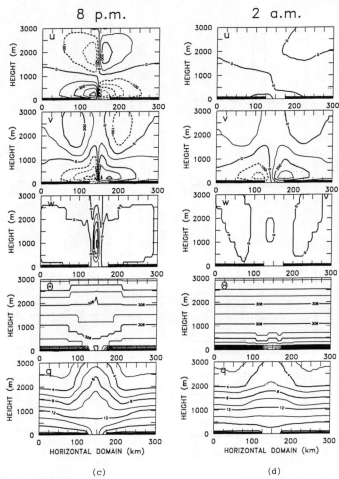

Figure 1. (continued)

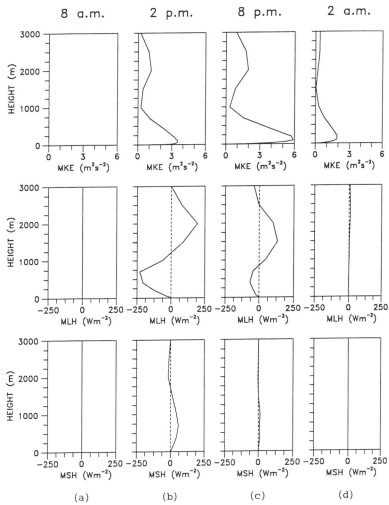

Figure 2. Vertical profiles of horizontally averaged i) mesoscale kinetic energy per unit mass (MKE); ii) mesoscale latent heat flux (MLH); and iii) mesoscale sensible heat flux (MSH), obtained in the planetary boundary layer that develops over a deforested region located in mid-latitude, during mid-summer time at a) 8 a.m., b) 2 p.m., c) 8 p.m., and d) 2 a.m.

particular, it is interesting to compare the extreme values of these profiles to typical values of kinetic energy and heat fluxes produced by turbulence. For instance, Yamada and Mellor[13] calculated a maximum turbulence kinetic energy of 1.18 m^2s^{-2} for Day 34 of the Wangara Experiment, and André et al. (1978)[14] calculated a maximum turbulent humidity flux of about 120 Wm^{-2} for Day 33 of the same experiment.

Of course, these experimental results cannot be compared directly to the mesoscale quantities calculated in the theoretical simulation presented here. Nevertheless, they provide an order of magnitude of the turbulent quantities and emphasize the relative importance of the mesoscale processes. It is important to remember that only the turbulent fluxes are parameterized in current GCMs (though assuming incorrectly horizontal homogeneity!), and that mesoscale fluxes are not.

Figure 3 depicts the diurnal variation of mesoscale kinetic energy, latent heat flux (positive only), and sensible heat flux (negative only) averaged over the entire PBL. There is also a clear relation between MKE and the heat fluxes. To be generalized, however, this relation needs to be investigated for various synoptic-scale background atmospheric conditions (i.e., wind speed, temperature lapse, humidity profiles) and various location on Earth.

CONCLUSIONS

The numerical deforestation experiment presented in this short paper emphasizes the importance of mesoscale dynamical processes generated by landscape heterogeneity. Heat fluxes associated with the mesoscale circulations may be more important than turbulent heat fluxes. Nevertheless, they are not represented in current GCMs.

Subgrid-scale clouds and precipitation are related to subgrid-scale latent heat flux. In return, this flux can be related to the mesoscale kinetic energy. Thus, one can derive a prognostic equation for the mesocale kinetic energy (similarly to the turbulent kinetic energy) that could be used to improve the representation of clouds and precipitation in GCMs. The author and colleagues have developed such an equation and investigated the impact of landscape heterogeneity and orography on the various terms of this equation. The potential of the approach is clear, but an intensive research activity is still needed to provide an appropriate parameterization.

ACKNOWLEDGMENTS

This research is supported by the National Science Foundation under Grants ATM-9016562 and EAR-9105059, by the National Aeronautics and Space Administration under Grant NAGW-2658, and by the U.S. Department of Energy under the CHAMMP Program.

REFERENCES

1. R. Avissar and M. M. Verstraete. *Reviews of Geophysics* **28**:35–52 (1990).
2. R. Avissar and R. A. Pielke. *Mon. Wea. Rev.* **117**:2113–2136 (1989).
3. D. Entekhabi and P. S. Eagleson. *J. Climate* **2**:816–831 (1989).

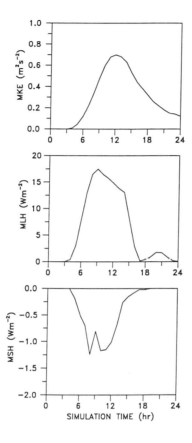

Figure 3. Diurnal variation of domain averages of i) mesoscale kinetic energy per unit mass (MKE); ii) mesoscale latent heat flux (MLH); and iii) mesoscale sensible heat flux (MSH), obtained in the planetary boundary layer that develops over a deforested region located in mid-latitude, during mid-summer time.

4. R. Avissar. In *Land Surface–Atmosphere Interactions for Climate Models: Observations, Models, and Analyses*, E. F. Wood, Ed., Kluwer, pp 155–178 (1991).
5. J. S. Famiglietti and E. F. Wood. In *Land Surface–Atmosphere Interactions for Climate Models: Observations, Models, and Analyses*, E. F. Wood, Ed., Kluwer, pp 179–204 (1991).
6. R. Avissar. *J. Geophys. Res.* **97**:2729–2742 (1992).
7. D. Collins and R. Avissar. *J. Climate* submitted (1992).
8. M. Segal, R. Avissar, M. McCumber, and R. A. Pielke. *J. Atmos. Sci.* **45**:2268–2292 (1988).
9. P. Mascart, O. Taconet, J.-P. Pinty, and M. Ben Mehrez. *Agric. For. Meteorol.* **54**:319–351 (1991).
10. R. A. Pielke, G. Dalu, J. S. Snook, T. J. Lee, and T. G. F. Kittel. *J. Climate* **4**:1053–1069 (1991).
11. R. A. Pielke, W. R. Cotton, R. L. Walko, C. J. Tremback, M. E. Micholls, M. D. Moran, D. A. Wesley, T. J. Lee, and J. H. Copeland. *Meteor. Atmos. Phys.* in press (1992).
12. R. Avissar and Y. Mahrer. *J. Appl. Meteor.* **27**:400–413 (1988).
13. T. Yamada and G. Mellor. *J. Atmos. Sci.* **32**:2309–2329 (1975).
14. J.-C. Andre, G. De Moor, P. Lacarrere, G. Therry, and R. du Vachat. *J. Atmos. Sci.* **35**:1861–1883 (1978).

Table I

Characteristics of the sandy loam soil type which was adopted for the simulations, as provided by the United States Department of Agriculture textural classes. The hydraulic properties are soil water content at saturation η_s, soil water potential at which water content η starts to be lower than saturation Ψ_{cr}, hydraulic conductivity at saturation K_{hs}, and b is an exponent in the function which relates soil water potential and water content.

Soil Property	Value
density	1250 kg m^{-3}
roughness	0.01 m
η_s	0.435 m^{3}m^{-3}
Ψ_{cr}	-0.218 m
K_{hs}	2.95 m day^{-1}
b	4.90
albedo	0.20
emissivity	0.95

Table II. Initial profiles of soil temperature and soil water content.

	Initial Profiles	
Soil Depth (m)	Temperature (K)	Wetness (m^{3}m^{-3})
0.00	300	0.04
0.01	300	0.06
0.02	300	0.11
0.03	300	0.15
0.04	300	0.18
0.05	300	0.20
0.075	300	0.20
0.125	300	0.20
0.20	300	0.20
0.30	300	0.20
0.50	300	0.20
0.75	300	0.20
1.00	300	0.20

Table III. Vegetation characteristics.

Property	Value
roughness	1 m
leaf area index	5
height	10 m
albedo	0.15
transmissivity	0.125
emissivity	0.98
initial temparature	300 K

Table IV. Plant root distribution

Soil Depth (m)	Root Fraction
0.00	0.00
0.01	0.00
0.02	0.01
0.03	0.01
0.04	0.02
0.05	0.07
0.075	0.12
0.125	0.14
0.20	0.18
0.30	0.16
0.50	0.15
0.75	0.14
1.00	0.00

Table V. Atmospheric input parameters

Input Parameter	Value
synoptic-scale wind	0.5 m s^{-1}
initial surface temperature	300 K
initial potential temperature lapse	3.5 K/1000 m

Table VI. Initial profile of atmospheric specific humidity

Height (m)	Humidity ($g\ kg^{-1}$)
10,000	0.5
8000	0.5
6000	0.5
5000	0.5
4000	0.5
3500	0.5
3000	1.0
2500	1.5
2000	2.5
1500	6.0
1000	8.0
700	10.0
400	10.0
200	10.0
100	10.0
60	10.0
30	10.0
15	10.0
5	10.0

THE REPRESENTATION OF LANDSURFACE-ATMOSPHERE INTERACTION IN ATMOSPHERIC GENERAL CIRCULATION MODELS

Dara Entekhabi and Peter S. Eagleson
Ralph M. Parsons Laboratory
Department of Civil and Environmental Engineering
Massachusetts Institute of Technology
Cambridge, MA 02139

ABSTRACT

In numerical models of global climate, there are a number of processes that cannot be resolved explicitly at the level of discretization that is computationally feasible. Such processes are *parameterized* according to simplified relationships and empirical formulae. Nevertheless the realism of the regional climates produced by the models is critically dependent on such processes. For illustration, we focus on landsurface atmosphere interaction and soil hydrology. In this paper we outline the influence of landsurface processes on climate and then follow to describe the various approaches to their modeling.

INTRODUCTION TO SOME BASIC ISSUES

The physical processes in the atmosphere act on scales ranging from the molecular to the planetary. Radiative processes such as selective absorption, scattering and transmission are dependent on the particular type of molecule encountered and are thus dependent on the atomic scale. On the other hand, differential heating and the planetary rotation set into motion waves whose lengths are on the order of the earth's radius.

Clearly no numerical model can explicitly capture the entire spectrum of scales evident in the climatic system. Once a particular discretization is selected, then the processes operating on scales larger than the resolution are part of an explicit solution to the governing state equations. Physical processes that occur on dimensions smaller than the discretization are therefore only parameterized. With typical numerical grids on the order of 10^4 to 10^6 km^2, radiative, turbulent, hydrologic and convective processes are all parameterized; the mass-flux general circulation of the atmosphere is, however, solved for explicitly in the climate model. Global circulation models (GCMs) resolve motions that only border on the synoptic scale (at most up to wave number 30 over the Globe). Weather systems are often not resolved well in the models. Local feature and regional climates are therefore not represented adequately in climate models. Increasing the model resolution will significantly improve the capability of the model climate to capture sub-synoptic scale features of the atmosphere.

In this report, we examine aspects of the computational issues involved in the numerical simulation of the climatic system. We argue that the dynamic computations involved in climate simulation may undergo significant improvement with the introduction of new machine architectures and coding techniques. The accuracy and quality of the simulated climate will nevertheless always remain constrained by the physical parameterizations included in the model. For illustration, we will focus on one of these physical parameterizations: the landsurface hydrology scheme used in specifying the interaction of the atmospheric fluid with its lower boundary.

LANDSURFACE HYDROLOGY

The atmospheric fluid, whose general circulation features define the regional distribution and seasonal march of temperature and moisture, is mostly forced through its lower boundary at the earth surface.

Surface roughness, at the large scale through mountain ranges and elevations, determines the magnitude of angular momentum transfer between the rotating solid earth and its enveloping fluid. At smaller scales, the roughness introduced by vegetation and smaller scale topographic features has strong influence on the turbulence structure of the atmospheric boundary layer.

Much of the forcing of thermally-direct circulations is also attained at the landsurface boundary. The atmospheric region that contains most of the general circulation and weather features that affect regional climate is by large transparent to solar radiation. It is the surface that partially absorbs the radiation and makes it available to the atmospheric fluid as heat. The differential deposition of solar energy due to the spherical geometry, diurnal rotation and seasonal orbit of the earth results in heating gradients that force instabilities and thermally-direct circulations. Thus the landsurface and the manner in which it partitions and stores the available solar radiation is a component of the climatic system that must be modeled in detail.

A third function of the landsurface, besides momentum and heat transfer, is related to its role in the hydrologic cycle. Water is deposited at the landsurface as either rain or snow. The landsurface partitions the incoming precipitation into storage and loss as runoff into the oceans and other water bodies. The fraction that is stored as soil moisture is evaporated with some delay. The landsurface thus plays an active role in modulating the seasonal cycle of moisture availability over large regions. The evaporation of soil moisture and soil water loss through transpiration take up an amount of energy equal to that required for the phase change of water from liquid to vapor. Thus the surface moisture balance and the energy balance are coupled together; this in turn results in feedbacks, physics and dynamics whose representations are crucial in the proper simulation of climate.

None of these landsurface-atmosphere interaction processes (momentum, heat and moisture transfer) may be modeled explicitly over the numerical grid defined for the integration of the model dynamics. They involve mechanisms and processes that occur at scales considerably smaller than that adequate for the simulation of atmospheric flows. Soil moisture diffusion processes, plant stomatal control of water loss and the turbulence introduced by roughness characteristics are all necessary components of the proper physical modeling of the climate.

The land surface parameterization in GCMs is concerned with the simultaneous water and energy balance at the surface. The incoming precipitation is partitioned into runoff according to a dimensionless runoff ratio R. The bare soil evaporation and transpiration rates are fractions β of the potential rate that occurs under conditions of unlimited surface moisture supply. The imbalance between the net precipitation input and evapotranspiration loss over any period of time is added to the soil water storage.

The net surface shortwave and thermal radiative forcing are balanced by ground heating and turbulent heat fluxes. These latter fluxes include the conducted sensible heat and the latent heat flux; the ratio of the two is the Bowen ratio that is sensitive to the surface hydrologic conditions.

The chief task of land surface hydrology parameterizations is to provide adequate models of R and β. Most GCMs have simple linear or broken-linear expressions for R and β as a function of soil water depth normalized by its saturated value (the relative soil saturation s). The NASA Goddard Institute for Space Studies (GISS) GCM[1], for example, uses

$$R = \frac{1}{2}s \tag{1}$$

and

$$\beta = s \tag{2}$$

These functions have empirical forms and their constants are calibrated until the model continental values for hydrologic balance matches the observed. Some GCMs such as the Geophysical Fluid Dynamics Laboratory (GFDL) GCM does not partition any rainfall to runoff until the soil is thoroughly saturated. These simple approaches to the surface hydrology are referred to as bucket-type. The treatment of ice and snow are also different for each GCM.

It is commonly recognized that the treatment of soil and especially vegetation properties described above is far too simplistic. The role of vegetation in controlling the surface water and heat balance is considered to be an important factor in most climates. Modeling the transpiration process is, however, a difficult task. Dickinson et al.[2] and Sellers et al.[3] developed biosphere models (*B*iosphere-*A*tmosphere *T*ransfer *S*cheme, BATS ; and the *S*imple *B*iosphere Model, SiB). These models are based on dividing the soil and canopy into several layers that are coupled through a series of resistances. These models are generally computationally intensive and have to be simplified in order to become part of operational GCMs. They are also characterized by a large number of parameters such as stomatal resistance for various vegetation types, etc.

In both the bucket-type formulation and the biosphere models, it is assumed that the grid value for soil wetness, precipitation, runoff or any other variable is the same everywhere within the grid. These models assume that the grid area acts as a uniform hydrologic unit. This assumption is problematic given typical grid size is on the order of 10^4 to 10^6 km^2. This is the case especially if one chooses to use physically based equations for the fluxes. Most surface hydrologic fluxes contain thresholds. Under the severe space-time averaging inherent in coarse mesh GCMs, the values of forcings such as precipitation are rarely high enough to exceed a realistic infiltration rate. If one is to incorporate realistic soil conductivity data and vegetation characteristics into GCMs in order to improve the hydrology, then it is necessary to include subgrid scale spatial variability into the model parameters. This principle was

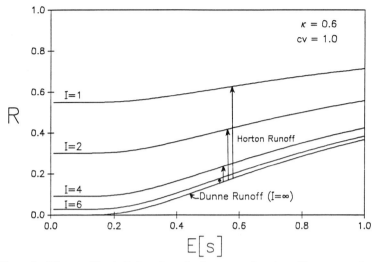

Figure 1. The runoff ratio R for a large land area as a function of its mean surface moisture state (E[s]) and the normalized rainfall intensity I. The fractional coverage of the rainstorm is κ and cv represents the coefficient of variation for the subgrid values of soil moisture.

recognized early on by Eagleson[4] and Warrilow et al.[5] and incorporated as GCM parameterizations by Entekhabi and Eagleson[6] and later by Famiglietti and Wood[7].

The GCM value for the grid precipitation intensity is taken to be the areal mean of a subgrid-distributed process covering a fraction κ of the entire grid. Observations support the use of an exponential distribution for convective-type rainfall[6]. The soil moisture state for the entire grid is also considered to be subgrid spatially variable due to redistribution under topography and heterogeneous soil properties. It is assigned the flexible gamma distribution with mean corresponding to the grid mean and the dimensionless coefficient of variation cv as a parameter. Runoff is generated by the interaction of subgrid scale distributed rainfall intensity and infiltration rate dependent on the local soil saturation. Figure 1 illustrates the derived distribution for the grid total runoff ratio (R) as determined by Entekhabi and Eagleson[6]. Horton runoff refers to runoff due to the excess of rainfall intensity over the local infiltration rate, and Dunne runoff refers to runoff due to precipitation directly over areas that have been saturated from below.

Figure 2 demonstrates similar results for bare soil exfiltration. In this case, however, the grid must be divided into areal fractions governed by water-limited evaporation regime (evaporation less than potential) and climate-controlled regime (evaporation rate equal to potential value). Entekhabi and Eagleson[6] also include a β function for vegetal transpiration based on a simple root-zone soil water extraction

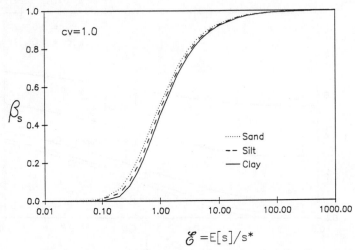

Figure 2. The bare soil evaporation efficiency β as a function of a dimensionless parameter ε incorporating areal mean soil moisture and soil texture.

model. This statistical-dynamical approach to land surface hydrology parameterization allows the incorporation of realistic hydrologic threshold functions (such as the infiltration rate). It has a relatively low number of parameters and may be implemented in GCM models without significant costs in computations.

CONCLUDING REMARKS

The capability of GCMs in representing regional climates and important features of the land-ocean-atmosphere system is critically dependent on the accuracy and reliability of its physical parameterizations. Here we have focused on the role of landsurface on the overlying atmosphere. The numerous ways in which the roughness, heat capacity and moisture storage characteristics of the near-surface soil and vegetation may influence the regional climate have been outlined. We next described the various approaches to the modeling of these interactions. The landsurface hydrologic parameterizations in current GCMs range from simple empirical formulae, to detailed transfer models based on resistance formulations and newer approaches that are based on physical principles and incorporate the heterogeneity of some key variables at spatial scales smaller that the GCM numerical grid.

REFERENCES

1. J. Hansen, G. Russel, D. Rind, P. Stone, A. Lacis, S. Lebedeff, R. Ruedy, and L. Travis. 1983. *Mon. Wea. Rev.* 111(4):609–662.
2. R. E. Dickinson, J. Jager, W. M. Washington, and R. Wolski. 1981. *Boundary subroutine for NCAR global climate model*. NCAR Technical Note 173TIA.
3. P. J. Sellers, Y. Mintz, Y. C. Sud, and A. Delcher. 1986. *J. Atmos. Sci.* 43(6):505–531.
4. P. S. Eagleson. 1978. *Water Resour. Res.* 14(5):705–712.
5. D. A. Warrilow, A. B. Sangster, and A. Slingo. 1986. *Modelling of land surface processes and their influence on European climate*. U.K. Meteorological Office, DCTN 38.
6. D. Entekhabi and P. S. Eagleson. 1989. *J. of Climate* 2(8):816–831.
7. J. S. Famiglietti and E. F. Wood. 1991. In E. F. Wood, Ed., *Land surface-atmosphere interactions for climate modeling*. Kluwer Academic Publishing, 179–204.

SPATIAL DISTRIBUTION OF PRECIPITATION RECYCLING IN THE AMAZON BASIN

Elfatih A. B. Eltahir and Rafael L. Bras
Ralph M. Parsons Laboratory
Department of Civil and Environmental Engineering
Massachusetts Institute of Technology,
Cambridge, MA 02139

ABSTRACT

Precipitation recycling is the contribution of evaporation within a large region to precipitation in that same region. The rate of recycling is a diagnostic measure of the coupling of land surface hydrology and regional climate. Here we describe the spatial and seasonal variability of the precipitation recycling process over the Amazon basin. The results are based on data of evaporation and water vapor fluxes from the European Center for Medium Range Weather Forecast (ECMWF). We estimate that 25% of all the rain that falls in the Amazon basin is contributed by evaporation within the basin. The contribution of recycled water vapor increases westward and southward with significantly different spatial distributions in the different seasons.

INTRODUCTION

Hydrology affects climate in many different ways. Evaporation provides the water vapor necessary for precipitation processes. Latent heat fluxes associated with evaporation and condensation provide an important energy transport mechanism in the Earth's atmosphere. Because land surface hydrology plays such a significant role in maintaining the equilibrium of regional climate, many recent studies[1,2,3] suggest that anthropogenic changes in surface hydrology, e.g., deforestation of the Amazon basin, may result in serious impacts on climate. The precipitation recycling rate is a diagnostic measure of the current degree of coupling and the potential interactions of land surface hydrology and regional climate.

Previous studies suggested different ways for computing precipitation recycling. Budyko[4] provides a spatially lumped estimate of precipitation recycling. It describes the seasonal but not the spatial distribution of the recycling rate. Lettau[5] describes precipitation recycling along a single streamline. We study both the spatial and seasonal variability of the recycling process.

We consider two species of water vapor molecules; those which evaporate outside the region and molecules which evaporate within the region. The definition of the word 'region' includes all the area under study which is the Amazon basin. It is not restricted to the area of a single grid point. For a finite control volume of the atmosphere, conservation of mass requires the following relations.

$$\frac{\partial N_w}{\partial t} = I_w + E - O_w - P_w$$
$$\frac{\partial N_o}{\partial t} = I_o - O_o - P_o \qquad (1)$$

where w denotes molecules which evaporate within the region and o denotes molecules which evaporate outside the region. N is the number of water vapor molecules. I and O are the inflow and outflow, respectively. P is precipitation and E is evaporation. N, I, O, P, and E are variable in space and time. It is assumed that the two species are well mixed which implies that

$$\rho = \frac{P_w}{(P_w + P_o)} = \frac{O_w}{(O_w + O_o)} = \frac{N_w}{(N_w + N_o)} \qquad (2)$$

ρ is defined as the precipitation recycling ratio. Mixing of water vapor in the atmosphere is primarily achieved by dry thermal convection near the surface of the Earth. It is observed that the change in total precipitable water is small relative to fluxes over an appropriate time period, like a month. We will then assume that both derivatives in Equation (1) are zero. Substituting from Equation (2) into Equation (1) and rearranging we then get

$$\rho = \frac{I_w + E}{I_o + I_w + E} \qquad (3)$$

The spatial resolution of the data should be small enough to resolve significant spatial variability in evaporation and fluxes. The temporal resolution should be large compared to the travel time across the basin.

The evaporation and fluxes data for the Amazon basin are part of the ECMWF global data set. The data assimilation system at the ECMWF combines data from surface meteorological stations, upper air observations and satellite data. Although the frequency and spatial coverage of the observations in the Amazon basin is limited, the ECMWF global data set is regarded as one of the best available data sets[5]. The data covers the period 1985–1990 inclusive. It has a twelve-hour resolution for the flux data and six-hour resolution for the evaporation data. It has a spatial resolution of 2.5° latitude by 2.5° longitude.

Figure 1 shows the spatial distribution of the precipitation recycling ratio in the Amazon basin. The annual recycling map indicates that ρ increases westward and southward. These directions are consistent with the directions of water vapor fluxes in the basin. Previous studies of recycling in the Amazon assumed that the variability of the process is mainly in one direction from east to west[6]. The results in Figure 1 suggest that ρ has a significant southward gradient. This gradient dominates the recycling map in the month of December consistent with the southward migration of the Inter-Tropical Convergence Zone (ITCZ) and the resulting southward flux of moisture over most of the Amazon basin in that month.

Figure 2 shows the estimated precipitation recycling ratio averaged over the basin and for the different months of the year. The seasonal variability of the averaged recycling ratio is small. The values for the Southern Hemisphere summer are slightly smaller than those for winter. The explanation of that result is in the high levels of rainfall in summer at the eastern subregion of the basin. Due to its location next to the

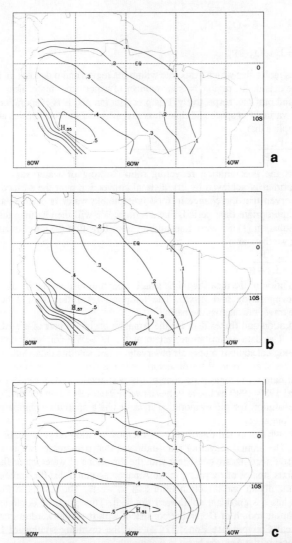

Figure 1. Spatial distribution of precipitation recycling ratio. a) annual, b) June, and c) December. The recycling ratio at each point in the basin is the ratio of recycled precipitation to total precipitation at that point. Recycling in a) is estimated by a weighed average of the recycling maps of all the months. The weighing function is the amount of precipitation for each month normalized by annual precipitation. This procedure is carried for each point in the basin.

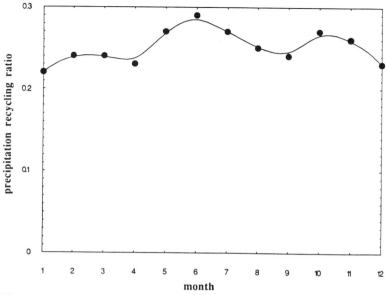

Figure 2. Seasonal distribution of precipitation recycling ratio. Recycling is estimated by a weighed average of the recycling spatial distribution. The weighing function is the amount of precipitation at each location normalized by the entire basin precipitation. This procedure is carried for each month of the year.

Atlantic Ocean, this subregion has a smaller ρ and hence the weighted average recycling ratio in summer is slightly smaller than that of winter.

The annual recycling ratio for the Amazon basin is estimated to be 25%. This ratio is significantly smaller than the frequently quoted estimate of around 50%[7,8]. The higher estimate of recycling was based on an inaccurate picture of atmospheric moisture flux over the Amazon which assumes that flux out of the basin is negligible. For a closed basin with no atmospheric water vapor outflow, the recycling ratio is given by the ratio of total evaporation to total precipitation. It was estimated that evaporation represents about 56% of precipitation in the Amazon[9], and hence it was concluded that the recycling ratio in the Amazon is about 56%[7]. But the Amazon is not a closed system, the data shown in Figure 3 indicate that the atmospheric water vapor flux out of the basin is significant. The flux out of the basin accounts for almost 70% of the flux into the basin. Although this ratio is significantly smaller than that for other rivers, e.g., 95% for the Mississippi basin[10], it is evident that the atmosphere above the Amazon basin is far from being a closed system. Most of the outflow occurs through

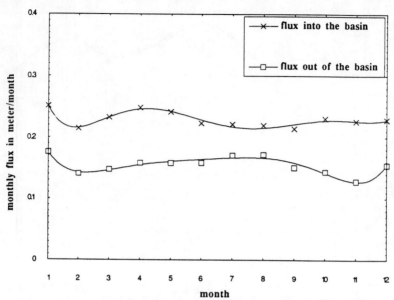

Figure 3. Atmospheric water vapor fluxes above the Amazon basin. Units correspond to depth of water over the basin area.

the Northern and Southern borders. This migration of moisture is confirmed by isotopic studies[11].

The east-west gradient of ρ as estimated by this study agrees with the results of Lettau et al.[6] ρ is comparable with $(1 - 1/\gamma)$ which is defined in their study. Our estimate of annual recycling of 25% is very close to a recent estimate of recycling in the Amazon using a modified Budyko's model[12]. We conclude by emphasizing that estimates of precipitation recycling provide a diagnostic measure of the coupling between hydrology and climate. It is inappropriate to make any prognostic statements regarding the effects of deforestation or any other disturbance of surface hydrologic parameters based solely on estimates of precipitation recycling and before studying the equilibrium of the disturbed climate system.

ACKNOWLEDGMENT

We acknowledge the support of the National Aeronautics and Space Administration (NASA). The views, opinions, and/or findings contained in this report are those of the authors and should not be constructed as an official NASA position, policy, or decision, unless so designated by other documentation.

REFERENCES

1. R.E. Dickinson and A. Henderson-Sellers. *Q.J.R. Meteorol. Soc.* **114**:439–462 (1988)
2. J. Lean, and D. A. Warrilow. *Nature* **342**:411–413 (1989)
3. J. Shukla, et al. *Science* **247**:1322–1325 (1990)
4. M.I. Budyko. *International Geophysical Series* **18**, Academic Press (1974)
5. K.E. Trenberth, and J. G. Oslon. *NCAR Technical Note* NCAR/TN-300+STR (1988)
6. H. Lettau, et al. *Mon. Weather Rev.* **107**:227–238 (1979)
7. L.C.B. Molion. Ph.D. thesis, University of Wisconsin, Madison (1975)
8. E. Salati and P. B. Vose. *Science* **225**:129–138 (1984)
9. R.E. Oltman. *U.S. Geological Survey Circular* **552** (1968)
10. G.S. Benton, et al. *Trans. Amer. Geophys. Union* **31**:61–73 (1950)
11. E. Salati, et al. *Water Resour. Res.* **15**:1250–1258 (1984)
12. K.L. Brubaker, et al. M.I.T. Tech. Report **333** (1991)

GLOBAL ESTIMATION OF RAINFALL: CERTAIN METHODOLOGICAL ISSUES

Witold F. Krajewski
Iowa Institute of Hydraulic Research
The University of Iowa, Iowa City, IA 52242

ABSTRACT

The paper discusses certain methodological and practical aspects of global rainfall estimation on the climatological scale. Estimation of monthly rainfall accumulation over a 2.5° by 2.5° grid is investigated, using methods based on in-situ observations and satellite-measured radiances. It is shown that validation of satellite-based estimates of rainfall requires uncertainty analysis. Such methods are described for climatological rainfall estimation. Uncertainty analysis for the methods based on point observations can be accomplished using concepts of spatial interpolation of random fields. For the satellite-based methods, two approaches are recommended: cross-validation and self-consistent uncertainty propagation using the framework of optimal estimation. The paper also presents a new method for measurement and estimation of open ocean rainfall using acoustic underwater sensors.

INTRODUCTION

There is a growing recognition that in order to make long term predictions of climate variability one must understand the global behavior of circulation patterns. An important component of energy sources that control circulation is latent heat released in the process of condensation which then leads to rainfall. Quantitative assessment of rainfall intensities and amounts is the only feasible method of estimating this energy component. Since 70% of the earth is covered by oceans, land-based rainfall-measuring networks are not adequate for global rainfall pattern evaluation; satellite-based techniques are the only realistic means. The temporal scale considered adequate for climate analysis is one month. Evaluation of monthly rainfall patterns for the whole globe is a challenging task which requires large amounts of data, appropriate methodologies for data processing, and considerable insight into the physical processes causing and controlling rainfall occurrence.

Until recently, it was possible to evaluate rainfall over land only. With the advent of space-born sensors, global estimation of rainfall, both over lands and oceans, has become possible. However, one of the main problems facing the developers of satellite-based rainfall estimation techniques is the lack of suitable data sets for validation of the new methods. This is a particularly important problem in the tropical zone, which shapes the global atmospheric circulation patterns. In this zone ground-based rainfall measuring networks are sparse and, as a consequence, available data are not adequate for climate studies.

There is great interest within federal agencies such as the National Aeronautics and Space Administration (NASA) and the National Oceanic and Atmospheric Administration (NOAA) in global monitoring of rainfall[1,2]. Its culmination will be the

Tropical Rainfall Measuring Mission—a US-Japan satellite project devoted to rainfall measurements from space[3]. International activities such as establishment of the Global Precipitation Climatology Project within the World Climate Research Program of the World Meteorological Organization and the Global Precipitation Climatology Center in Offenbach, Germany attest to the importance of the problem.

Over land, the availability of rainfall data varies from place to place. Rainfall estimates used for climatologic purposes are derived predominantly from raingage network data[13,15]. Radars are too few, too sparse, and with unresolved methodological problems of rainfall estimation to currently provide a reliable source of data over large areas. The future potential, however, of radar networks used for rainfall estimation is very significant and will be realized soon with advent of the NEXRAD system in the United States and European efforts to establish a common network.

In this paper two issues associated with global rainfall estimation are discussed. These are: 1) validation of satellite-based methods of rainfall estimation; and 2) estimation of uncertainty associated with rainfall estimates. Under the first issue a new in-situ method is proposed for observation and estimation of rainfall over open ocean.

PROBLEM FORMULATION

Of interest is estimation of areally averaged rainfall accumulations over the whole globe (with exception of regions north of 80° N and south of 80° S) at a grid of 2.5° by 2.5°. An example of such a grid is shown in Figure 1 for Florida. The use of the term "estimation" signifies the fact that rainfall is not observed at the desired scale and needs to be estimated from the available observations. These observations may be measurements of rainfall itself but at a different scale than the scale of interest or, may comprise other, rainfall-related variables. The direct observations of rainfall come from raingage readings. Raingages provide quite accurate observations of temporal accumulations of point rainfall but spatial estimates rely on the use of interpolation techniques. Indirect observations of rainfall come from radars and satellites.

Radar is capable of providing rainfall estimates over areas as large as 100,000 km^2 with resolution of 1° in azimuth and 1 km in range, but these estimates are subjected to errors which are quantitatively not well understood. This fact, combined with limited availability of radars throughout the world, makes contributions of radar rainfall estimates to global climatology of rainfall negligible. This situation is bound to change soon with the establishment of operational radar networks in Japan, England, and deployment of the NEXRAD system in the United States and organization of a European radar network.

Satellites are perhaps the most attractive means of estimating global rainfall. Not only can they see over the vast oceanic cloud systems, but they also collect global data through a single system. The relationship between satellite-measured radiation intensity and rainfall, which contributes only partially to this intensity, is much more complex than in the case of radar. Theoretical methods of radiative transfer are of limited use because of infeasibility of solutions for large scale problems. Therefore, validation of the empirically based methods for estimating precipitation from satellite measurements against ground-based estimates is of critical importance.

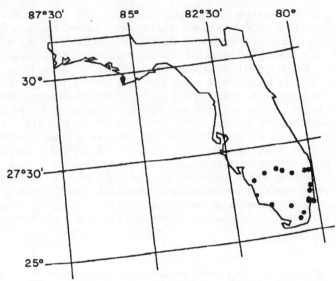

Figure 1. Climatological rainfall estimation grid (2.5° by 2.5°) superimposed over Florida. A network of eighteen raingage stations is also shown.

In brief, the problem of global rainfall estimation, at the scales stated above, can be formulated as the problem of multisensor rainfall estimation. For many locations (oceans) only a single sensor will be available, for other locations two or three sensors may be combined for optimal (in minimum variance sense) climatological rainfall estimation. Below, several issues related to the problem are discussed. First, a discussion of the validation problem is given. It is followed by estimation of uncertainty of rainfall estimates from raingage networks. Finally, methodologies of uncertainty estimation which do not use land-based data are considered.

VALIDATION OF SATELLITE-BASED METHODS OF RAINFALL ESTIMATION

The main problem in using satellites to estimate rainfall is that satellite sensors do not measure rainfall directly. They measure radiation reflected and/or emitted by clouds and rain drops. The relationship between satellite-measured radiation and rainfall-related atmospheric processes is very complex. This complexity and the lack of measurement networks over vast ocean areas makes it particularly difficult to validate the satellite-based methods. On the other hand, the knowledge of accuracy of these estimates of global rainfall is of paramount importance because they will affect, through a calibration process, the performance of the Global Climate Models (GCM), which in turn, will be instrumental in important political and economic decisions[6].

The above discussion makes it clear that there is a need for methodologies which can provide not only global rainfall estimates but also the error bounds associated with them without resorting to independent observations (which are unavailable over oceans, anyway). Existing satellite-based methods (see Lee et al.[7] for a recent summary) of rainfall estimation are deterministic, i.e. are not capable of producing uncertainty measures.

Comparison with Raingage-Based Estimates

For the areas where independent estimates of rainfall are available, as is the case over many land locations with raingage networks, uncertainty of satellite-based estimates can be assessed through comparison with raingage-based estimates.

Denote the true mean areal rainfall over a grid box as R, and the same rainfall estimates based on raingage observations and satellite data as R_g and R_s, respectively. We are interested in evaluating the uncertainty of the satellite-based estimate, which can be measured as the following variance of the difference from the true rainfall:

$$E\left[(R-R_s)^2\right] = E[RR] - 2E[RR_s] + E[R_s R_s] \qquad (1)$$

and, similarly:

$$E\left[(R-R_g)^2\right] = E[RR] - 2E[RR_g] + E[R_g R_g] \qquad (2)$$

where E[•] denotes expectation operator. The variance of the differences between the raingage and the satellite-based estimates can be calculated as:

$$E\left[(R_g - R_s)^2\right] = E[R_g R_g] - 2E[R_g R_s] + E[R_s R_s] \qquad (3)$$

If one writes

$$E\left[(R_g - R_s)^2\right] = E[R_g - R + R - R_s]^2 \qquad (4)$$

it can be shown, that after some simple derivations

$$E\left[(R_g - R_s)^2\right] = \sigma_g^2 + \sigma_s^2 + 2\sigma_{gs}^2 \qquad (5)$$

where σ_g^2 is the error variance of the raingage-based estimate, σ_s^2 is the error variance of satellite-based estimate, and σ_{gs}^2 is the error covariance between the raingage and the satellite estimates. The unknown σ_s^2 can be expressed as the following function of the known, or easy to calculate, terms:

$$\sigma_s^2 = E\left[(R_g - R_s)^2\right] - 2\sigma_{gs}^2 - \sigma_g^2 \qquad (6)$$

The first term in Eq. (6) can be calculated from the historical record of satellite and raingage estimates for the location (grid box) of interest. This record is not very long, about 10 years, which implies, if the annual cycle effects are taken into account, only 10 data points. This very small sample can be enlarged if four seasons are considered, or, especially if all the months are used together. In such a case, probably justified for the tropical zones, the sample size would be about 120 data points—adequate for meaningful evaluation of the term. The error variance of the raingage-based estimation can be calculated following the methodology outlined in the next section. Finally, it seems reasonable to assume independence between the two sensor errors which leads to cancellation of the error covariance term.

The above method can be used only for the land locations where raingage observations are available. Although its application for those locations may give an idea about the performance of the satellite-based methods over open oceans, there is a risk that due to a significant difference in the environmental factors controlling both the rainfall process and its satellite-based sensing, these results are not representative for the open ocean. A new method to observe and estimate open ocean rainfall from inexpensive in-situ sensors is proposed in the next section. A different approach is necessary when no independent data is available. For such situations a self-consistent methodology based on optimal estimation theory[8] is proposed later in this paper.

ESTIMATION OF UNCERTAINTY

Estimation of uncertainty associated with global rainfall patterns is important for at least two reasons. First, meaningful use of estimated rainfall patterns in diagnostics of General Circulation Models is difficult without knowing their quality and reliability. Second, according to the theories of static estimation[8], optimal combination of rainfall estimates coming from several independently operating sources (sensors) can be based only on measures of their relative errors.

Estimation, as the term implies, is associated with errors. These errors include sampling errors due to discrete coverage of the estimated space-time domain, measurement error (or instrument error), and methodology-related errors. Since for most land locations rainfall is estimated using raingage data, first consider a methodology for estimation of uncertainty of raingage-based space-time rainfall accumulations.

Raingage-Based Estimation

The general framework for raingage-based estimation of mean areal rainfall and its uncertainty is discussed by Bras and Rodriguez-Iturbe[9]. Below, some practical aspects of the methodology will be emphasized. The purpose is estimation of σ_g^2 in Eq. (6). First, let us establish the notation.

Suppose that the true but unknown mean areal rainfall is

$$R = \frac{1}{|A|} \int_A P(\mathbf{u}, T) d\mathbf{u} \qquad (7)$$

where A is the area of interest (a 2.5° by 2.5° grid box), $\mathbf{u} = (x,y)$ is a two-dimensional space location, and P is the true process of rainfall accumulation at the time horizon T. In the following derivations this time index will be dropped to simplify the notation. It will be assumed that both the estimated areal rainfall and its observations are given at the same time scale (monthly accumulations). Therefore, Eq. (7) denotes, say, monthly mean area rainfall. The task at hand is to estimate R given n point raingage observations $G_i(\mathbf{u},T)$, $i = 1,2,...,n$. A linear estimator of the following form can be proposed

$$\hat{R} = \sum_{i=1}^{n} \lambda_i G_i(\mathbf{u}) \qquad (8)$$

The weights λ_i, for $i = 1,2,...,n$, can be found by requiring that the estimator be unbiased and characterized by minimum variance. Depending on the assumptions concerning the first and second moment statistics of the monthly rainfall process for the particular location (grid box), minimization of the variance expression leads to the solution of a linear system of equations with coefficients being values of the spatial covariance (or variogram) function. For example, if it is assumed that the mean rainfall is constant over a box, the solution for the optimal weights comes from the system

$$-\frac{1}{|A|}\int_A \text{cov}(\mathbf{u}-\mathbf{u}_i)d\mathbf{u} + \sum_{j=1}^n \lambda_j \text{cov}(\mathbf{u}_i - \mathbf{u}_j) + \mu = 0 \quad \text{for } i = 1,2,\ldots,n \quad (9)$$

$$\sum_{i=1}^n \lambda_i - 1 = 0 \quad (10)$$

The solution of this system yields the optimal set of weights denoted with λ_i^* which, when substituted into the variance expression, give

$$\sigma_R^2 = \sigma_g^2 = \frac{1}{|A|^2}\int_A\int_A \text{cov}(\mathbf{u}_1 - \mathbf{u}_2)d\mathbf{u}_1 d\mathbf{u}_2 - \frac{1}{|A|}\int_A \sum_{i=1}^n \lambda_i^* \text{cov}(\mathbf{u}-\mathbf{u}_i)d\mathbf{u} - \mu \quad (11)$$

In order to use the above method one needs to estimate the spatial covariance function of the rainfall process. Of course the true covariance, which appears in all the double integral terms, cannot be inferred from the noisy data and is substituted by a model obtained by fitting a theoretical covariance function to the data.

There are many functions which could be used as models for covariance. For a discussion of these models and the conditions they have to meet see Journel and Huijbregts[10]. Before a covariance model is selected, the empirical (or raw) covariance values need to be inspected. In Figure 2 the raw correlations (normalized covariances) are shown for all the pairs of raingages for monthly rainfall accumulations in the southern Florida grid box shown in Figure 1. The size of the box is about 250 km by 275 km. Although exponential-type decaying of the correlation function is evident, it is also clear that estimation of the function is based on highly variable data. The data plotted are for all the months pooled together. To account for seasonal effects it may be prudent to perform the estimation based on correlation function developed separately for each month of the year.

In a case of orographic effects within a grid box of interest, estimation of the mean areal rainfall cannot be based on the assumption of the constant mean and variance. Still, a method used should be simple to implement and robust. Perhaps the best choice is a separate estimation of the spatial trend (the mean) followed by estimation of the residual correlation. The mean can be modelled using a low order polynomial to capture the major features of the spatial variability.

This requirement for simplicity of a method becomes obvious if one realizes that there are roughly 3,000 grid boxes over the land areas where rainfall is estimated from raingage observations. Due to various factors affecting many of these locations (orography, coastal effects, nonuniform raingage network coverage, poor quality data, low density network, etc.), full automation of the estimation procedure is a dangerous approach. The process should be semi-interactive, but with minimum degree of subjectivity involved.

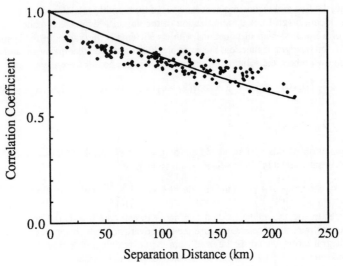

Figure 2. Spatial correlation function for monthly rainfall calculated from the eighteen raingages in Florida (shown in Figure 1).

The above rainfall estimation process will result in two maps: one for the mean areal rainfall, and one for the associated estimation variance. The variance values will reflect the effects of local network density as well as variability of rainfall. This variance can subsequently be used in validation of the corresponding satellite estimates of rainfall and/or in optimal merging of the two estimates.

<ins>Satellite-Based Estimation</ins>
For those situations where uncertainty of a rainfall estimation method can not be assessed by comparison with an independent estimate a different approach is necessary. The optimal estimation framework as described, for example, by Schweppe[8] offers such a possibility. The basic idea of this approach is propagation of uncertainty through the systems of equations (model) which describe the relationship between the actual measurements (radiances), the model parameters, and the estimated variable of interest, i.e., rainfall. As an example consider the model proposed by Wilheit et al.[11,12] for rainfall estimation based on passive microwave observations. Wilheit and colleagues assume a particular, and fixed, cloud configuration. The model involves the cloud base level, thickness, dropsize distribution, temperature lapse rate, etc. This model, when combined with a simple one-dimensional radiative transfer model, constitutes the basis for solving the inverse problem of rainfall estimation at a pixel scale, and is used in climatological rainfall estimation in the approach of Wilheit et al.[12].

In order to construct a radiative transfer model, information about several factors is required, including surface temperature and emittance, water droplet size and distribution at the surface (rainfall rate) and throughout the atmosphere, atmospheric temperature profile, and cloud-base height and thickness. From this information, the radiative transport equation is solved and the radiant energy arriving at the satellite radiometer is computed. By iterative changing of the model parameters which control both the computed brightness temperature and rainfall rate, one can make the computed and observed signals to agree with each other. The rainfall rate that corresponds to such a set of model parameters is the estimated value. The information required for description of atmospheric processes controlling radiative transfer cannot be known exactly. Errors in description of the atmospheric parameters are propagated through the model and result in erroneous estimates of rainfall despite agreement of the calculated and observed radiances. This approach represents a deterministic estimation framework, the accuracy of which can be assessed only by comparison with a known reference estimate.

The new method proposed here is a stochastic extension of the physically-based estimation described above. The formal framework of the method is that of optimal estimation of static systems. According to this framework, the model with its parameters and inputs is combined in an optimal way with measured quantities to produce a minimum variance estimator. The associated variance is used as an accuracy measure of the values produced. The solution is difficult if the model linking the observations, inputs and parameters is nonlinear. Worse yet, in the case of atmospheric radiative transfer, the model is given only in terms of a numerical scheme making analytical manipulations virtually impossible. Still, in principle, it should be possible to numerically linearize the model and apply it within a linear estimation framework. The linearization involved has to be performed in such a way that statistical expectations of nonlinear functions involved can be evaluated. This is referred to as statistical linearization[9].

Mathematically, a radiative transfer model describing the relationship between satellite-measured radiances and rainfall rate can be described as a nonlinear function of the form:

$$\hat{r} = f(\mathbf{T_V}, \mathbf{T_H}, \mathbf{a}) \quad (12)$$

where $\hat{r}$ is rainfall rate estimate over a footprint of satellite measurements, $\mathbf{T_V}$ is a vector of brightness temperature observations at vertical polarization, $\mathbf{T_H}$ is a vector of brightness temperature observations at horizontal polarization, and $\mathbf{a}$ is vector of parameters describing the vertical structure of the atmosphere. These include cloud height and thickness, rain drop size distribution parameters, cloud ice crystal characteristics, emittance of the background, parameters describing temperature profile, and others. Since, as pointed out earlier, the precise knowledge of these parameters is impossible, our lack of knowledge, or uncertainty, can be described in terms of probability distributions which will characterize the range of variability of the parameters. For the vectors of observations $\mathbf{T_V}$ and $\mathbf{T_H}$, the uncertainty source is called measurement error and it can be determined based on design considerations of the satellite system. Now, (12) can be rewritten as a problem of optimal estimation:

$$\min \rightarrow \text{Var}\{\hat{r}\} = g(\mathbf{T_V}, \varepsilon_V, \mathbf{T_H}, \varepsilon_H, \mathbf{a}, \varepsilon_a) \quad (13)$$

where $\text{Var}\{\hat{r}\}$ is variance of estimate $\hat{r}$, $g(\cdot)$ is a function similar to $f(\cdot)$ in (12), and ε_V, ε_H, and ε_a, are error terms associated with observations and model parameters. Solving the above optimization problem is a difficult although achievable task. The main difficulties stem from the complex, nonlinear, and non-explicit nature of function $f(\cdot)$ in (12).

In order to estimate rainfall at temporal and spatial scales of interest to climate modeling, the estimates obtained according to (13) could be averaged and the variance of the averaged quantity R_s estimated.

The proposed method is original and has never been applied in satellite estimation of rainfall. If developed, it can provide a self-consistent approach to rainfall estimation giving satellite-based methods more credibility and allowing for multisensor rainfall estimation and meaningful intercomparisons of various methods.

OTHER METHODS

Among other approaches to validation of satellite-based methods of global rainfall estimation are: 1) extensive instrumentation of selected sites around the world[2]; and 2) numerical simulation of the measurement and estimation processes[13,14]. In this paper a new method is proposed for measurement and estimation of rainfall over open ocean.

A New Method of Estimating Oceanic Rainfall at Climatological Scale

The new method is based on underwater acoustic signal measurements and relationship between mean monthly rainfall and fractional time of rainy periods within the same month. Let us begin by discussing the second feature of the method. Doneaud et al.[15] showed, using radar data, that mean areal rainfall is highly correlated with fractional area covered with rain above a certain threshold. This fact stimulated considerable interest and is the basis for the area threshold method of rainfall estimation[16,17,14]. Extending the original idea of Doneaud et al.[15] to temporally averaged rainfall, results in a strong linear relationship between fractional time of rainy periods and mean rainfall. For example, in Figure 3 the results are given for mean monthly rainfall and fractional time both calculated from raingage observations at a Florida station. The fractional time was calculated from 15 minutes data. The correlation coefficient is 0.93. Such high correlation is a consistent result checked at many locations using 15 minute data from the National Climatological Data Center, Ashville, North Carolina. The reasons for this high correlation can be easily explained following an analogy to the approach used by Kedem et al.[16] and Kedem and Pavlopoulos[17]. They propose a simple statistical model of point rainfall in which both unconditional and conditional (on rainfall occurrence) statistics are used. An interested reader is referred to their papers for details.

The significance of this high correlation between the fractional time of rain and temporally averaged rainfall lies in the possibility of estimating temporal rainfall at climatological scales by measuring time of rain only, and not the intensity itself. This can be accomplished using underwater acoustic sensors. The advantage of acoustic sensors as compared to a traditional raingage is its lower cost. Traditional raingages do not work well when placed on buoys in open ocean environment. Severe conditions,

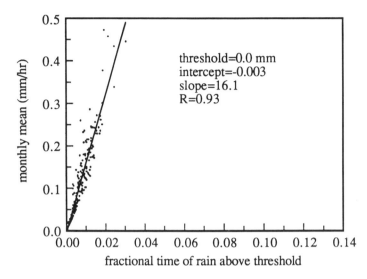

Figure 3. Mean monthly rainfall vs. fractional time in rain. Based on 15 minutes data from Melbourne, Florida, for the period 1972–1989. R is the correlation coefficient.

high winds, and wobbling of the buoys cause logistic difficulties and significant errors in measurements of rainfall intensities. An improvement of this situation can be offered by using optical raingages which have no mechanical parts and, due to the principle of measurement they use, are not susceptible to wobbling. However, their deployment and maintenance is rather expensive. An acoustic sensor is inexpensive, and since it is placed under water it avoids problems associated with surface wave action.

Franz[18] conducted an extensive experimental study on the spectrum of underwater sound from the splashes of droplets and sprays. He then used the scaling laws to estimate the spectrum levels of underwater sound from the splashing of rain on the surface of an ocean in terms of the rate of rainfall. In general, the rain sound spectra are relatively flat, with broad peaks near the middle of the audible frequency range. These distinctive spectral characteristics allow their separation from other sound sources[19]. The shape of the sound spectrum clearly indicates the presence of rain (Figure 4), while the spectral levels may be used to estimate quantitative rainfall rates. However, the problem of converting this spectrum to rainfall rate at the ocean surface has not yet been solved. Also, no mathematical description of underwater rainfall-generated acoustic signal has been developed. The problem is very difficult due to complex hydrodynamical and acoustic interaction of the involved processes of raindrop impact

190   Global Estimation of Rainfall

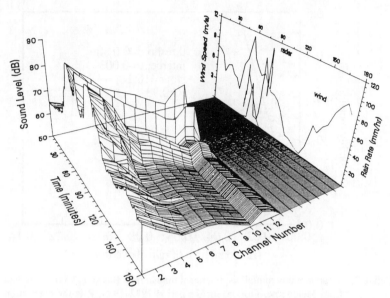

Figure 4. Acoustic signal and the corresponding rain rate and wind velocity. Data were collected by NOAA off the coast of Florida.

and sound propagation. To add to this complexity, wind generated waves break and the resulting spray contaminates the acoustic spectrum.

However, it seems that rather than to estimate rainfall intensity from the acoustic signal, one can, without much difficulty, determine whether it rains or not. This can be accomplished much easier as it is clear that the relevant information is contained in the acoustic signal. One approach to calibration of the methods to distinguish rain from no-rain is the use of a neural network.

The practical implementation of this combined approach, using measurements of fractional time, would work in the following estimation scheme. A climatological grid box would be covered with several inexpensive buoys equipped with the acoustic sensor. Its measurements of the acoustic signal would be converted to a rain/no-rain time series with high temporal sampling frequency (on the order of one minute sampling interval). The calculated (at the end of the month) fractional time in rain would be converted, via a linear regression, to an estimate of mean monthly rainfall. This will establish several (as many as the number of acoustic sensors) point estimates of monthly rainfall. From these points mean areal rainfall for the month would be calculated using the methodology outlined for the raingage networks. The linear regression necessary for the calculations needs to be calibrated locally. Assuming statistically homogeneous rainfall over the box, a single optical raingage could be used

statistically homogeneous rainfall over the box, a single optical raingage could be used for such purposes. Initial analysis of the regression in question justifies such assumption. These findings confirm a theoretical analysis by Kedem et al.[16] for the case of spatial estimation. This new approach will allow, for the first time, for truly open ocean estimation of rainfall by in-situ sensors.

CLOSING REMARKS

Several practical issues of estimating global rainfall were discussed. In particular, the author attempted to justify the importance of estimating uncertainty associated with different rainfall estimation methods. It was shown that uncertainty in one method can be estimated either by comparison with another, independent method, or by propagation of uncertainty in the observed quantities (measurements) and parameters through the model used in estimation. General examples are given to illustrate both approaches.

A new methodology is proposed for measurements and estimation of monthly mean areal rainfall from simple acoustic sensors placed under water.

It is hoped that the issues raised in this paper will serve as an aid in designing and planning activities of global rainfall estimation by national and international organizations such as NOAA in the United States, World Meteorological Organization, and the Global Precipitation Climatology Center in Germany. Other related problems not explicitly discussed, concern rainfall estimation from multiple sensors, climatological raingage network design, and estimation of rainfall from radars.

Acknowledgments. The author would like to acknowledge the support of the National Oceanic and Atmospheric Administration through grant NA16RC0444-01. Discussions with Phillip Arkin, Bruno Rudolf, and John Wilkerson were particularly useful in shaping the described ideas. The help of Gualiang Lin in preparation of the manuscript is appreciated.

REFERENCES

1. P. Arkin and P. Ardanuy. 1989. Estimation of climatological scale precipitation from space: A review. *J. of Climate* **2**(11):1229–1238.
2. J. Wilkerson, (Ed.). 1988. Validation of satellite precipitation measurements for the Global Precipitation Climatology Project. Proceedings of an international workshop, Washington, D.C., 1986, World Meteorological Organization, WMO/TD—No. 203.
3. J. Simpson, R. F. Adler, and G. North. 1988. A proposed tropical rainfall measuring mission (TRMM). *Bulletin of the American Meteorological Society* **69**:278–295.
4. D.R. Legates. 1987. A climatology of global precipitation. *Publications in Climatology* **40**(1), Newark, Delaware.
5. B. Rudolf, H. Hauschild, M. Reiss, and U. Schneider. 1991. Operational global analysis of monthly precipitation totals planned by the GPCC. *Dynamics of atmospheres and oceans.*
6. EOS, Transactions of the American Geophysical Union. 1991. Congress probes climate change uncertainties, **72**(43), October 22.

7. T.H. Lee, J. E. Janowiak, and P. A. Arkin. 1991. Atlas of products from the Algorithm Intercomparison Project, 1: Japan and surrounding oceanic regions, June–August 1989. University Corporation for Atmospheric Research, p. 130.
8. F. Schweppe. 1973. *Uncertain dynamic systems*. Prentice Hall.
9. R.L. Bras, and I. Rodriguez-Iturbe. 1985. *Random functions and hydrology*. Addison-Wesley.
10. A.G. Journel and Ch. J. Huijbregts. 1978. *Mining geostatistics*. Academic Press.
11. T.T. Wilheit, A. T. C. Chang, M. S. Rao, E. B. Rodgers, and J. A. Theon. 1977. A satellite technique for quantitatively mapping rainfall rates over the oceans. *J. Appl. Meteorol.* **16**:551–560.
12. T.T. Wilheit, A. T. C. Chang, and L. S. Chiu. 1991. Retrieval of monthly rainfall indices from microwave radiometric measurements using probability distribution functions. *Journal of Atmospheric and Oceanic Technology* **8**(1):118–136.
13. R.F. Adler, H.-Y. Yeh, M. N. Prasad, W.-K. Tao, and J. Simpson. 1991. Microwave simulations of a tropical rainfall system with a three-dimensional cloud model. *J. Appl. Meteor.* **30**:924–953.
14. W.F. Krajewski, M. Morrissey, J. A. Smith, and D. T. Rexroth. 1992. On the accuracy of the area threshold method: A model-based simulation study. *Journal of Applied Meteorology* (in press).
15. A.A. Doneaud, S. I. Niscov, D. L. Priegnitz, and P. L. Smith. 1984. The area-time integral as an indicator for convective rain volumes. *J. Clim. Appl. Meteor.* **23**:555–561.
16. B. Kedem, L. S. Chiu, and Z. Karni. 1990. An analysis of the threshold method for measuring area-average rainfall. *Journal of Applied Meteorology* **29**(1):3–20.
17. B. Kedem and P. Pavlopoulos. 1991. On the threshold method for rainfall estimation: Choosing the optimal threshold level. *Journal of American Statistical Association* **86**:626–633.
18. G.J. Franz. 1959. Splashes as sources of sound in liquids. *J. Acoust. Soc. Am.* **31**:1080–1096.
19. J.A. Nystuen. 1986. Rainfall measurements using underwater ambient noise. *J. Acoust. Soc. Am.* **79**:972–982.

METHODS AND PROBLEMS IN ASSESSING THE IMPACTS OF ACCELERATED SEA-LEVEL RISE

Robert J. Nicholls, Karen C. Dennis
Claudio R. Volonte, and Stephen P. Leatherman
Laboratory for Coastal Research, 1113, LeFrak Hall
University of Maryland, College Park, MD 20742

ABSTRACT

Accelerated sea-level rise is one of the more certain responses to global warming and presents a major challenge to mankind. However, it is important to note that sea-level rise is only manifest over long timescales (decades to centuries). Coastal scientists are increasingly being called upon to assess the physical, economic and societal impacts of sea-level rise and hence investigate appropriate response strategies. Such assessments are difficult in many developing countries due to a lack of physical, demographic and economic data. In particular, there is a lack of appropriate topographic information for the first (physical) phase of the analysis. To overcome these difficulties we have developed a new rapid and low-cost reconnaissance technique: "aerial videotape-assisted vulnerability analysis" (AVA). It involves: 1) videotaping the coastline from a small airplane; 2) limited ground-truth measurements; and 3) archive research. Combining the video record with the ground-truth information characterizes the coastal topography and, with an appropriate land loss model, estimates of the physical impact for different sea-level rise scenarios can be made. However, such land loss estimates raise other important questions such as the appropriate seaward limit of the beach profile. Response options also raise questions such as the long-term costs of seawalls. Therefore, realistic low and high estimates were developed. To illustrate the method selected results from Senegal, Uruguay and Venezuela are presented.

INTRODUCTION

Accelerated sea-level rise is one of the more certain responses to global warming and presents a major challenge to mankind.[1] Rising sea level causes erosion, submergence, salinization, higher water tables and a greater risk of flooding and storm impacts.[2] Coastal scientists are increasingly being called upon to make an analysis of these physical phenomenon so that economic and societal impacts of sea-level rise can be assessed and appropriate response strategies can be investigated. Subsequently to the development of the approach described in this paper, the IPCC formalized a seven step procedure for vulnerability analysis (VA) to accelerated sea-level rise—termed the Common Methodology (CM).[3] While our method was developed independently of the CM, it addresses the same problems and is one approach towards VA in the context of the CM.

The ability to conduct VA depends critically on the availability of a range of physical, demographic and economic data. In developed countries such as the Netherlands, sophisticated analyses have been possible[4] including the development of new physical and modelling concepts.[5,6] A long observational database on coastal evolution was an important prerequisite for the successful completion of this analysis. However, in the developing world, the necessary data often does not exist, or is found in inappropriate forms for VA and many analyses are essentially qualitative.[7] There is a clear need to provide more quantitative VA for such developing countries. Furthermore, these VA should be completed in a reasonably short period (less than a year).

A feasibility study on the Chesapeake Bay demonstrated that the AVA method is unbiased and suitable for reconnaissance surveys. A pilot study of Senegal demonstrated the utility of the method at a country scale.[8] Other national studies are in progress or completed, including Uruguay[9] and Venezuela.[10] These results give: 1) national assessments of vulnerability to sea-level rise; and hence, 2) allows the attention of future studies to be directed to those areas most "at risk" from sea-level rise.

Other important problems emerged while preparing these assessments: 1) quantifying land loss; and 2) the cost of response measures. On sandy coasts, land loss due to beach erosion is estimated using the Bruun Rule. However, several uncertainties necessitate the construction of realistic high and low erosion estimates. The cost of responses is also difficult to assess due to a lack of experience with the range of options. Again, realistic high and low estimates were developed.

SEA-LEVEL RISE AS A COASTAL HAZARD

Sea-level rise is a hazard which operates over a long timescale (decades to centuries). As such sea-level rise can be seen as an insidious hazard within the coastal zone. While its impacts are real and costly, they are only apparent when viewed in a long-term context.

In general, hazards within the coastal zone operate on a number of different time (and space) scales: the larger the time (and space) scale of the hazard, the larger the number of people affected by the hazard. A conceptual model to describe this is presented in Figure 1. (This model can be considered as a third dimension of the Large Scale Coastal Evolution Concept[5,6]) There are a number of existing hazards in the coastal zone that do not involve any rise in sea level: most particularly tropical and extra-tropical storms. These temporarily raise water levels causing flooding while the associated waves may destroy coastal property. However, any single storm is localized and short-lived and the number of people affected on a global scale is relatively small. On a longer timescale (months to years), changing weather patterns have more widespread impacts and hence affect larger numbers of people. For instance, an El Niño has consequences throughout the Pacific, including raising sea level up to 0.6 meters along the coast of North America for several months, combined with a high storm frequency.[11] Sea-level rise acts over decades to centuries and impacts the entire population of the coastal zone, directly or indirectly. Its negative consequences include: 1) raising the baseline condition and hence, causing higher water tables, salinization and land loss due to erosion and inundation; and 2) exacerbating extreme events such as storm damage and flooding. People more readily perceive the extreme event than the long-term sea-level rise that progressively enhances the destructive consequences of

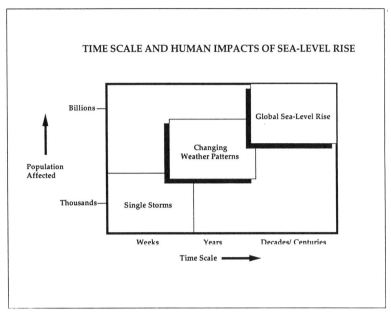

Figure 1. Time scale and coastal hazard.

such events. For instance, in the Chesapeake Bay, a number of islands have been abandoned by their inhabitants in the last century.[12,13] The underlying cause appears to be relative sea-level rise, but extreme events were important in triggering island abandonment.

Although implicitly acknowledged in most studies, this conceptual model explicitly acknowledges the large time and space scales that must be considered in any analysis of the impacts and responses to sea-level rise. Often these scales are considerably larger than available experience and considerable judgment is required.

SEA-LEVEL RISE

The most recent and comprehensive analysis suggests that in the last 100 years (1880 to 1980), the rate of sea-level rise was 1.8 mm/yr.[14] Thus, without any change in present trends we already face significant problems in the coastal zone.[15] Best estimates are that by the year 2100 there will be a rise in the range 0.33 to 1.10 meters, with a most likely rise of 0.66 meters.[1] Thus, while there is uncertainty about the magnitude of future sea-level change, there is a consensus that it will rise in response to global warming. Simply knowing the direction of change greatly assists planning appropriate response strategies to such change.

In our impact analysis, we considered three scenarios of accelerated eustatic sea-level rise by the year 2100: 0.5, 1.0 and 2.0 meters, plus the present trend of 0.2 meters.

COASTAL CHARACTERIZATION USING AVA

One of the major problems with studying the impacts of sea level rise in developing countries is the lack of detailed information on coastal geomorphology and the existing pattern and scale of development.[7,16,17] Most fundamentally, coastal elevations are poorly known. For instance, the best topographic maps of Senegal have 5-meter contours, but these only have very local coverage (around the capital Dakar). Most maps have a 40-meter contour interval. This is of limited value when analyzing land loss for any reasonable sea level rise scenario. Similar problems are found in many other developing countries. New topographic surveys using aerial photography and satellite imagery are prohibitively expensive and time-consuming, and in the case of satellite imagery a 20-meter contour is the highest accuracy possible.[18]

Given these constraints, our solution is "aerial videotape-assisted vulnerability analysis" (AVA). (This has also been termed videomapping). This characterizes the coastal development and geomorphology, including rough estimates of coastal elevations, by obliquely videotaping the coastline at low elevation from a small airplane just off the coast.[19] Position is determined with a hand-held global positioning system (or GPS). This information is analyzed in conjunction with existing and limited new ground-truth information to determine physical and hence economic impacts of sea level rise. The video record gives us the following information on the coastal zone, including those areas affected directly by sea-level rise:

a) <u>Index of terrain and relief changes</u>. We have a cross sectional view along the coastline which enables us to subjectively estimate the relative topography of the coastline.
b) <u>Types of coastal environments</u>. We are able to identify the size and location of wetlands, sandy beaches and dunes, hard rocky coastlines, etc. This is useful for determining the dominant land loss mechanism (erosion or inundation).
c) <u>Land use practices</u>. We can identify coastal land use such as agricultural plots and aquaculture facilities.
d) <u>Infrastructure</u>. We have a record of coastal infrastructure.
e) <u>Population</u>. We have quantitative indicators of the population living in the coastal zone.

The most important limitation of this method is that, unlike traditional procedures involving aerial photographs, parallax is not used to estimate elevations. Instead, researchers use the video record to classify the coastal landforms and subjectively interpolate elevations between occasional ground-truth data (either reliable spot heights or surveyed transects). Clearly, this procedure would not provide an accurate contour for anyone who needed to know whether a particular parcel of land would be inundated. However, validation experiments in the Chesapeake Bay, Maryland, have demonstrated that this method is unbiased and reasonably accurate when estimating land loss at a large scale.

This technique is especially attractive because it allows us to survey large expanses of coastline quickly with a limited budget. A similar procedure is already utilized in

Louisiana, where the coastal zone is retreating so rapidly that a quick and inexpensive method for determining coastal geomorphology is essential.[20,21] The flexibility of the method allows rapid mobilization, for instance after the passage of a hurricane.

In outline, the fieldwork procedure is as follows: 1) divide the coast into a working geomorphic classification using published information and expert knowledge and obtain the best maps; 2) videotape the coastline at a low and (sometimes) a high altitude of about 70/100 meters and 300 meters, respectively, including a recorded commentary and accurate position; 3) view the videotapes and update the geomorphic classification as appropriate; 4) visit as many representative coastal types as possible and collect topographic data and information concerning land and property values, agriculture, future plans for development, etc.[19]

The first stage of the analysis is to use the video record to develop an inventory of the coastal zone, including coastal geomorphology, coastal land use and development, and estimates of coastal elevation. Thus, the coastline can be classified into sections with similar characteristics, which can be considered homogeneous units. Subsequent analysis, such as land loss estimates, utilizes procedures which are described in the following sections. The video record is still important to this analysis: 1) infrastructure losses are estimated directly from the video record, and any available maps, by simply overlaying the predicted recession and estimating the number of buildings that would be destroyed; while 2) the lengths of coastline requiring protection are measured from the video record.

LAND LOSS ESTIMATES

Erosion. The Bruun Rule[22,23,24] is commonly used to calculate erosion due to sea-level rise. We used the following form, after Hands[23]:

$$R = G \times S \times (L/(B + h^*)) \tag{1}$$

where:
R is the shoreline recession due to a sea level rise S;
h^* is the depth of closure;
B is the dune height;
L is the active profile width from the dune to the depth of closure;
G is an overfill ratio.

These parameters were estimated from field data and available charts. G was assumed to be unity (all eroded material is sand), except where material is not erodible (hard rock), so these recession estimates are minimum values. The depth of closure, h^*, (and hence the active profile width, L) is the variable that is most difficult to estimate.[23] In particular, it depends upon timescale: the longer the period of interest, the larger the depth of closure.[5] Beach erosion (and the cost of beach nourishment) is sensitive to the value chosen, as a larger value of h^* generally implies a proportionally larger active profile width (L) and hence, a larger value for the term $L/(B + h^*)$. This is due to the exponential form of beach profiles (Figure 2). There are no accepted methods for estimating depth of closure over the long timescales of interest. Therefore, for our calculations we constructed a low and high estimate of closure depth (d_{L1} and d_{L100}, respectively). d_{L1} is the same definition as d_l of Hallermeier,[25] which is the

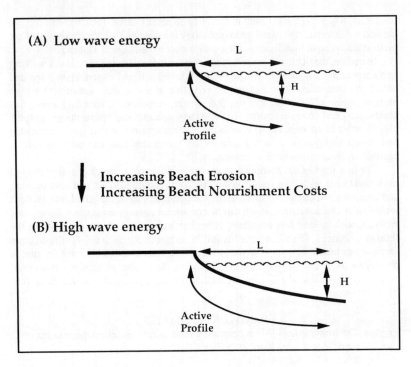

Figure 2. Active profile width (and hence the ratio (L/H)) increases with wave energy. Thus, sub-aerial beach recession and the costs of beach nourishment increase with wave energy. (Note H = 8+h*—Equation 1).

greatest depth where intense on/offshore transport and significant alongshore transport will occur within a typical year. d_{L1} can be approximated:

$$d_{L1} = 2H_s + 11s \qquad (2)$$

where:
H_s is mean significant wave height;
s is the standard deviation of wave height.

Experience suggests that d_{L1} is a good estimate for a typical year and it provides a robust low estimate. Our high estimate is a new depth definition termed d_{L100}. It extends Hallermeier's original concept and has the same definition as d_{L1} except that the timescale is a century. Based on the very limited experience (e.g., Hands[23]) and some new calculations a simple approximation was developed:

$$d_{L100} = 1.75 \times d_{L1} \qquad (3)$$

Further work to better define how h* increases with timescale would be useful.

Both values of h* are referenced to a datum one meter above low water. The corresponding low and high active profile widths were determined at different locations along the coast using the best bathymetric charts. The coastline was divided into segments of similar beach width and dune height and the Bruun Rule was applied.

It is worth noting that the availability of data on wave climate was a major constraint on these analyses. A comprehensive and accessible global database on wave climate is urgently required if the objectives of IPCC[16] are to be attained.

Inundation. In the case of wetlands and other coastal lowland we applied a direct inundation or "drowning" concept. Inundation is most significant in deltas and in wetlands around estuaries. For these low-lying areas, the video record is insufficient to define an estimate of elevation and it was integrated with available maps, plus expert judgment. This is not to undervalue the video record as it a) provided a check on the validity of the maps, and b) helped to define the present extent of wetlands and mangroves.

Two, and where possible, one meter contours (above high water), plus the area of existing wetlands or mangroves were estimated. We considered sedimentation in the inundated area as best as possible. Coastal marshes[26] and mangroves[27] can accrete vertically in response to slow rates of sea level rise and are only inundated above a certain threshold value of sea level rise. Deltas may or may not be receiving sediment from the river catchment. In most cases the availability of sediment is poorly understood and assessments of wetland evolution due to relatively small amounts of sea-level rise (up to 0.5 meters) is difficult, and often our best estimate of land loss was linear interpolation of land loss from the higher scenarios of sea-level rise.

RESPONSE OPTIONS

Four response options were considered as follows:
1. No protection.
2. Present protection. Existing protection is maintained.
3. Developed areas protection. Medium to highly developed areas are protected (i.e., cities, tourist beaches, factories). Using the video record, medium to high development is defined as areas with more than 15 percent of the ground covered in structures. For tourist areas, beach nourishment is assumed. Elsewhere, seawalls are utilized.
4. Total protection. All coastal areas with a population greater than 10 people/km^2 are protected (c.f. IPCC[16]). All additional protection above 3. Developed Areas Protection is seawalls.

Nourishment and Seawall Designs

The protection options were developed as follows using in-country costs, where possible. For seawalls and groins, we used the cost of purchasing, transporting, and placing rock, plus design costs (10%) and maintenance costs (20%), to determine a total cost. The cost of beach fill (sand) was estimated to be U.S. $5/m^3, unless alternate cost estimates were available. No protection costs were determined for baseline conditions (0.2 m rise).

Beach Nourishment. The volume of beach fill was calculated by raising the entire active profile by the sea level scenario. However, as with beach erosion, the active width is difficult to determine.[28] In our case it is logical to use the low and high active widths already defined for the Bruun Rule as low and high cost designs, respectively.

An additional cost on long open beaches is longshore loss of sand out of the nourished area.[29] To stop such losses we propose large terminal groins at the ends of each nourishment project which, in effect, converts the open beach into a pocket beach. The terminal groins extend from the dune to the seaward limit of the profile, which can be several kilometers offshore. Costs do not rise uniformly with sea-level rise as the groins are a large but declining proportion of the total cost as the sea-level rise scenario is raised. While this procedure raises the costs of beach nourishment significantly, the resulting costs are more realistic than simply considering the cost of a single beach fill (c.f. Titus et al[30]). From an environmental perspective, regular renourishment may be preferred, but the costs will again be much greater than a single beach fill. More work on the optimal application of beach nourishment would be useful.

Seawalls. In order to calculate the cost of erecting a seawall we developed three simple seawall designs, whose application depends on wave environment. These are a low and high cost seawall for open (or wave-exposed) coasts (LCOC and HCOC) and a seawall for sheltered coasts (SC) (Figure 3). They all utilize 1:2 slopes and a two meter berm. A seawall design for erodible cliffs was also developed.

The LCOC and HCOC designs are necessary due to the uncertain response of the beach in front of the seawall to sea level rise. In the best case (lowest cost), the beach would not erode at all, while in the worst case (highest cost) the beach would be completely lost. Actual behavior will lie somewhere in between these extremes, with total beach loss being more likely with higher sea level rise scenarios. The LCOC design assumes no beach erosion and only considers the effects of a more severe nearshore wave climate. The HCOC design assumes total beach loss and the consequent need to prevent undermining of the wall. The size and hence cost of the HCOC design increases with wave climate. Thus, the LCOC and HCOC designs reflect the two cost extremes. The SC design is lower than the LCOC design as the sheltered coast would have limited problems with a more severe nearshore wave climate. In general, existing seawalls are uncommon in the countries being considered and the problems of costing upgrades of existing structures was relatively minor.

The 1:2 slope was chosen instead of the 1:4 slope that is used for the cost estimates of IPCC[16] since the 1:2 slope is more economical and the standard in the United States. Using the IPCC[16] designs which utilized a 1:4 slope and a 4 meter berm width would double the cost of seawall construction. However, the range of our costs for wave-exposed coasts embrace the IPCC costs due to the HCOC design.

RESULTS

Some results from AVA for Senegal,[18] Uruguay[9] and Venezuela[10] are given in Tables I, II, and III. The high and low estimates give a large possible range of land

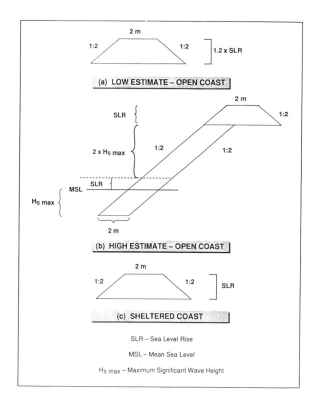

Figure 3. Seawall designs: (a) Low estimate - open coast (LCOC); (b) High estimate - open coast (HCOC); (c) Sheltered coast (SC).

Table I. Land Loss Estimates (in km^2) Assuming No Protection

Rise (m)	0.2	0.5	1.0	2.0
Uruguay	15–20	36–51	72–103	145–206
Senegal	349–356	1,947–1,963	6,042–6,073	6,494–6,546
Venezuela	1,138–1,147	2,844–2,866	5,686–5,730	11,277–11,358

Table II. Protection Costs (in millions of U.S. dollars)

	Rise (m)	Developed Areas Protection	Total Protection (this study)	Total Protection (IPCC[16])
Senegal	0.5	146–575	407–1,422	n.a.
	1.0	255–845	973–2,156	1,596
	2.0	492–1401	2,792–4,269	n.a.
Uruguay	0.5	2,068–6,871	2,162–7,505	n.a.
	1.0	2,903–8,578	3,144–9,437	1,805
	2.0	4,758–11,802	5,432–13,188	n.a.
Venezuela	0.5	454–960	718–1,613	n.a.
	1.0	999–1,517	1,717–2,634	3,155
	2.0	1,921–2662	4,118–5,172	n.a.

n.a.: not available

Table III. Relative Vulnerability For Developed Areas Protection and a 1-Meter Rise

Relative Cost		Land Loss	(%Total)
1. Uruguay	(6.5 to 19.2%)	1. Senegal	(3.08)
2. Senegal	(0.9 to 3.0%)	2. Venezuela	(0.62)
3. Venezuela	(0.1 to 0.2%)	3. Uruguay	(0.04)

Note: Relative Cost is based on 1987 gross investment and assumes the investments occur over 50 years.

loss/cost, but we feel that this embraces the uncertainty inherent in analyses of this type.

Table I gives land loss assuming no protection. It is interesting to note the significant land loss due to a 0.2 meter rise (the present case). Thus, present rates of sea-level rise are already a significant hazard. Land loss, under all scenarios, is one to two orders of magnitude smaller in Uruguay, than Senegal and Venezuela, because inundation is not an important process in Uruguay. In Senegal and Venezuela inundation is the dominant land loss process. Most of the inundated land is wetland or mangroves. However, erosion causes a high proportion of the economic and societal impacts in all three countries.

In Table II, costs of protection are given with the cost estimates of IPCC.[16] Uruguay has high protection costs due to the importance of beach-based tourism and hence, a large requirement for beach nourishment (over 98 per cent of costs for Developed Areas Protection). IPCC[16] appear to have greatly underestimated the cost of sea-level rise on Uruguay. Senegal has smaller protection costs in absolute terms, but they are still significant. Again beach nourishment is a large proportion of the costs. There is a large increase in cost between Developed Areas and Total Protection. It is questionable what benefits Senegal would gain for this increase in expenditure as much of the additional protection is for low value desert areas. In general, we believe that Developed Areas Protection, as assessed in these studies, is the most likely protection option. Venezuela, also has large protection costs, but given the large length of its coastline, they are surprisingly low. This is due to the low level of coastal development in Venezuela, occupying only 13 percent of the coastline. For Venezuela, the IPCC[16] cost appears to be a significant overestimation.

In Table III these three countries are ranked by: 1) their ability to pay for protection (using gross investment); or 2) land loss (as a percentage of total land area) (Table III). For the purposes of this analysis it is assumed that the investment in protection occurs over 50 years (2050 to 2100). Uruguay has the highest relative costs, consuming a significant proportion of the available investment (at 1987 levels), but it has the lowest land loss. Senegal has both high costs and high land loss. These results support the need for a vulnerability profile, as opposed to a vulnerability index, as recommended in the Common Methodology.[3]

DISCUSSION

The AVA approach is only one method of obtaining the information necessary for a vulnerability analysis (VA). This first-step reconnaissance allows one to focus upon the more critical areas of a coastline. More detailed follow-up studies using more traditional approaches are recommended. However, the video record provides much useful information on land use, infrastructure, and "brings the coast to your desk." Therefore, it may be useful even if high quality topographic data is available, as is the case in Louisiana.[20,21] AVA is least valuable in deltaic areas due to their low topography. However, as already noted the video record still provides a useful inventory of many coastal characteristics for VA.

AVA as described in this paper has been constructed as a series of modules, whereby individual elements of the overall procedure can be improved, while the remaining elements of the approach remain robust. Therefore, as our understanding improves, VA can be easily and rapidly repeated to generate new and improved estimates of the impacts of sea-level rise.

By the nature of reconnaissance surveys, many possible improvements to the procedures described are possible and would help to more precisely define impacts and appropriate responses to sea-level rise. In particular, it would be useful to improve our understanding of beach erosion and hence the costs of beach nourishment and seawalls. In this regard, better wave data is essential. In addition, a more explicit description of the time element of the problem is required, including factors such as economic growth and development.

CONCLUSION

AVA is a rapid and low-cost method to conduct reconnaissance vulnerability analyses at large scales. It provides a focus on which areas can be most usefully studied with more conventional and expensive techniques and hence helps to maximize the results from any limited national budget for such an analysis. Aerial video techniques are also useful for more detailed studies when high quality topographic data already exists. However, other problems remain such as the appropriate seaward limit of the active beach profile. These are independent of the data source utilized. Procedures have been developed to provide a realistic range of values for these estimates. Further work is required on these problems, including the establishment of observational databases for the world's coastlines, most particularly some meaningful description of wave climate.

ACKNOWLEDGMENTS

This work was funded by the Office of Policy Analysis, United States Environmental Protection Agency. Mr. Jim Titus (Project Officer) and Mr. Joel Smith gave much useful advice. The authors would like to thank our in-country collaborators, Dr. Isabelle Niang, Senegal, and Mr. Jose Arismendi, Venezuela, without whom this paper would not be possible. Dr. Robert J. Hallermeier and Mr. Edward Fulford gave important technical advice. Dr. Marcel Stive made useful comments on an earlier draft of this manuscript.

REFERENCES

1. R. Warrick and J. Oerlemans. 1990. *Climate change: The IPCC scientific assessment.* Cambridge Univ. Press.
2. National Research Council. 1987. *Responding to changes in sea level: Engineering implications.* National Academy Press, Washington, D.C.
3. IPCC. 1991. *Assessment of the vulnerability of coastal areas to sea level rise: A common methodology.* Revision No. 1. IPCC Coastal Zone Management Subgroup.
4. E. B. Peerbolte, J. G. de Ronde, L. P. M. de Vrees, M. Mann, and G. Baarse. 1991. *Impact of sea level rise on society: A case study of the Netherlands.* Delft Hydraulics,.

5. M. J. F. Stive, D. J. A. Roelvink, and H. J. De Vriend. 1990. *Proc. 22nd Coastal Eng. Conf.*, ASCE, N.Y.
6. M. J. F. Stive. 1990. *Proc. 3rd European Workshop on Coastal Zones*. National Technical University, Athens, p. 4.1.
7. J. G. Titus, ed. 1990. *Changing climate and the coast*, 2 Vols. U.S. Environmental Protection Agency, Washington, D.C.
8. I. Niang, K. C. Dennis, and R. J. Nicholls. 1992. *The Rising challenge of the sea*. U.S. Environmental Protection Agency, Washington, D.C.
9. C. R. Volonte and R.J. Nicholls. 1992. *The rising challenge of the sea*. U.S. Environmental Protection Agency, Washington, D.C.
10. J. Arismendi and C. R. Volonte, *The rising challenge of the sea*. U.S. Environmental Protection Agency, Washington, D.C.
11. P. D. Komar and D. B. Enfield. *Sea-level fluctuation and coastal evolution*. SEPM Special Publication No. 41, Tulsa, Okla.
12. S. P. Leatherman. In press. *The regions and global warming: Impacts and response strategies*. Oxford Univ. Press.
13. M. Kearney and J. C. Stephenson. 1991. *J. Coastal Res.* 7:403.
14. B. C. Douglas. 1991. *J. Geophys. Res.* 96(C4):6981.
15. P. Vellinga and S. P. Leatherman. 1989. *Climatic Change* 15:175.
16. IPCC. 1990. *Strategies for adaptation to sea level rise*. IPCC Coastal Zone Management Subgroup, Rijkswaterstaat, The Netherlands.
17. J. G. Titus, ed. 1992. *The rising challenge of the sea*. U.S. Environmental Protection Agency, Washington, D.C.
18. E. I. Theodossiou and I. J. Dowman. 1990. *Photogram. Eng. and Remote Sensing* 56:1643.
19. K. C. Dennis, R.J. Nicholls, and S. P. Leatherman. 1991. *Sea level rise and coastal management*, No. 1. NOS/NOAA and Rijkswaterstaat.
20. M. R. Byrnes, S. Penland, R. A. McBride, K. Debusschere, K. A. Westphal, and D. W. Davis. 1990. Workshop Summary, Great Lakes Shoreland Management Workshop Series, Cleveland, Ohio.
21. K. Debusschere, S. Penland, K. A. Westphal, P. D. Reimer, and R. A. McBride. 1991. *Proc. Coastal Zone "91*, ASCE, N.Y.
22. P. Bruun. 1962. *J. of Waterways and Harbors Div., ASCE* 88:117.
23. E. B. Hands. 1983. *Handbook of coastal processes and erosion*. CRC Press, Boca Raton, Fla. p. 167.
24. S. P. Leatherman. 1991. *Prog. Phys. Geog.*, 14:447.
25. R. J. Hallermeier. 1981. *Coastal Eng.* 4:253.
26. R. Gehrels and S. P. Leatherman. 1989. *Sea level rise: Animator and terminator of coastal marshes*. Vance Bibliographies, Monticello, Ill.
27. J. C. Ellison and D. R. Stoddart. 1991. *J. Coastal Res.* 7:151.
28. M. J. F. Stive, R. J. Nicholls, and H. J. De Vriend. 1991. *Coastal Eng.* 16:147.
29. R. G. Dean. 1983. *Handbook of coastal processes and erosion*. CRC Press, Boca Raton, Fla.
30. J. G. Titus, et al. 1991. *Coastal Management.* 15:39.

AUTONOMOUS AEROSONDES FOR METEOROLOGICAL SOUNDINGS IN REMOTE AREAS

Tad McGeer
The Insitu Group, 224 Robin Way, Menlo Park, California USA 94025

ABSTRACT

Meteorology historically has been handicapped by lack of in situ data in remote and oceanic regions of the globe. Often only satellite observations are available over such areas, and in many instances these remain inadequate for both forecasting and research. However completion of the Global Positioning System (GPS) and developments in satellite-based communication have created an opportunity to relieve the problem. It is now possible for atmospheric soundings to be made by aircraft no larger than hobbyist's models, flying autonomously over long ranges and durations, from sea level to high altitude, and regularly relaying data back to a monitoring site. The cost of soundings with such aircraft, when made on a wide scale, might well be quite comparable with that of balloon-borne radiosondes. Balloon sondes, however, are tied to fixed sites. "Aerosondes" could reach anywhere on earth.

MOTIVATION

The most valuable data set in contemporary meteorology is provided by balloon-borne radiosondes of the World Weather Watch. These are flown at least twice daily from about 800 sites worldwide[1], sampling pressure, temperature, humidity, and wind from the surface to 20 km altitude or higher. (About 800 more sites fly wind balloons only.) The data are used to initialize numerical models for routine forecasting, and also compose much of the archival record for meteorological and climatological research. However balloons provide effective coverage only near populated areas. A few remote stations and weather ships are maintained to fill problematic gaps elsewhere, but high costs force most of the globe to be left more or less unsampled. Satellites, buoys, airliners, and ships provide some information in these areas[2], but their reports are an inferior substitute for full in situ data. The result is that the quality of forecasting suffers, with diffuse economic consequences over much of the world, and occasionally more acute losses due to poor warning of tropical cyclones and other severe weather. Consequently the World Meteorological Organisation has an ongoing search for economical methods of improving the global sounding network.

Meanwhile technological developments have brought unmanned aircraft into consideration for atmospheric studies. In particular Boeing's *Condor*[3] and the proposed Aurora Flight Sciences *Perseus*[4] have raised the unprecedented possibility of over three days endurance and 20,000-km range at altitudes up to about 20 km. The cost of these (relatively large) aircraft limits their potential to major research efforts[5,6] but they illuminate the possibility of more general application if similarly attractive performance can be achieved at a smaller and more economical scale. Thus emerges the opportunity for an "Aerosonde": a small aircraft which, at a real cost comparable to that of radiosondes, could make measurements of pressure, temperature, humidity and wind from sea level to high altitude, but which at the same time had sufficient range to carry the instrument set from a convenient base into remote areas, and to remain there for a useful period of time. Here we review the design and applications of such an aircraft; more detail can be found in the references[7,8].

ENABLING TECHNICAL DEVELOPMENTS

The idea of using small aircraft to gather meteorological data is not new; they were used for boundary-layer work in the 1970s[9] and perhaps before. Nor is it new to imagine long flights in small aircraft; the endurance record for the Fédération

Table I. An Aerosonde's "Payload"

component	weight (g)	power (W)	remarks
computer	300	2	80386 class
GPS navigator	250	2	5-channel or better; 100 m, 1 m/s accuracy
300 baud VHF modem	250	< 2	available 1994 through *Orbcomm* constellation; full duplex
2-axis magnetometer	120	<1	0.5° heading accuracy. Allows wind component measurement to ≈ 1 m/s by comparison with airspeed and GPS velocity
yaw rate gyro	110	2	RC model type for lateral direction stabilization
2 × P, T, H transducers	160	<1	standard sonde sensors
microphone	50		for low-altitude altimetry by detecting the engine echo
servos (7)	250	< 2	RC mode type for control surfaces, propeller, and powerplants adjustments
pitot/static pressure	10	<1	
engine monitoring	20	<1	
wiring, packaging, fusing, etc.	200		
total electronics	1720	< 10	
DC generator and gearing	100	20	modified brush motor
batteries	200	5 W-h	16 × AA NiCad for transmitter bursts and generator transients
Total	2020		

Aéronautique Internationale model class (< 5 kg take-off weight) is over 20 hours[10]. These aircraft, however, were radio-controlled by pilots watching continuously from the ground. Consequently flights were confined to visual range both horizontally and vertically, and limited in endurance. The novelty today is that radio-control flying has become unnecessary; instead a small aircraft can be left to fly itself over long distances and durations. The necessary equipment is listed in Table I.

Notice that satellite-based navigation and communications are essential elements. Navigation can now be done using the Global Positioning System, which has been building up over the 1980s and will be complete in 1993. Lightweight receivers first marketed in autumn 1991 can calculate position and velocity anywhere on the globe, with respective accuracies of 100 m and 1 m/s in three dimensions. Meanwhile Motorola, Inmarsat, and a number of other firms have proposed satellite constellations for global telephone and data relay through pocket-sized terminals[11]. These networks are projected to become available by mid-decade (e.g. mid-1994 for Orbital Sciences's *Orbcomm*) and will be ideal for communication with Aerosondes.

Adding the remaining components listed in the table brings the weight of a complete electronics set to about 2 kg, and the power requirement to about 10 W. This constitutes a complete Aerosonde payload, and allows design at a very economical scale.

SIZING AND PERFORMANCE

Figure 1 illustrates a candidate aerosonde sized for missions of about 3 days' duration and 8000 km range. Table II lists its specification. Here we note a few salient features of the design, details of which are given in our 1991 engineering study[7].

Powerplant. The powerplant traditionally is the biggest item in an aircraft development program, and the Aerosonde program is no exception to the rule. It need not rely on any exotic technology; in fact the powerplant will be based on a four-stroke gasoline engine commonly used by aero modelers. However modifications will be necessary for extended autonomous operation and reduced fuel consumption. Appropriate techniques have been developed in "Mileage Marathon" competitions, which use model four-stroke engines to power lightweight road vehicles. Fuel consumption still will be high by large-engine standards (because of high thermal losses), but quite adequate to the purpose.

The powerplant design problem does not end with the engine itself. Many applications involve flight at high altitude, for which three additional components are needed: 1) a compressor, whose ratio can be varied with altitude; 2) a heat exchanger, to cool the intake air after (nearly adiabatic) compression; and 3) a variable-ratio exhaust expander, to recover energy through the pressure drop from the exhaust manifold to ambient. (Without the expander a large fraction of engine power would be absorbed just in running the compressor.) At small scale, design of these components is relatively simple because it is appropriate to use positive-displacement pumps rather than turbomachinery. On the other hand unusually close tolerances are required, so the design and manufacture must be done with some care.

Table II. Prospective specification for the Aerosonde of Figure 1

Weights (kg)
electronics	2.0
powerplant	2.7
airframe	1.2
empty weight W_e	5.9
gasoline	6.1
oil	0.3
gross weight $W_{t/o}$	12.3

Dimensions
wing span	3 m
wing area	0.5 m^2
aspect ratio	18

Powerplant (based upon the O.S. FS-70 four-stroke)
engine displacement	11.5 cc
compressor displacement	60 cc
speed	11,000 rpm
power	685 W
specific fuel consumption c	0.48 kg/kWh
propeller diameter	0.9 m
propeller efficiency η_p (with variable pitch)	0.83

Performance
service ceiling	16 km
optimum lift/drag ratio	
at sea level	23 @ 20 m/s
at 16 km	18 @ 50 m/s
maximum speed	
at sea level	40 m/s
at 16 km	70 m/s
range	8300 km
time to 16 km at gross weight	1.3 hr
climb/descent cycle time	4 hr
maximum soundings/flight	≈ 20

Structural design. The primary structure will be in carbon and kelvar composite, which offers advantages over more traditional metal structure in both performance and manufacturing economics. However Aerosonde design will benefit much more significantly from fundamental scaling effects, which allow unusually low structural weight fractions in small aircraft. The structural weight listed in Table II is only 10% of gross weight; nevertheless the aircraft will be sufficiently strong to survive even severe turbulence and, moreover, landing without an undercarriage. (In fact "belly" landing is common practice even at tenfold larger scale.) Hence the aircraft will not need a runway; any bit of smooth open space should offer a satisfactory base of operations.

Aerodynamics. Aerodynamic performance suffers in scaling to small size; viscous losses, and hence drag, become disproportionately large. Consequently the lift-to-drag ratios listed in Table II are modest but, nevertheless, quite acceptable. As the table indicates, range (which depends on aerodynamic, structural, and propulsive efficiency) works out to just over 8,000 km.

Choice of altitude. Range is insensitive to altitude, so the 8,000 km could be covered at any level between the surface and 100 hPa. However higher altitude implies

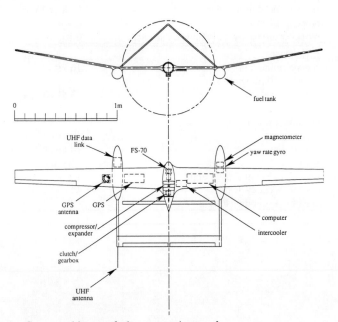

Figure 1. Conceptual layout of a long-range Aerosonde.

higher speed, and so reduced endurance; it would be about 4.5 days at sea level, but only 2 days at altitude. A meteorological mission might involve a continuous series of soundings between these extremes; in that case the flight would last about 3 days.

PROSPECTIVE ECONOMICS

Exciting opportunities would arise if missions of this type could be flown at a per-sounding cost comparable with that of radiosondes, i.e., (according to Australian Bureau of Meteorology figures) about US$200. Roughly half of this is in the cost of the staff, site, and communications. We anticipate that similar staff and facilities will be required for operating Aerosondes, so this element of cost should be roughly the same. That leaves US$100 as the target for per-sounding hardware cost, which in general can be reckoned as

$$\text{hardware cost per sounding} = \frac{\text{unit price}}{\text{flights/aircraft} \times \text{soundings/flight}}$$

US$10,000 is a reasonable round figure for unit price. (Obviously this will depend on production rate.) Twelve, on average, is a practical number of soundings per flight; this will allow for transit time and a 6-hour (rather than 4-hour) sounding cycle. The per-sounding cost target then would be achieved if the average lifetime were 10 flights. This seems reasonable, considering the potential for losses to weather (particularly icing) and equipment failure. Experience will be necessary before a better estimate can be made, but the possibilities are certainly sufficient to justify a very thorough set of trials.

APPLICATIONS

Global upper-air observations. The first opportunity suggested by the Aerosonde's economics is to increase routine upper-air observations over oceans and remote areas. The new in situ data not only would be tremendously useful in themselves, but moreover would allow ongoing calibration of satellite retrievals which otherwise can harbor serious biases[12,13]. Observations could be maintained by regular out-and-return missions to specific stations, or alternatively by rotation through bases arranged to take advantage of prevailing winds. For example much of the southern oceans could be covered by a grand circuit through Australia, the Society Islands, South America, southern Africa, and back to Australia.

Tropical cyclone reconnaissance. In the near term a higher-priority application is likely to be severe weather reconnaissance, which would remain attractive even at much higher cost. Hurricanes coming ashore are especially strong drivers: on the one hand "overwarning" of populated shoreline is commonly reckoned to cost of order $100,000 per kilometer in the U.S.[14] and Australia[15], while on the other hand "underwarning" (or warning without credibility) can have much more serious consequences. Thus even expensive methods for improving forecasts become economically justifiable. Aircraft reconnaissance is demonstrably in this category[16,17,18,19,20,21], and the U.S. consequently spends about $15~million (Figures for aircraft reconnaissance were obtained from the Office of the Federal Coordinator for Meteorological Services and Supporting Research, Washington, D.C.) annually to maintain a *Hercules* squadron for the purpose off its Atlantic coast[22,23]. Yet even at that price the level of operations is thought to be less than optimal, in that improvements in forecasting which could be obtained through more intensive in situ measurements would justify the additional expense involved[15].

A difficulty, however, is that the high costs of maintaining a reconnaissance program are much more obvious, and so harder to accept, than the high costs of *not* maintaining a reconnaissance program. Therefore in practice the optimal economic level is never achieved. In fact in the Pacific there is no systematic reconnaissance at all. The USAF operated a *Hercules* flight in the area until 1987, but withdrew it because the $20-million annual cost was considered unsustainable. The very sensitive Bay of Bengal is similarly exposed. The WMO considers this situation to be seriously inadequate, and has made improvement a high priority under the United Nations International Decade for Natural Disaster Reduction[24].

Aerosondes offer the prospect of materially improving the situation without requiring impolitic levels of funding. They could both complement present aircraft observations where they are available, and fill gaps elsewhere. Missions would probe both the near environment and the core of the storm[8], including the high-altitude region which at present is poorly observed. Because of the Aerosonde's slow speed flight planning would have to take account of winds; in particular entry would be made in the low-level inflow, and exit in the high-level outflow. Entry to the storm core probably would be from above to avoid the dangers of the eye wall. The attrition rate on such missions will be higher than in more routine service, but we anticipate that it will be justified by the results obtained.

DEVELOPMENT PROGRAM

In early 1992 work began on a program which could permit regular missions in three to five years. The Wackett Aerospace Centre at Royal Melbourne Institute of Technology is responsible for powerplants, and The Insitu Group is developing airframes, avionics, and software; the Australian Bureau of Meteorology is providing scientific support. The first step is to build a prototype for demonstration of autonomous operation and data gathering. It will be used mainly in engineering trials and so will be limited to low altitude and only a few hundred kilometers range. However these capabilities should be sufficient for a small meteorological experiment, for example in a sea breeze front. Following the prototype we plan a design suitable for operational trials in reconnaissance of shallow cold fronts off the south Australian coast[8]. This will involve flights of about 36 hours' duration, but still at low altitude. Meanwhile parallel work will proceed on a complete high-altitude powerplant, leading to initial tests to the 100-hPa level in about three years' time.

However while the technical plan is in place, the necessary funding is not. Whether it becomes available of course will depend upon the weight of scientific opinion as expressed to funding agencies. Moreover the significance of community input at this stage extends beyond simple competition for money. Autonomous aircraft also raise issues concerning air traffic control[8]; technically satisfactory solutions exist, but the momentum necessary to adopt them will accumulate in direct proportion to the push provided by atmospheric scientists. We have outlined the potential of autonomous aircraft; when, or whether, that potential is realized will depend upon the participation of the community as a whole.

REFERENCES

1. Secretariat of the World Meteorological Organisation. 1988. *Annual Report*. Geneva, WMO No. 713.
2. C. H. Dey. 1989. *Weather and forecasting*. pp. 297–312.
3. A. M. S. Goo, N. Arntz, and R. D. Murphy. 1989. *Aerospace America* **27**(2).
4. J. Langford. 1990. In *Proceedings of the 17th Congress*, International Council of the Aeronautical Sciences, Stockholm.
5. D. L. Albritton, F. C. Fehsenfeld, and A. F. Tuck. 1990. *Science* **250**(4977):75–81.
6. K. A. Emanuel and J. G. Anderson. 1991. In *Proc. 19th Conference of Hurricanes and Tropical Meteorology*, 435–436, American Meteorological Society.
7. T. McGeer. 1991. *Autonomous aerosondes for meteorological measurements in remote areas*. Technical Report, Aurora Flight Sciences, Alexandria, Va.
8. G. J. Holland, T. McGeer, and H. Youngren. 1992. Submitted to *Bulletin of the American Meteorological Society*.
9. D. Martin. 1980. In *The Boulder low-level intercomparison experiment Report No. 2*, J. G. Kamal, H. W. Baynton, and J. E. Gaynor, eds., 18–25. NOAA/NCAR Boulder Atmospheric Observatory, Boulder, Colo.
10. M. L. Hill. 1982. *Astronautics and Aeronautics* **11**(20):47–54.
11. American Institute for Aeronautics and Astronautics. 1992. *Proceedings of the 14th International Satellite Communications Conference*, Washington, D.C.
12. G. Kelly, E. Anderson, A. Hollingsworth, P. Lonnberg, J. Pailleux, and Z. Zhang. 1991. *Monthly Weather Review* **119**:1866–1880.
13. E. Anderson, A. Hollingsworth, G. Kelly, P. Lonnberg, J. Pailleux, and Z. Zhang. 1991. *Monthly Weather Review* **119**:1851–1864.
14. U.S. Special Interagency Task Group on Airborne Geoscience. 1989. *Airborne geoscience: The next decade*. Airborne Geoscience Applications Program Office, NASA, Washington, D.C.
15. Bureau of Meteorology. 1986. *Proposal for operational aerial reconnaissance of tropical cyclones in the Australian region*. Melbourne.
16. G. J. Holand and R. T. Merrill. 1984. *Quarterly Journal of the Royal Meteorological Society* **110**:723–745.
17. R. T. Merrill. 1985. In *Proc. 16th Conference on Hurricanes and Tropical Meteorology*. American Meteorological Society, Boston.
18. J. Molinari and D. Vollaro. 1989. *Journal of the Atmospheric Sciences* **45**:1093–1105.
19. J. L. Franklin. 1990. *Monthly Weather Review* 118:2732–2744.
20. J. L. Franklin, M. DeMaria, and C. S. Velden. 1991. In *Proc. 19th Conference on Hurricanes and Tropical Meteorology*, 87–92, American Meteorological Society.
21. R. L. Elsberry. 1990. *Bulletin of the American Meteorological Society* **71**:1305–1315.
22. Office of the Federal Coordinator for Meteorological Services and Supporting Research. 1991. *National hurricane operations plan*. Washington, D.C., 18th ed., FCM-P12-1991.

23. Office of the Federal Coordinator for Meteorological Services and Supporting Research. 1991. *National winter storms operations plan*. Washington, D.C., 18th ed., FCM-P13-1991.
24. International Council of Scientific Unions. 1991. *Third report of the special committee for International Decade for Natural Disaster Prevention*. Vienna.

INDUSTRY, CLIMATE CHANGE, AND NATURAL DISASTERS

CLIMATE CHANGE
AND NATURAL DISASTERS

GLOBAL WARMING AND THE INSURANCE INDUSTRY

G. A. Berz
Munich Reinsurance Company, Munich, Germany

ABSTRACT

In the last few decades, the international insurance industry has been confronted with a drastic increase in the scope and frequency of great natural disasters. The trend is primarily attributable to the continuing steady growth of the world population and the increasing concentration of people and economic values in urban areas. An additional factor is the global migration of populations and industries into areas like the coastal regions which are particularly exposed to natural hazards. The natural hazards themselves, on the other hand, have not yet shown any significant increase.

In addition to the problems the insurance industry has with regard to pricing, capacity and loss reserves, the assessment of insured liabilities, preventive planning and the proper adjustment of catastrophe losses are gaining importance.

The present problems will be dramatically aggravated if the greenhouse predictions come true. The increased intensity of all convective processes in the atmosphere will force up the frequency and severity of tropical cyclones, tornados, hailstorms, floods and storm surges in many parts of the world with serious consequences for all types of property insurance. Rates will have to be raised and in certain coastal areas insurance coverage will only be available after considerable restrictions have been imposed, e.g., significant deductibles and/or liability or loss limits.

In areas of high insurance density the loss potential of individual catastrophes can reach a level where the national and international insurance industries run into serious capacity problems. Recent disasters showed the disproportionately high participation of reinsurers in extreme disaster losses and the need for more risk transparency if the insurance industry is to fulfill its obligations in an increasingly hostile environment.

INCREASE IN THE SCOPE AND FREQUENCY OF GREAT NATURAL DISASTERS

The scope and frequency of loss incurred by great natural disasters have experienced a drastic increase worldwide over the past decades. This observation is based on the statistical analysis of the comprehensive loss data that the "Geoscience Research Group" of Munich Re has been continuously collecting over the last 30 years and has evaluated in connection with the preparation of its "World Map of Natural Hazards," which is now in its second edition. Each month brings data on 20 to 50 new loss events. Fortunately, however, only a few of these have to be added to the list of great natural disasters; that is, only those involving damage that considerably exceeds what the country or region affected can handle itself and for which the country requires substantial national or international aid. This is generally the case if the number of casualties runs into the hundreds or thousands, if tens or hundreds of thousands of people are rendered homeless and if the total economic loss is in excess of US$100 million.

Fairly comprehensive data exist on these great disasters, even those that occurred farther back in the past as seen in Figure 1. They numbered only 14 in the 1960s, and 70 in the 1980s, which means that they increased within three decades by a factor of 5. The trend is also very clear when we consider the total economic loss which, adjusted for inflation, increased during the same period from an annual average of US$3.7 billion per year to US$11.4 billion, i.e., by a factor of 3.1. The insured losses increased by an even larger factor of 4.8. At the present time, the insurance industry worldwide must count on an annual average burden from natural catastrophes of US$5 billion. The

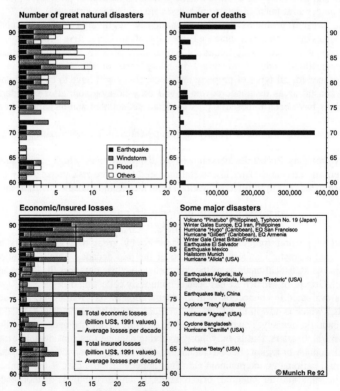

Figure 1. Great natural disasters, 1960–1991

fact that the total of insured losses in certain years can by far exceed these amounts was proven in 1990 by the winter storms in Europe, which amounted to approximately US$10 billion. At the same time, this must certainly also be taken as an indication for the fact that catastrophic losses are continuing to climb sharply.

This upward trend is mainly attributable to the persisting unbridled growth of the world population paralleled by a general improvement in living standards and the increasing concentration of people and economic values in urban areas. A UN prognosis states that the number of cities with populations in the millions will increase from 200 to 400 within 20 years and, in the same time period, cities with populations in the tens of millions from 10 to 25. An additional factor is the global migration of populations into coastal regions, which are generally more highly exposed to natural catastrophes. And, finally, there is industry which, to some extent following the movement of the people and usually seeking new resources or transport routes, is nowadays venturing increasingly into regions that are subject to elemental perils, without always making proper allowance for this by implementing appropriately safe technologies. Examples of this range from the hotels concentrated along the hurricane-exposed coasts of Florida and the Caribbean to the offshore industry operating in the North Sea and in the Arctic, or to nuclear power plants located in extremely hazardous earthquake-prone areas or even on top of active tectonic faults. In addition to this, the earthquake of October 1989 in California showed how extremely vulnerable highly-developed regions are today to any disruption of the infrastructure, especially communication networks.

At the same time, the density of insured natural hazards in certain geographic areas has greatly increased, and of course particularly so in highly exposed countries. This not only applies to California, a region exposed to earthquakes where approximately every third house is insured against earthquakes today in comparison with a few years ago when only every fourteenth house was. In the Federal Republic of Germany, in the time period since 1976, the year of the "Capella" gale for example, the number of homeowner's policies with windstorm coverage has increased by 65%; furthermore during the same period the repair-cost index for residential buildings increased by 70%. The rapid spread of all-risks and loss of profits covers is also considerably increasing the loss burdens to insurers from natural catastrophes.

REPERCUSSIONS FROM THE GREENHOUSE EFFECT

Natural hazards themselves have not yet shown any significant increase, although today there are more and more alarming factors indicative of a gradual change in worldwide environmental conditions. Let us consider for example the extreme intensity of super-hurricanes "Gilbert" in 1988 and of "Hugo" in 1989, as well as the extraordinary series of winter gales in 1990, and the very intense typhoons in Japan and the Philippines in 1991. They can be regarded as a further indication that the warming of the global climate is gradually beginning to have some noticeable effects.

Assuming that all the predicted greenhouse effects become reality, we will be faced with consequences like the following:

- The number and intensity of natural meteorological catastrophes are growing because the atmospheric heat engine is operating at a faster speed than ever before. The result is increased circulation in the atmosphere and the oceans. Tropical cyclones are possibly becoming more frequent and more severe and are extending their paths towards the moderate latitudes (e.g., towards Western Europe and California). Winter storms are penetrating further inland, as 1990 showed. There is also more water vapor in the atmosphere, generally causing more precipitation and flooding, thunderstorms, hailstorms and tornados.
- All over the world the risk of storm surges along the densely populated coastlines is increasing. Probably the situation is most extreme in Bangladesh, but many other areas, e.g., in Western Europe, North America and the Far East, are also becoming more exposed. Billions of dollars will be needed to protect the coastlines.
- Agricultural and forestry conditions will change. While they will presumably become better in some moderate and sub-polar regions—not only because of the warmer and wetter climate, but also due to the larger amount of carbon dioxide in the air—other regions will be confronted with more and more severe droughts. This applies particularly to regions along the dry sub-tropical zone, such as the entire Mediterranean, the Midwest USA and the south of the former USSR. It would appear to be only a matter of time until these changes trigger serious social and political crises.

It is too early yet for a reliable forecast of regional effects. However, one thing is certain: the transition to a new stable climatic situation, if this occurs at all, will be accompanied by an exceptional frequency of anomalies which will generally give rise to catastrophes.

EFFECTS ON THE INSURANCE INDUSTRY AND COUNTER-MEASURES

The insurance industry will feel the effects of this in many different ways, most obviously in its agricultural policies but also in the other classes of property insurance providing coverage against windstorm, hail, snow pressure, frost, lightning, flood, landslide and subsidence.

The effects on motor, marine, health and life insurance are less obvious, although they certainly do exist. In the latter two cases, there may well be some consequences if the infamous ozone "hole" expands over more densely populated areas as recently observed, and if the incidence of skin cancer increases as a result of "hard" ultraviolet radiation.

What should the insurance industry's response be to changed environmental conditions? This is a question that needs to be faced most urgently by the reinsurers because the individual losses resulting from any major natural catastrophe tend to accumulate in their accounts as a result of their international treaty commitments. In order to realize how large a reinsurer's share of the losses caused by a major natural catastrophe can be, let us recall the 1985 Mexico City earthquake in which only 2% of the total insured loss of US$275 million had to be paid by the local market leaving the prodigious "remainder" of 98% to be borne by the foreign reinsurers. The situation was much the same in the aftermath of hurricane "Gilbert" in Jamaica in 1988. And, as revealed in a recent analysis, even in an insurance market as strong as that of the USA more than half of the total loss of roughly US$14 billion caused by a hypothetical hurricane would have to be paid by domestic and foreign reinsurance companies. Only by ensuring a worldwide geographical spread of their liabilities can international reinsurers bear such extreme burdens. It is reinsurance that makes it at all possible for direct insurers to offer insurance against natural catastrophes. Such insurance coverage would be unthinkable on a strictly national basis.

Insurers and reinsurers basically have two options for overcoming the increased loss potential from natural catastrophes or, more generally, a change in risk conditions: either to adjust their own attitudes, or to prevent or limit the consequences.

The latter would imply, for example, the "ultima ratio" of partially or entirely excluding certain hazards from insurance coverage, or, alternatively, either requiring the policyholder to make a substantial contribution to the cost of losses through deductibles or other methods of coinsurance or limiting the insurer's liability to a certain agreed maximum. It is, however, a never-ending challenge for the insurance industry, and especially for the industry's reinsurers, to guarantee by means of the appropriate conditions that virtually every risk remains insurable.

The insurer cannot provide unlimited capacity blindly. Like every private enterprise, an insurance company must ensure that the capital it makes available is suitably protected. Our "product" is basically a promise to assume liability for the future. The consequences of any changes taking place between the time that liability is assumed and the end of the policy period therefore have to be borne by the insurer and reinsurer. Our duty as conscientious businesspeople to exercise the necessary care means that we must constantly check and evaluate the extent of our obligations. In the first place, a reinsurer must know what liabilities s/he has assumed for each type of hazard in each region that is exposed to that particular hazard. On this basis s/he can calculate the size of his/her potential loss; that is, the maximum loss s/he will have to bear as a result of the most extreme event that is likely to affect each of the zones and classes of insurance in which s/he operates.

To this end the concept of accumulation control was developed. This enables us to gain a general picture of the current status of our liabilities, broken down into the relevant accumulation zones. These zones are defined by scientists and underwriters in such a way that they allow a simulation of various catastrophe scenarios for each market with sufficient exactitude in a way that will not create too much administrative work. Usually this is done by using statistical methods to calculate the geographical extent and intensity distribution of hypothetical extreme events; that is, major earthquakes, storms, etc., as a function of various probabilities of occurrence. The liabilities existing in the zones involved are multiplied by the loss ratios that correspond to the calculated intensity and probability of occurrence in each liability category. The resulting loss amounts per zone are then added together to obtain the total loss potential for each catastrophe scenario.

The most "costly" catastrophe scenarios today are still repetitions of the earthquakes of 1906 in San Francisco and 1923 in Tokyo, which would burden the insurance industry with claims worth tens of billions by cautious estimate. Similar sums would mount up as a result of a "hundred-year" windstorm in the USA or Central Europe. That these figures are not mere fabrications was proved by the 1989 hurricane "Hugo," the 1990 winter storms in Europe and the 1991 typhoon "No. 19" in Japan, which cost the insurance industry record sums of US$4.5, 10 and 5 billion.

In view of these sums, if both the frequency and the intensity of catastrophic events continue to increase, the question may arise as to what the insurance industry can do in order to avoid to whatever extent possible undesirable limitations of liability. How long the insurance industry will be able to continue fulfilling its function depends on the speed and agility with which it can adjust to a changing environment.

This can be achieved by applying loss reduction measures, like introducing adequate deductibles and, most important, by charging premiums that are truly commensurate with the risk. Unfortunately, this is still today exclusively based on loss experience of the past and, under competitive pressure, corners are often even cut further.

It is becoming more and more important for insurance markets exposed to catastrophe to create an organizational structure in anticipation of problems which predictably arise in the wake of a natural catastrophe for assessment and adjustment of what might be millions of individual losses incurred simultaneously.

Authorities are also challenged to set more stringent construction regulations and limit land use in highly exposed regions. Industry can contribute significantly by applying more strict criteria in its choice of a location for particularly vulnerable or hazardous facilities.

Reference should also be made in this connection to the "International Decade for Natural Disaster Reduction," as proposed by the United Nations for the last decade of this century to combat the increasing danger of natural disasters to mankind. This programme is of great significance to the insurance industry, as it includes the following topics:

- mapping of hazard zones
- assessment of loss potentials resulting from different catastrophe scenarios
- land use regulations/restrictions
- standardization of building codes
- standardization of loss information
- forecast and warning services
- public information/awareness/motivation.

The insurance industry can expect significant long-term benefits from the programme on a national and global scale and should, therefore, become a major partner in this programme.

If the annual fluctuations in results increase as a consequence of more frequent and greater catastrophes, it will also become necessary to set aside more funds for catastrophe reserves. The issue here is to convince tax authorities that premiums paid for the coverage of natural catastrophes cannot be considered as "earned" at the end of the year, but rather only after very long periods.

In summary it can be said that by properly applying all instruments of insurance and working together with science, industry and authorities, the repercussions of the increase of catastrophic events need not necessarily be negative for the insurance industry and there is every reason to be optimistic that these risks will remain insurable for insurance clients worldwide.

REFERENCES

1. Munich Reinsurance, Windstorm (Munich Re, Munich, 1990).
2. Munich Reinsurance, World Map of Natural Hazards (Munich Re, Munich, 1988).

NATURAL HAZARDS AND CLIMATE CHANGE: THE RESPONSE OF THE ENERGY INDUSTRY

David C. White
Ford Professor of Engineering
Department of Electrical Engineering and Computer Science
Energy Laboratory
Massachusetts Institute of Technology
Cambridge, Masschusetts 02139

ABSTRACT

An extensive review of the energy industry is presented. It is argued that the only strategy to reduce both costs and emissions of greenhouse gases is to increase the efficiency with which energy is used and also shift the energy systems more toward electric energy that is produced by more efficient generating systems using lower carbon-based fuels and eventually to non-carbon-based energy sources. The non-mobile energy uses in the residential, commercial and industrial sectors can shift to electrical energy, albeit requiring extensive changes in capital stocks of energy-using equipment and major changes in industrial process technology. This cannot occur rapidly and will require well designed incentives to stimulate the innovation needed to bring forth the cost competitive new technologies that will be needed.

INTRODUCTION

Natural hazards can occur anywhere, but they tend to be characteristic for given regions. Hurricanes normally start in the sub-tropical oceans (except for 1991 where a Nor'easter started in August in northern waters); dust storms in desert areas; lightning almost everywhere affecting electric power lines, starting forest fires, killing people or animals; blizzards in the northern climates; etc. The natural hazards are the result of complex and chaotic thermodynamic phenomena driven by the interaction of the energy from the sun and the earth's ecosystem of air, water, and land masses.

The potential exists for major climate change due to man's changing the ecosystems by increased emissions to the atmosphere, destroying tropical forests, changing wetlands, building cities, or other activities that are a consequence of a growing use of the ecosystem to support a rapidly growing world population. Regional activities influence the global environment and climate. Similarly, global climate and environment influence regional weather patterns and natural hazards.

The Energy System and the Energy Industry supplying the requirements of today's industrial economy have much in common with the world's climate system. They are a worldwide system interconnected through a number of international corporations, both private and governmental. At times these corporations, with widely varying business goals, produce chaotic events in energy markets, as amply demonstrated by energy markets in the '70s and early '80s.

Primary energy resources—whether they be coal, crude oil, natural gas, bitumen, biomasses, uranium, water, wind or sun—are distributed worldwide, but with significant regional differences. Coal is found on most continents with major resources being exploited in North America, Northern Europe and Asia. Crude oil has been found and exploited in North America, South America, Australia, Northern Europe, Asia and Africa. Particularly large resources (about 2/3 of the world's proven reserves) of light-weight premium grade crude oil exist in the Middle East (Africa).

Another widely distributed hydrocarbon is natural gas, which consists of methane and other light hydrocarbons such as ethane, pentane. North America has large reserves of natural gas and a well-developed network of pipelines to distribute the gaseous product to centers of demand. Northern Europe, particularly the new Commonwealth of Independent States, Great Britain, Norway and Denmark, has large natural gas resources connected to a network of pipelines to deliver it throughout Europe. Africa, South America and Australia also have major reserves of natural gas, although the pipeline network between reserves and end use is relatively undeveloped. With the growing importance of CO_2 emissions from fossil fuel combustion the advantage of methane (CH_4) over petroleum (C_2H_2) and coal ($CH_{0.7-0.9}$) in terms of CO_2 produced per unit of heat energy during combustion is environmentally a major plus for methane.

Over 85% of today's world's energy demands[1] are supplied by fossil fuel reserves. The remaining 15% or less comes from hydro power, nuclear and biomass. The demand for energy in industrially developed nations is about equally divided between the industrial, residential and commercial, and the transportation economic sectors. The various economic sectors in any specific nation have different use patterns for primary energy and each sector responds differently to constraints on energy use. Predicting how the energy industry will respond to the threat of global climate change due to burning fossil fuels is not a question that can be addressed easily or with any assurance that any predictions will have any significant validity.

One strategy that is technically feasible, at least with today's knowledge, is to increase the efficiency with which energy is used and also shift the energy systems more toward electric energy that is produced by more efficient generating systems using lower carbon-based fuels and eventually to non-carbon-based energy sources. The non-mobile energy uses in the residential, commercial and industrial sectors can shift to electrical energy, albeit requiring extensive changes in capital stocks of energy-using equipment and major changes in industrial process technology. This cannot occur rapidly and will require well designed incentives to stimulate the innovation needed to bring forth the cost competitive new technologies that will be needed. The transportation sector cannot, with today's technology, shift to electric energy and will for a long time be based on liquid hydrocarbon fuels.

The current energy use patterns in various nations and projections of how these might change to reduce the potential for climate change due to CO_2 emissions are discussed in the following sections. Natural hazards do not play a significant role in energy use pattern but could be significant in determining where and how energy supply and distribution systems are sited and operated.

WORLD ENERGY DEMAND

In 1987 the world consumed approximately 322 quads of energy (Figure 1) (1 quad = 10^{15} Btu = 1.05×10^{15} KJoules). Of this energy fossil fuels supplied 88%. The

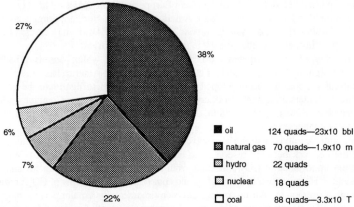

Figure 1. World energy demand 1990[1.]

various fossil fuels used as a percentage of total world energy demand are given in Figure 1[1].

Fuel	% World Demand	Quantity
Petroleum	39%	(23×10^9 Bbl)
Coal	27%	(3.3×10^9 T)
Natural Gas	22%	(1.9×10^{12} m^3)
Non-Fossil Fuel	13%	(40 quads)

The breakdown by major consuming regions is shown in Figure 2. The U.S., Western Europe and the new Commonwealth of Independent States (CIS) account for 65% of the total energy demand. At these annual consumption rates the proven reserves and estimated total recoverable resources listed in Table I show that oil and gas will probably last into the 22nd century. Coal resources are larger than either oil or gas both in proven reserves and total estimated recoverable resources. Coal can meet demand for centuries, in fact well beyond our ability to project anything about future society. Running out of fossil fuels is not a concern for today's society. The major concern is how to prevent global climate changes and possibly ecological disaster caused by burning these fossil fuels[3].

In 1987, the consumption of fossil fuels released approximately 6 GT (1 GT = 10^9 metric tons) of carbon into the atmosphere. About one half of the released carbon is believed to remain in the atmosphere, the other half appears to be absorbed by the oceans[3]. The total amount of carbon in the form of CO_2 in the atmosphere is currently approximately 735 GT. Table II shows that burning all the estimated reserves of petroleum and natural gas might release an additional 400 to 800+ GT of carbon. If these resource estimates are correct and the oceans keep absorbing about half of the released carbon, these two fuels alone could raise the atmospheric CO_2 content by 50%.

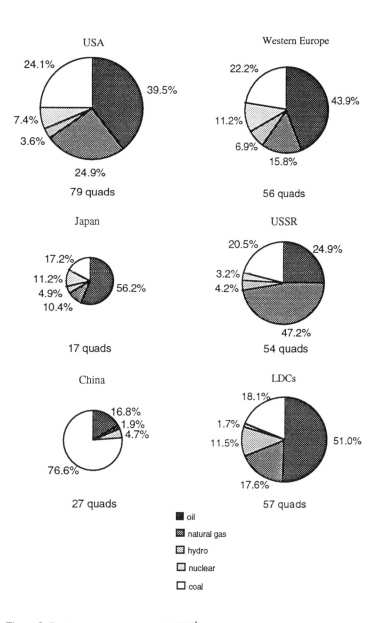

Figure 2. Regional energy demand (1990)[1].

Table I. World Fossil Fuel Reserves and Estimated Recoverable Resources (1990)[1]

	Proved reserves[1]	R/P ratio	Estimated recoverable resources
Oil	1051×10^9 bbl	42 yrs	$2–4 \times 10^{12}$ bbl
Natural gas	119×10^{12} m^3	58 yrs	$300–1000 \times 10^{12}$ m^3
Coal	1.6×10^{12} T	238 yrs	$10–100 \times 10^{12}$ T

The world's resource base for coal exceeds by 1 to 2 orders of magnitude the carbon currently in the atmosphere. World shale oil resources are also about an order of magnitude larger in carbon content than current atmospheric CO_2.

Clearly fossil fuel resources of solid fuels—depending upon the rate they are used and how the oceans and biota on land masses remove the carbon from the atmosphere—have the potential of doubling, tripling or even producing an order of magnitude increase of CO_2 in the atmosphere. Currently it is assumed that the CO_2 increase is linearly related to the CO_2 emitted, an assumption which is not proven at this time. Such a large-scale increase of the atmospheric CO_2, if it occurs, could lead to incalculable changes in the global climate and eco-system. Even though we cannot reliably predict the CO_2 induced global changes, the large risks make it prudent to start now to implement a worldwide energy usage policy that seeks to prevent such potentially large global changes.

To address how world energy needs might be met in the future, some information is required on the current use of energy in the various industrial sectors. Figure 3 shows the total and percentage energy use in the pre-1991 non-communist world for transportation, industrial, residential and commercial sectors in 1985. The total and percentage use by sector is shown in Figure 4 for the U.S. and OECD countries. The major consumer of primary energy is the industrial sector (estimated 35–40% for the pre-1991 non-communist world, 36% in the U.S., and 39% in OECD countries). The energy demand in the residential/commercial sector is approximately equal to the industrial sector. These two sectors are users of all energy forms including almost all the electricity produced, of which approximately 2/3 is used by the residential/commercial sector and 1/3 by the industrial sector in the U.S. Electricity uses about 30% of all primary energy in the world (37% in OECD countries), and electricity demand is expected to grow in the future. Electricity is discussed more fully in Sections 3 and 5, including ways to reduce the use of coal which today is the dominant fuel for producing electricity.

Table II. Carbon released by burning estimated world recoverable fossil fuel resources

Fossil fuels	Estimated recoverable resources	Gigatons carbon
Oil	$1–4 \times 10^{12}$ bbl	240–480
Natural gas	$300–1000 \times 10^{12}$ m^3	150–500
Coal	$10–100 \times 10^{12}$ T	7000–70,000
Shale oil	$10–20 \times 10^{12}$ bbl	2500–5000

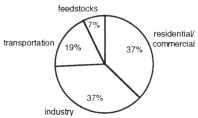

Figure 3. Estimated energy demand by consuming sector for the non-communist world[9].

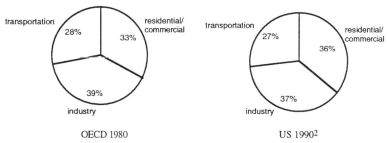

Figure 4. Energy demand by consuming sectors for the U.S. and OECD.

Approximately 90% of the energy demand in the industrial sector is supplied, by fossil fuels (including electricity produced by fossil fuels). Industrial use of fossil fuels is divided between petroleum (50%), coal (25%), and natural gas (25%). The data shown in Table III on process heat demand by the industrial sector are for the U.S., but are somewhat representative of industry worldwide. If these energy requirements are to be supplied by other than fossil fuels, then it is quite apparent that the energy must come from electricity produced by non-fossil sources or from process heat and steam obtained using biomass, solar, geothermal, or nuclear sources. It is also evident that major capital investments would be required to make these kind of energy use changes. If such changes in energy sources were to occur, there would be a major incentive to modify the industrial process used to produce the goods or products, both to minimize the energy used and to match the energy form to the industrial process selected.

WORLD ELECTRICITY PRODUCTION AND CONSUMPTION

World-wide environmental concerns and societal ideologies have influenced the planning and operation of electric power systems. In the U.S. the past decade has seen a major focus on conservation in electricity end use with relatively small additions to new electric supply. The Clean Air Act of 1991 may result in increased emphasis on

Table III
U.S. "heat" requirements by industrial sectors (note steel and iron),
1974 data 10^{12} BTU

Industry/Segment	Hot water <212°F	Steam 212–350°	Steam >350°	Direct heat/Hot air <212°	Direct heat/Hot air 212–350°	Direct heat/Hot air >350°	rounded totals
Aluminum			38.4			63.1	101.5
Automobiles and trucks	13.0	1.4		21.3	10.0	0.9	46.6
Cement						490	490
Ceramics						280	280
Concrete block and brick		9.7	4.6				14.3
Gypsum			31.9		14.0	45.9	
Chemicals (inorganic)		212		4.1	10.9	173	400
Coal mining and cleaning						11	11
Copper	14.2			2.4		55.4	72
Food processing	15.2	68	25.3	2.7	23.5	3.9	139
Glass	8.0	9.8			9.7	310.6	338
Lumber	3.5	13.1	2.5	67.5	2.6	44.7	134
Mining (Frasch sulfur)		43.7					43.7
Paper and pulp		465				94	559
Petroleum refining		120	380			2600	3100
Plastics/selected polymers		7.6	9.4				17
Rubber/SBR manufacture		5.4					5
Steel and iron		65				1712	1777
Textiles	19	191	4.3		69.3	12.5	296.1
Rounded totals	72.9	1212	496	98.0	140	5851	7870
Percent of total	0.9	15.4	6.3	1.2	1.8	74.3	

Table IV[6]
1986 Total Primary Energy and Electricity Primary Energy Quads
(10^{15} BTU or 10^{15} KJ)

	Total energy (quads)	Primary energy for electricity (quads)	Electricity as % total
World	323	94	30.5
N.A.	86	31	35.3
U.S.	77	26	34.1
OECD	161	60	37.2
W. Germany	12	7	35%
USSR	60	16	26.7

upgraded and new supply equipment. A broad brush look at the rest of the world finds that the northern European countries have programs where technology improvement in fossil based plants is underway while France's nationalized industry has followed a path of major additions of nuclear power plants. Great Britain is going through a transition of a government owned and operated electric power industry to the transfer of this industry to the private sector. Eastern Europe has old inefficient and environmentally dirty coal based power plants. The developed Asian countries have relatively efficient and environmentally clean electric power systems. The developing countries are all increasing their electricity supply as rapidly as they can generate capital to do so. This often takes the form of large hydro plants, even before the end use infrastructure is in place to utilize the power effectively.

The production of electricity from primary energy in various regions and countries of the world is shown in Table IV[6]. The OECD countries use about 37% of their energy for electricity, the U.S. about 34%, West Germany 35%, the Commonwealth of Independent States and Baltic nations 27%, and the total world about 30%. The production and use of electricity is a major and growing source of the demand for primary energy in all forms. Table V[6] shows the breakdown between primary sources for world electricity production: fossil 64%, nuclear 16%, hydro 20%, and geothermal less than 1%. For the U.S., 80% of the fossil energy used to produce electricity is supplied by coal, or equivalently, coal is used to produce 57% of the U.S. electricity (see Table VI)[6]. For OECD countries, coal produces approximately 43% of the electricity. This is followed by nuclear 22%, hydro 19%, natural gas 9%, and oil 8%. The role of electricity in meeting the energy needs in all sectors of the economy, except for transportation, has grown steadily throughout the 20th century and there is strong evidence that it will continue to expand in the 21st century. With electricity's growing importance, a major problem for the future will be to supply this electricity demand in ways that are acceptable to society and which have the least local, regional and global impact on the environment. Some indication of how this problem can be approached can be obtained by reviewing the regulation and management of the Electric Power Industry in the U.S.

U.S. Electric Power Industry: Historical Trends and Today's Issues

Table V[6]. World Electricity Production 1986 (10^9 KWhr)

Region	Thermal	Hydro	Nuclear	Geothermal	Total
N. America	2059	642	485	15	3202
S. America	83	281	6	—	370
Europe	2621	664	832	6	4128
Africa	181	49	4	0.3	234
Mid East	170	22	—	0.04	191
Far East	1115	326	229	6	1676
Oceania	120	39	—	2	161
World total	6349	2027	1556	30	9962
%	63.7	20.3	15.6	.3	

Table VI[6]. U.S. Electricity by Fuel (1987)

Coal	Petroleum	Natural gas	Nuclear	Hydro	Geothermal
56.9%	4.6	10.6	17.7	9.7	0.5

Total primary energy to produce electricity 26 quads

Prior to the 1970s, the U.S. electric utility industry actively marketed electric power to increase system size and also revenues. For over half a century a combination of low fuel prices and increased efficiency of thermal engines and electrical equipment brought forth reductions in electricity costs as new electric generation and transmission facilities were added to the system. By approximately the 1960s the rate of increase in efficiency gains in electrical equipment had slowed (the Second Law of Thermodynamics limits the attainable thermal efficiency) such that increased efficiency did not always offset capital costs of new generating equipment. The 1970s added another disruption to the cost of electricity with the price increases in world oil markets which concurrently increased prices for all fuels used by the Electric Utility Industry.

The 1970s brought major changes to the U.S. energy industry and particularly the electric power sector. The public encountered increased electricity prices at a time when demand dropped rapidly. Many energy activists, such as Amory Lovins[7] and others, questioned the validity of the historic pathways taken by various sectors of the energy industry. Soft energy paths, decentralized energy systems, conservation in end use equipment, were all part of a rethinking of how the energy industry and particularly the electric sector, should function in the future. Much of this turmoil was captured and many regulatory activities started (either rightly or wrongly) by the National Energy Acts PL 95-617, 618, 619, 620, and 621 dated Nov. 9, 1978 which included the Public Utility Regulator Policy Act PL 95-617 (known typically as PURPA).

There is little to be gained for our purposes to review all the activities that the above public laws stimulated. Today the major activities that are being actively pursued in the U.S., and particularly in New England and California, is the active participation of electric utilities in funding end use conservation programs to reduce electricity demand. The particular method of paying for end use conservation programs, which (at least in New England) involves having all rate payers fund the conservation programs that may benefit directly only part of them, has produced much contentious debate among regulatory economists and energy conservation activists.

In the electric industry the major conservation activities are still focused on end use efficiency improvements in all sectors—industrial, commercial and residential, with residential being the most difficult to change in a cost effective manner. A major article in a recent issue of Business Week, September 16, 1991, entitled "Conservation Power: The Payoff in Energy Efficiency,"[8] discussed only end use efficiency gains for electricity—electric motors, lighting, refrigeration, heat pumps/air conditioners, etc. For predominantly non-electric applications, such as transportation, increased efficiency in airplanes, autos, trucks, are listed, and potentials for efficiency gains discussed. There was no similar discussion of equally important potential efficiency gains in electric supply equipment.

Since the middle seventies when the activities in conservation for the electric industry became a major force in public regulatory policy-making, there has been great myopia shown by all actors, including Electric Utility leaders. Energy efficiency for electric power has been almost exclusively focussed on the consumer side of the electric meter (e.g., end-use efficiency). Even though the major emphasis has been on the demand side, there have been efficiency gains on new equipment for the supply side, such as: the gas turbine used singly and in combined cycles, the integrated gasifier combined cycle, the new humid air turbines, etc. These technologies bring forth much reduced environmental emissions as well as improved thermal efficiencies for prime movers.

For additions to the supply side of electric power systems, these new generating systems are being installed. However the major focus on conservation activities and currently the current economic downturn have greatly reduced or eliminated major new supply side additions to the electric power systems. Also, because of perceived advantages, at least in the short term, to reducing capital expenditures, various programs to extend lifetimes of existing generation equipment have become the norm in most utilities, often aided by programs developed at EPRI. The net result in the United States has been an aging of the electrical generation infrastructure, reduced capital expenditures, retarded R&D on new generating systems, and reduced electric power equipment manufacturing capability. Taken together those changes have directly added to environmental pollution produced by the electric power industry.

A few regulatory economists (e.g., Joskow and Schmalensee)[9] foresaw these effects and have argued that the present regulatory trend of using monopoly power resulting from natural physical monopolies (electric utility systems) plus regulatory decree, either to push conservation programs or artificially stimulate competition in electric power generation systems, could have less than optimal results. Their concerns, no matter how learnedly presented (using essentially a top down economic viewpoint) have been largely ignored. Another group at MIT's Energy Laboratory (using a bottoms-up, more engineering based methodology) have studied in detail the New England Electric Power Systems. They used multiple simulations of many possible future configurations of the electric system over a twenty year period. End use conservation, various new supply technologies, life-extension of existing plants, under several load growth and fuel price uncertainties were investigated in multiple studies of how the system would be operated under an economic dispatch logic using modeling tools and methodologies currently practiced by the electric power industry. Some of these results have been summarized in three papers by members of this group[10,11,12].

The results from the New England Project: Analyzing Regional Electricity Alternatives[7] will be discussed here, but the simple conclusion is: *Energy efficiency of both supply and demand will produce the most economic electric power while substantially improving the environment.* Neither strategy alone can yield the best results. Understanding these results and acting upon them aggressively will change many future activities of the Electric Power Sector in the U.S. and World-wide.

A Short Review of the New England Project: Analyzing Regional Electricity Alternatives

As part of their ongoing work with THE NEW ENGLAND PROJECT: Analyzing Regional Electricity Alternatives, the AGREA research group at the M.I.T. Energy Laboratory has evaluated the performance of New England's electric power system for a broad range of strategies[13].

Using information obtained from individual New England utilities, NEPOOL, EPRI and other sources, and using EPRI's EGEAS production costing model, the NEPOOL Load Forecasting Model, as well as a capacity planning module AGREA developed by them, the M.I.T. research team has simulated the operation of New England's electric power system for a broad range of strategies. In all, 288 individual strategies were evaluated across fifteen combinations of load growth and fuel prices (for a total of 4320 simulations). For each simulation, data was recorded that tracked the costs, emissions, fuel consumption, reliability, and load and capacity growth for a twenty-year period (1990–2009). By comparing the performance of these strategies, least-cost, least-emissions strategies can be identified.

Each strategy is a combination of five separate components:
1) a choice of new generation technologies (technology mix),
2) a choice of the level and relative contribution on conservation and peak load management (DSM),
3) the persistence of existing capacity in the system, and how—if applicable—it is replaced (existing capacity treatment),
4) the choice of fuel in existing residual oil-fired units (boiler fuel switching), and
5) the desired level of extra capacity to have available (reserve margin).

Table VII[13] shows the options available within each component. Exhaustively combining all the components' options yields 288 separate strategies. No effort was made beforehand to determine which strategies perform better than others although the range of strategies include those technologies most likely to be chosen. Simulation results themselves communicate which strategies were less expensive and had lower emissions. Examination of these strategies in detail can inform decision makers of which options to stress or discourage in actual resource procurement.

Figure 5[14] (referred to as a "tradeoff graph") shows the performance of these strategies for the measures of direct cost, and cumulative sulfur dioxide (SO_2) emissions, for a future with a pessimistic economy/low load forecast (adapted from NEPOOL Resource Adequacy Report), and medium oil and natural gas prices (the "PM" future). This future has been selected for discussion because, of the fifteen futures used in the simulations, it most closely matches the long-term load growth trend contained in the 1991 NEPOOL CELT Report.

Figure 6[14] compares the PM future with a one based on the NEPOOL Resource Adequacy Report's "Normal" economy ("NM"), and the peak load trajectories from the last two CELT reports. The variability of the PM and NM futures comes from the reintroduction of uncertainties associated with short-term economic fluctuations and weather.

Table VII[13]. Winter 1990/91 Scenarios for the New England AREA Project

Number of Scenarios	4,320*		
No. Strategies	288	No. Futures	15
New capacity mixes	4	**Economic growth**	3
Gas dependent	1	Base	1
50/50 gas & coal	1	Hight	1
Coal dependent	1	Low	1
50/50 gas & nuclear	1		
Fixed PV & gas	†	**Fuel prices**	5
Fixed PV & gas/coal	†	Migh oil/high gas	1
		Medium oil/medium gas	1
Existing capacity mixes	3	Low oil/low gas	1
Life extension	1	High oil/low gas	1
Scheduled retirement	1	Low oil/low gas	1
Repower existing	1	Discontinuity	*
Un-grandfather existing	†		
		DSM costs	1
Operations	2	Base	1
Current procedures	1	High	‡
Low sulfur oil	1		
Sulfur dispatch	†	**Supply tech costs**	1
		Base	1
Reserve margin	2	High	‡
Default 23%	1		
Higher 30%	1	**Supply timing**	1
		Base	1
DSM option sets	6	High	‡
No conservation	1		
Current utility progs.	1	*An exhaustive combination of all*	
More conservation & peak mgt	1	*the options and uncertainties*	
A lot more conservation	1	*listed yields a total of 124,416*	
A lot more peak mgt.	1	*scenarios*	
Technical potential	1		

*—Each scenario will require multiple EGEAS production costing runs in order to include electricity price and demand interactions, increasing computational time
†—Option-set or Uncertainty deferred to the next scenario set
‡—Uncertainties that will be evaluated through sensitivity analysis

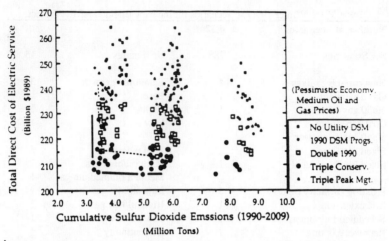

Figure 5[14]. Direct costs vs. sulfur dioxide emissions tradeoff graph for five levels of demand-side management.

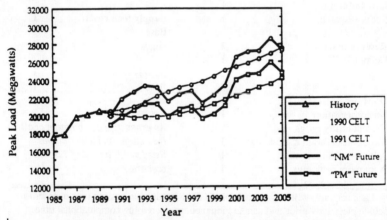

Figure 6[13]. Comparison of the PM and NM load growth futures with the last two NEPOOL CELT reports.

Direct costs are defined as the total cost stream associated with the provision of electric service in New England, discounted at inflation to 1989 dollars. The external, societal cost of emissions are not included in the direct cost calculation, although they may be added later. Direct costs include customer as well as utility expenditures for conservation measures. With this information, the change in direct costs and emissions between any two strategies can be used to calculate the cost incurred to avoid a given level of pollutant emissions.

The strategies in Figure 5 are keyed by their level of demand-side management. No strategies for the Technical Potential level of DSM appear in Figure 5 since the level of DSM program costs required to obtain even a conservative estimate of Technical Potential were not estimated. The difference in costs and emissions among strategies of like DSM is due to the combinations of the other four strategy components: New Technology Mix, Existing Capacity Treatment, Boiler Fuel Choice-Residual oil, and Target Reserve Margin identified in Table VII. As can be seen, increased levels of DSM tend to reduce the overall level of direct costs. It is important to note that while it is possible to simulate a given level of demand-side management, such as Triple Conservation, whether such an option can be implemented, at the assumed costs, is not assured.

In Figure 5, the least-cost, least-SO_2 emission strategies on the left and lower edges of the strategy "cloud" have been connected together with a solid line. These strategies, referred to as the "Decision Set," are those strategies for which society must raise direct costs to obtain additional reductions in SO_2. The line that connects them is referred to as a "tradeoff curve," and graphically represents a "cost-effectiveness function" for avoiding SO_2 emissions.

Due to the uncertainty associated with attainable levels of DSM, tradeoff curves for two levels of demand-side management have been drawn in Figure 5; the solid line is the tradeoff curve for Triple Conservation, and the dashed line for the Double 1990 Programs level of DSM. Levels of DSM in this set of strategies were based on 1990 utility DSM filings. Conservation programs proposed by utilities this year (1991) are roughly one-third larger than 1990's on an electricity savings basis. The Double Programs level of DSM therefore represents approximately a fifty percent increase in electricity savings over 1991 programs.

The three "clouds" in Figure 5 represent particular combinations of options for treating existing capacity and choice of boiler fuel. The cloud on the right (highest SO_2 emissions) results from life extension and 1990 choice of boiler fuel. The middle cloud is life extension and 0.5% sulfur oil 6 or it is 1990 boiler fuel and either scheduled retirement or repowering. The lower emission cloud (far left) are scheduled retirement or repowering plus 0.5% sulfur oil 6. It is important to observe that it is possible to move from the 1990 DSM program and life extension (upper part of right cloud - the strategy of least resistance) to lower cost lower emission strategy by replacing existing capacity in a planned program and substituting a lower sulfur residual oil as the boiler fuel. A combination of higher efficiency in both end-use and electrical generating equipment can save both money and emissions.

Results Drawn from the New England Project

Strategies identifying the cost effective ways to reduce emission in New England are discussed in reference 14. Figure 7[14] identifies one important characteristic of the New England system. These data show that the medium construction year of existing capacity is 1965 and today more than 10% of the regions operating fossil fuel-fired power plants have already exceeded their "book" lives. New England's historical pathways for meeting electricity demand have focused on life extensions of existing plants and electric energy imports, especially from Hydro Quebec. Figure 8[14] shows SO_2 emissions for four simulated extensions of historic practices for the next 20 years. The dotted lines are estimates of the SO_2 limits of the new clean air act taking place in the year 2000 or alternatively 1995 (current Massachusetts regulation). It is clear that historic pathways for expanding the New England electric power system cannot meet the new Federal Clean Air Act. It is equally clear from the reduction in SO_2 by conservation programs (Figure 5) that conservation alone, while it may stabilize or even reduce electricity demand, cannot meet long term clean air regulations for SO_2. The underlying reason for this result is rooted in the current mix of generating plants and the type of conservation programs that are available in the region. This occurs because SO_2 comes predominantly from the oil 6 fired generating plants and these make up approximately one-third of the base load capacity. Almost all residential conservation programs and much of commercial conservation reduce peak load and upper intermediate loads. As such, it has little impact on the dispatch of the base load oil 6 fired power plants. Only activities that affect the base load plants can reduce sulfur emissions. One can, in fact, reduce SO_2 by shifting to low sulfur oil 6 (either 0.5% or 1.0% sulfur oil). This strategy will reduce SO_2, but does not help NO_x whose new regulations are unknown at this time. The current directions of environmental regulations are pointing to a new direction for planning generation expansion in the future. An alternative direction for long-term electricity generation planning is shown in Figures 9, 10, and 11a,b,c[11,13] where unit cost of service is plotted against SO_2, NO_x, and CO_2 for supply technology options that includes life extensions, scheduled retirement and repowering of existing plants and the range of DSM options and fuel choices shown in Table VII. The three tradeoff curves Figure 11a,b,c, identify the DSM, fuel choice and technology choice options for CO_2. These five trade-off graphs show that new generating technology can reduce these three important environmental emissions with very small changes in cost of electric service. This result occurs because of the economic dispatch logic under which all New England plants are operated within the New England Power Pool (NEPOOL). Economic dispatch uses only operating costs to determine which plants run. The higher efficiencies of new generation reduces the fuel required per kwh generated. Fuel costs are the dominant factor in operating costs. For highly efficient plants this results in their being operated low in the loading order and hence having very high capacity factors. These new clean plants reduce emissions per kwh (even CO_2) and also cost less per kwh because of fuel savings. These cost savings often offset the increased capital charges of the new plants, hence the relatively modest increases in total cost in a properly planned system. This somewhat counter-intuitive effect has not been given sufficient attention in the regions power planning. The new clean air act and future environmental activity, in particular

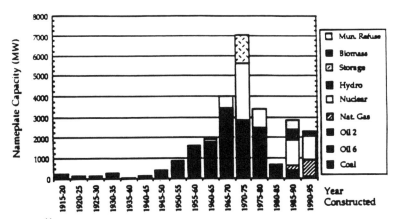

Figure 7[11]. New England capacity additions by fuel type.
Data Sources: ECNE (1988)w EIA (1990b), and NEPOOL (1990a).

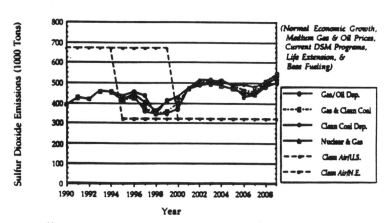

Figure 8[11]. SO$_2$ emissions for four new capacity strategies.

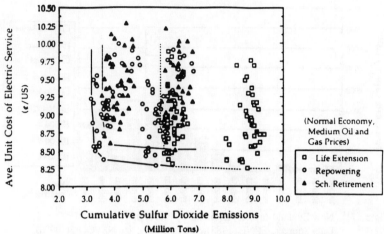

Figure 9[11,13]. Unit cost of electric service vs. sulfur dioxide by existing capacity treatment.

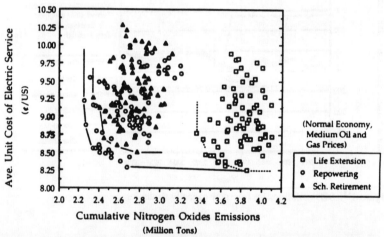

Figure 10[11,13]. Unit cost of electric service vs. NO_x by existing capacity treatment.

David C. White   241

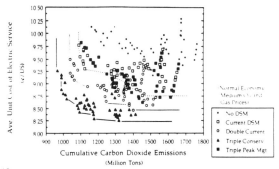

Figure 11a[11,13]. Unit cost of electric service vs. carbon dioxide by DSM strategies.

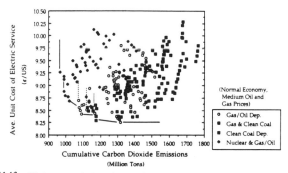

Figure 11b[11,13]. Unit cost of electric service vs. carbon dioxide by new technology choice.

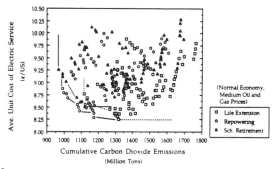

Figure 11c[11,13]. Unit cost of electric service vs. carbon dioxide by existing capacity treatment.

global warming, will require a careful re-thinking of both supply and demand side electric system power planning. In fact, if the gains from DSM are combined with the gains from new supply, considering both cost and emissions, it is possible to both reduce costs and emissions.

The studies of the New England system discussed above cannot be transferred automatically to other regions of the U.S. or other countries. However, economic dispatch of power systems is essentially used worldwide. Thus, insights from the New England project are useful in identifying important issues that electric power planners in any country or region should consider. For an existing system which has a mix of plants, some which are environmentally clean and others which are not, it is important that the power planning strategy addresses how to improve the environmental characteristics of the existing system. Simultaneously, activities to reduce energy use are also helpful. The most effective strategy is one which both reduces energy demand and improves the generating systems environmental characteristics. This can be summed up by a general rule as follows:

Environmental improvements of existing electric power systems will occur best through a balance of increased efficiency in both end use and electric generation. For electric generation the increased efficiency must be obtained using environmentally acceptable technology

Fortunately there are opportunities in both end use and generating equipment that meet these requirements, and further equipment improvements are possible.

THE ATMOSPHERE: REVIEWING STUDIES ON CONSTRAINING CO_2 EMISSIONS

There have been a number of studies by environmental economists on the cost of controlling CO_2 emissions. Typical studies are those of Edmonds-Reilly[15], Manne-Richels[16], Jorgensen-Wilcoxen[17] and Nordhaus[18]. All of these studies predict a rising cost (usually imposed by a tax on carbon) to reduce CO_2 emission both over time and as the level of emissions increase. A typical carbon tax versus carbon reduction is shown in Figure 12 developed by Nordhaus[18]. A recent study on Mitigating Climate Change[19] by the National Research Council has compared those economic studies (top down) with more engineering based studies (bottom up). This study found significant reductions in CO_2 emission were possible while also reducing the costs of energy services by using engineering analysis and technologies which are currently commercially available. The projected carbon tax versus emission reduction data developed in the engineering studies, are shown in Figure 13 from the Mitigation Study[19]. Similar results have been found in other engineering analyses such as "Energy Conservation in the Industrial Sector" by R. U. Ayres[20] and the New England Study of Electric Alternatives discussed previously.

The principal reason for the major differences in the economic versus engineering studies can be summarized as follows. Typically bottom-up studies look at the newer technologies which use energy more efficiently than current technologies, and make some assumptions about the extent to which these new technologies can be and will be

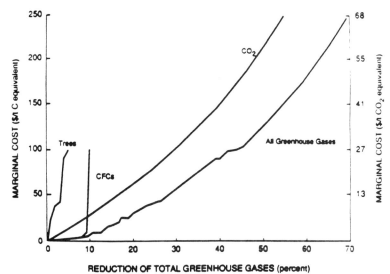

Figure 12. The Nordhaus study of marginal cost of greenhouse gas reduction.

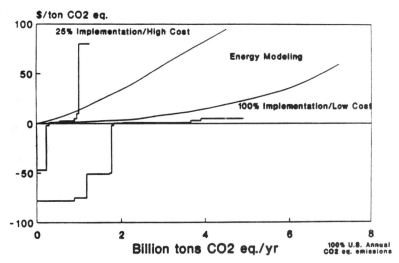

Figure 13. Comparison of technology costing and energy modeling methods of mitigation costs.

used. Usually, the degree to which these new technologies are assumed to be commercialized and used is taken to be quite large. Inherent in most bottom-up analysis is that there are no problems with actually implementing new technologies on a broad scale. The critical component in such studies is that the newer more cost effective new technologies will be installed and used. In some cases bottom-up analysts advocate new policies such as energy efficiency standards for appliances, rewriting building codes, raising CAFE standards, requiring utilities to use BACTs, etc. Bottom-up analysts often propose changing regulations which favor existing technologies (i.e., regulations which are preventing the penetration of new technologies). Implicit in a bottom-up approach is the assumption that current markets are not operating efficiently.

Typically, top-down analysis take their reference case (how ever they define it) as a competitive solution. That is, they assume that their reference case is a close approximation for what a competitive economy would be, even though it may not be an optimization. This perspective assumes that there are a whole set of institutions that define how close a system will get to an optimization, and that to move closer to the optimal solution will incur hidden costs. For instance the top-down approach is concerned with issues like "take back" problems, and consumer preferences. While the bottom-up approach implicitly assumes that take back problems are small, the top-down approach assumes that if you reduce the cost of a service people will want more of it. Another concern of the top-down analyst is the administrative costs of actually running a regulatory or incentive program. Implicit in the top-down approach is the assumption that if we try to regulate or use incentives to encourage new technologies, then administrative costs will overwhelm the savings gained from using the more efficient technologies. Thus top-down analysis implicitly assume that it is very difficult to change institutional structures and procedures.

The real difference between the top-down and bottom-up perspectives stems from fundamentally different perspectives about the effectiveness of using policies to force technological change. Essentially, the bottom-up analyst assumes that policies can be effective, while the top-down analysts assumes that policies will not work. At this point it should be clear that both approaches are rather limited in scope. The bottom-up approach does leave out potentially important macro-economic effects (i.e., What happens in the labor market? What links are there between demand for energy, transportation, and other things?) On the other hand, the top-down approach typically leaves out potentially important information about how technology changes (i.e. How do new technologies penetrate over time? How close to optimal is the current mix of technologies?) What is really needed is a cross between the top-down (macro-economic) and bottom-up (engineering-technology) approaches.

The conclusion that can be drawn from the extensive work that has been done on energy supply and end-use systems is that it is possible to reduce CO_2 emissions per unit of services, even using the current mix of primary energy resources--including fossil fuels. This process of CO_2 reduction can be even more enhanced if some non-fossil energy sources are used to produce electricity. All this requires a carefully designed set of energy development strategies which are sensitive to particular requirements of each energy use situation. Considering the multitudes of energy use practices in the many nations of the world, each in a different state of economic development, it is apparent that identifying the right strategies and using the most efficient policy instruments to implement the strategies is a monumental and difficult assignment. Whether the world can develop and implement correct policies to reduce CO_2 emissions with minimum economic disruption is not obvious. The following section discusses this problem

CLIMATE CHANGE AND THE ENERGY INDUSTRY

The use of fossil fuels to meet the world's demand for energy is so large that reducing fossil fuel use to limit CO_2 emissions is not something that can be done easily or rapidly. The projections by the IPCC Working Group III "Climate Change, the IPCC Response Strategies"[5] project a growing population, particularly in developing nations, that—coupled with any improved standard of living in both developed and developing countries—leads to a growing demand for energy resources. In such circumstances, the following question must be addressed: Is it technically possible (i.e., does technology exist) to meet energy demand and also reduce CO_2 emissions? The answer to this question is yes, but not using end use technology and energy supply technology based on primary energy resources that are currently in use in every nation worldwide. Except for CO_2 sequestering technology[22] (which today is not proved to be scientifically feasible), all energy sources that are free of CO_2 emissions to the atmosphere are derived from nuclear, solar or geothermal sources. While these non-fossil energy sources can produce steam or process heat, most of these commercially available energy conversion technologies (based on these non-fossil energy sources) produce electricity. Thus, if major CO_2 reductions are required to mitigate the environmental and ecological effects of global warming, the use of electricity in all sectors of the economy will have to increase while direct fossil fuel use must decrease. It follows from the above conclusion that generating electricity at low cost becomes an imperative if energy costs in all sectors are to be managed without major detrimental effects on the economies of the world.

It is conceptually possible, using technology that is commercially available, to use electric energy to meet the energy requirements of the industrial, residential and commercial sectors of the economy. Such a change in the energy form used would result in major changes in industrial process technologies and some changes in the manufacturing sectors as well as residential and commercial energy use practices.

To meet the needs of the transportation sector, at least as it is today configured, it is not really feasible to use electric energy. Electric automobiles are under development and some commercial production will occur during the 1990s. However, a true substitute for liquid hydrocarbons to fuel the world's transportation system is not today available, either actually or conceptually. Gaseous hydrocarbons, including methane, could be substituted for liquid hydrocarbons, but the gain in reduced CO_2 emissions are relatively small (probably 20%). Hydrogen is a possible route to follow for a transportation fuel, but this requires a major source of inexpensive electricity as well as other major changes in the transportation fuel distribution system. It is technically feasible to propose a liquid hydrocarbon transportation fuel based on biomass which is produced on mega scale biomass farms in a closed cycle system producing no net CO_2

to the atmosphere. Such a massive change in the way the world's transportation fuels are produced is certainly not in any near term strategy to deal with CO_2 emissions. If it occurs at all, it undoubtedly will be a strategy of last resort after clearly establishing that avoiding ecological disaster from CO_2 emission is a worldwide accepted imperative.

In summary, the above discussion proposes that it is technically feasible, if undertaken in a long term planned strategic manner, to substitute non-fossil fuel derived electricity for most energy demands in stationary (non-mobile) applications in the industrial, residential and commercial sectors. The policy tools to undertake such a massive restructuring of the world's energy industry and the corresponding end use sectors using this energy are in a far greater state of disarray than are the technological products which would be needed to accomplish the transition. Some indication of how this process can be started, including some activities that today are economically viable, will follow. The previous discussion also indicates that analogous technical substitutes away from fossil fuels for the transportation sector are not available today and the prospects for future technology breakthrough are not very good. History shows that necessity is the mother of invention, and this could occur for the transportation sector, but to bet on this occurring as a major part of a CO_2 management strategy is a very high risk strategy indeed.

Electric Industry—How to Reduce CO_2 Emissions

Technologies are commercially available or under development for the generation of electricity that are environmentally clean. There are commercial proven technologies to control regional environmental air emissions, SO_2, NO_x, and particulates. The global environment pollutant CO_2 is more difficult to control if carbon based fuels (fossil fuels) are used. World-wide (and also for the U.S.) fossil fuels are used to produce approximately 2/3 of the electricity (Table 5). This means reducing CO_2 will require major changes in the world's electric generating system. Increased efficiency of power plants helps to reduce the CO_2 per unit of electricity generated. And, as shown in the New England Project discusson, new technology is an essential part of any near term strategy to reduce environmental emissions, both regionally and globally.

It may be possible to sequester CO_2 in the ocean, but this is still a research goal, not a proven approach[22]. Natural gas (CH_4) has less carbon per BTU than coal ($CH_{0.8-0.9}$) or petroleum ($CH_{1.8-2.0}$). Less carbon per BTU in natural gas and increased efficiency of well-designed natural gas fired systems can reduce CO_2 by about 50% over a well designed coal-based generation system. For old, less efficient coal based, generating equipment, the gains are even higher.

If CO_2 must be reduced well below today's level of emissions (approximately 6 Gigatons Carbon) then non-carbon-based energy sources will be required. Today non-carbon energy sources, if they are used, are almost exclusively used to produce electricity—nuclear, photovoltaic, wind, geothermal and even solar thermal. Any movement toward non-carbon-based energy systems almost certainly means a growth in electricity use.

Unfortunately the non-carbon-based generating technologies, at least today, are more capital intensive per kilowatt of capacity than fossil fueled equipment. Lower costs for the primary energy inputs to non-carbon-based generating technologies help to offset the capital costs and in some cases can result in lower cost electric power. If environmental emissions from fossil fueled plants are entered into the cost equation, this further helps the non-carbon-based technologies. Nevertheless, the worldwide markets for all fossil fuels, the worldwide regional distribution of coal, the worldwide knowledge base of the older less environmentally clean fossil fuel technology, and the expected growth in demand for electricity to come from the developing world, all point to the conclusion that fossil fuel technologies for electric power generation will remain the dominant source of electricity for many years into the 21st century.

If fossil fuels are the most likely choice for electric generating equipment in the next few decades, it is imperative that the transfer of knowledge of environmentally clean electric generating technology for both fossil fueled and non-carbon-based technologies be high on the agenda of the world's political and industrial leaders. Improving the generating systems in the developed world (of which many like New England are far from optimum) and installing the most environmentally acceptable technology in the developing world, requires a much longer term orientation in policy making both domestically and worldwide. Today there is some evidence, such as the International Panel on Climate Change (IPCC)[3,4,5], that longer-term environmental issues are being addressed. However, transforming this policy awareness into action plans which affect technology choice in the market place is far from being implemented. As policy actions occur for environmentally clean electricity supply, the need for new technology to replace old generating plants and also meet new electric load, will lead to an expanding market for the electric equipment industry and considerable development in the way electric power systems are operated, financed and managed by both the private and governmental sectors.

The previous discussion proposes that reducing emission of CO_2 can best be accomplished by a significant increase in the use of electricity for meeting non-transportation energy demands in industrial, residential and commercial sectors. This will not occur naturally in today's energy markets without the use of policy tools by government to change the factors which affect energy-related decisions. In the U.S. during the past fifteen years, there have been substantial intervention in electric energy markets, by regulators at both the federal and state levels. These activities, discussed in a previous section, give valuable insights against which to evaluate future policies which may be adopted to mitigate global climate change.

GLOBAL CLIMATE CHANGE—A CHALLENGE FOR THE ENERGY INDUSTRY

The Energy Industry in every nation and worldwide is predominantly a fossil fuel industry, with over 85% of the world's energy coming from hydrocarbon fuels. It is also an industry that is highly concentrated and international in scope. For example, worldwide petroleum companies have market shares as follows:

World-wide Petroleum Companies Market Share
  4 largest =   30%
  8 largest =   53%
 20 largest =   81%
 50 largest =   94%

The petroleum companies, particularly the larger ones, have significant energy resources in both natural gas and coal, in addition to petroleum, and are active in all energy markets. The larger companies also have very large and competent research and development laboratories with a long history of developing and commercializing new technology. The fossil fuel-based energy companies are, therefore, capable of responding to new market conditions for fossil fuels, and if given the proper incentives could be the major source of new technology for reducing CO_2 emissions from using fossil fuels.

The fossil fuel energy companies and the manufacturers of equipment for electrical power generation have worked in partnership on many occasions to bring forth environmentally improved and more efficient electrical supply technology. For example, the first prototype Integrated Gasifier Combined Cycle Electric Power Plant in Cool Water, California, used gasification technology from Texaco, gas turbines from G.E. and additional important components from other equipment manufacturers. There is an established history of the various energy companies combining technical capabilities to introduce new and innovative energy technology to the market place. The technical capability to respond to reducing CO_2 emissions exists and it needs to be motivated by realistic policy initiatives by national and international governmental bodies.

Some activities to deal with global change are currently underway, but there is no clear concensus among environmental scientists and governmental policy makers about the preferred responses to global change. The IPCC has produced a report on possible responses[5] but it isn't definitive nor approved by the respective governmental bodies involved in the IPCC Studies. The European Community has proposed setting a goal of reducing CO_2 emissions by 20% from 1990 levels by the year 2000 with further reductions by 2010. A tax of $10/bbl of crude oil equivalent is one policy tool under consideration. The U.S. has as yet not accepted any constraints on CO_2, but instead is assuming the reduction of other green house gases (CFCs) now constrained by the Montreal Protocol Agreement is a way to meet the 20% reduction target proposed by the European Community. Developing countries are concerned by the prospect of global change, but need cheap energy to fuel their economic development to meet demands of growing populations and desired improvements in their standard of living. Their position is typically that, since the developed world has caused the problems, the developed countries should be responsible for solutions, including financial aid to minimize CO_2 emissions in developing countries.

A response to the threat of global climate change raises the question, what actions are reasonable considering the immense quantity of fossil fuels used, the number of independent nations involved in supplying and using fossil fuels, the massive economic structure represented by the energy supply companies, and the extensive manufacturing companies producing goods that consume fossil fuels to produce energy-related services? Further, there are many oil exporting nations in Africa, Asia, South America, and North America whose total economy depend upon fossil fuel exports, particularly petroleum. One thing is clear, changing the energy use structure of every nation in the world is not going to be easy, it cannot occur rapidly, and it will take unprecedented cooperation among nations for an effective response to develop.

The one positive aspect of the climate change problem, considering only energy resources, is the existence of technical solutions, at least for non-transportation energy use, to substitute away from fossil fuels for the energy used to produce goods and services. As discussed previously, the technologies using non-fossil fuel energy sources that are the most developed, are those used to produce electricity. A strategy to reduce CO_2 emissions is, therefore, a strategy to increase electricity production and build an energy infra-structure based on electricity produced from non-fossil energy sources. Such a strategy is possible and electricity use has grown throughout the 20th century. Nevertheless an all electric economy for non-transportation energy use is a major change in energy use practices, and will require major changes in industrial process technology and in many instances, also require a significant restructuring of the electric power industry in most nations. The following section will propose how this change might occur.

Global Change—A Possible Response by the Energy Industry

The response strategy to global change that would most effectively use today's commercial electrical generating technology and stimulate the developments needed in the future, follows directly from previous discussions on the U.S. and New England Electric Power Industry. These sections show that increased efficiency in both electrical generation and end use technology reduces energy costs and improves the environment. A long-term perspective of slow, steady changes over 50 to 100 years is the only feasible way to change today's energy infrastructure, thus today's response strategy must start by improving the existing electrical generating system using the most cost-effective technology available and also improving the end-use efficiency of electrical consuming equipment. Besides efficiency improvements, the use of the lower carbon to hydrogen ratio fossil fuels (e.g., methane, CH_4) helps to reduce CO_2 emissions. The most efficient electrical generating equipment also involves the use of gaseous fuels in either gas turbines or fuels cells (such as Molten Carbonate Cells). Thus improved efficiency and lower CO_2 emissions comes from using methane, or in some instances other gaseous fossil fuels.

Eventually non-fossil energy resources will be needed to further reduce CO_2 so the strategy should also include the introduction of non-carbon sources in the energy mix used to generate electricity. There are issues that must be addressed for all the non-fossil based electrical generating equipment. Of these, nuclear technology generates most societal concerns, but there are several technologies with the potential to overcome these concerns. These include advanced LWRs, Modular HTGRs, and Modular Liquid Metal Reactors, such as the IFR. Over-riding all these technical options is the need for an *international management and monitoring structure* which can deal effectively with the safety, proliferation, and waste management issues of nuclear sources. The transition to a growing use of nuclear fuels for electric power generation must include the international management systems—so important to their effective and safe use in all nations.

The transition to lower carbon fossil fuels and non-fossil sources for electrical power generations cannot occur at the same pace in all nations. The developing nations have financial and resource constraints that are even more restrictive than those facing the developed nations. For example, China has large resources of coal and its energy infrastructure is predominantly coal based, including its rail transportation system. For China higher efficiency coal-based technology is the strategy which must be followed as the initial path towards long-term conversion to the non-fossil based energy services. International financial and technical aid will be required from the developed nations to aid in this transition to a more efficient use of indigenous resources.

The complexity of the responses needed by the many nations of the world with their mix of indigenous resources, economic conditions and living standards, makes any overall strategy extremely complex to outline in even a rudimentary way. General guidelines are proposed in Table VIII, which captures the essential factors that policy instruments should try to influence in stimulating reduction in CO_2 emissions to the environment. The simplest and most often proposed CO_2 reduction policy instrument is the carbon tax. However, other policy instruments could be even more effective in bringing about environmental improvements. For example, in the U.S., regulatory constraints on the electric power industry are sufficiently severe to offset many capital expenditures which today are cost-effective and also environmentally superior. A carbon tax will not address these constraints, but regulatory reform could. In summary, there is no simplistic single fix, either economic or technical, to bring about CO_2 reduction from energy use. Governments and Industry working together can improve the CO_2 emissions problem and the energy industry has the financial research and development capability to be a major contributor to the solutions in the U.S. and worldwide.

CONCLUSIONS

The Energy Industry could be a major contributor in any response strategy that develops to reduce CO_2 emissions to the atmosphere. The time required to change the energy infra-structure will be very long—many decades—and will not occur without enlightened policies being promulgated by national and international leaders.

The only strategy which is technically feasible, at least with today's knowledge, is to increase the efficiency with which energy is used and also shift the energy systems more toward electric energy that is produced by more efficient generating systems using lower carbon-based fuels and eventually to non-carbon-based energy sources. The non-mobile energy uses in the residential, commercial and industrial sectors can shift to electrical energy, albeit requiring extensive changes in capital stocks of energy-using equipment and major changes in industrial process technology. This cannot occur rapidly and will require well designed incentives to stimulate the innovation needed to bring forth the cost competitive new technologies that will be needed.

The energy industries have the financial, research and development capability to be major contributors to this change. However, the transition to a much lower

Table VIII. Proposed strategies to reduce CO_2 emissions

A. Time Frame 1992 to 2030
   1. Non-Transportation Energy use—Industrial Commercial and Residential:
      a) Improve efficiency using best available technology supply and end-use equipment,
      b) Shift electrical generation toward low carbon-based fossil fuels (methane CH_4),
      c) Re-establish nuclear electrical generating equipment as a cost-effective source of electrical power, including improved international controls on plant and waste management
      d) Continue development and commercialization of non-fossil-based electrical generating sources.
      e) Stimulate development of industrial process technology based on electricity and implement it as the electrical supply system shift toward non-fossil fuels

   2. Transportation Energy use:
      a) Improve efficiency of transportation vehicles and highways.
      b) Improve mass transit and rail systems to reduce auto and truck use in essential services.
      c) Continue development of electric vehicles and introduce them in markets where they can be effective, such as fleet vehicles, postal delivery, etc.
      d) Stimulate research and development on transportation fuels from managed biomass systems for a closed cycle net zero CO_2 emission system.

   3. Needs of Developing Nations:
      a) Encourage World Bank and other International Funding Agencies to support improved electrical technology in developing nations.
      b) Support development of non-fossil-based technology, such as hydropower, to reduce fossil fuel demand.
      c) Encourage the indigenous manufacture of the most efficient advanced technology systems for electrical supply and end-use equipment in the developing nations.
      d) Establish international programs to train the manpower and develop the infrastructure to manage and maintain electrical systems for efficient operation.

consumption of fossil fuels will almost certainly change completely the structure of the energy industry. Petroleum companies could well become Electrical Energy and Environmental Service companies selling their products to electric transmission and distributing companies and to individual large consumers of electrical energy, such as the primary metals industry.

Internationally managed corporations that can manage and operate the complex new energy supply facilities, particularly new nuclear facilities, are almost certain to arise in the reformulation of the current energy industries into those which will be the technical and financial leaders of the future. Many of these changes were visualized by the forward-looking leaders of the petroleum companies in the 1970s and 1980s. Unfortunately the drop in crude oil prices in the middle 1980s has led most petroleum companies to destroy much of the R&D capability established earlier and to withdraw from broad energy markets to the narrower fossil fuel products markets.

The past twenty years have graphically shown that a transition to new conditions in energy markets doesn't occur easily. Dealing effectively with global change will be far more difficult and risky than the energy markets of the 1970s. Unusual vision by industrial and governmental leaders will be required for the massive changes that responding to global change will require in energy use practices. Possible technical pathways can be identified which, if implemented, can be as cost effective as today's fossil fuel-based energy system. Can policies be identified that will encourage the technological potential to become the new energy system of the future? Here is the challenge for the Energy Industry, what will be its response?

REFERENCES

1. British Petroleum. *BP statistical review of world energy* British Petroleum Company, p.i.c. June 1991.
2. Annual Energy Review 1990—Energy Information Administration, DOE, May 1991.
3. IPCC Working Group I, J. T. Houghton, C. J. Jenkins, J. J. Ephraums, editors. "Climate Change, The IPCC Scientific Assessment." Cambridge University Press, July 1990.
4. IPCC Working Group II, W. J. McG. Tegart, C. W. Sheldon, D. C. Griffiths, editors. "Climate Change, The IPCC Impacts Assessment." Australian Government Publishing Services, October 1990.
5. IPCC Working Group III, F. Bernthal, chairman. "Climate Change, The IPCC Response Strategies." Island Press, July 1990.
6. D. C. White, D. S. Golomb. "Closing the Energy Cycle: A Challenge for Energy Intensive Industries." Kohle-Stahl Kolloquium, T.U.B. Berlin, Feb. 1989.
7. A. Lovins. "Soft Energy Paths: Toward A Durable Peace." Ballinger Publishers, to 1977. A. Lovins, L. H. Lovins, I. Krause, W. Burch. "Least-Cost Energy Solving the CO_2 Problem." Brick House Publishing Co, 1981.
8. *Business Week*, "Conservation Power" pp 86–92, September 16, 1991.
9. P. L. Joskow, R. Schmalensee. "Markets for Power: An Analysis of Electrical Utility Deregulation." MIT Press, Nov. 1983.

10. S. R. Connors, C. J. Andrews. "The Role of Demand-side Management in Strategic Emission Reduction: Integrating End-Use Efficiency Improvements in the Electric Power Sector." Demand-side Management and the Globe Environment Conference, Arlington, Virginia, April 22–23, 1991.
11. C. J. Andrews, S. R. Connors. "Existing Capacity—The Key to Reducing Emission." Energy Laboratory, MIT May 17, 1991.
12. C. J. Andrews. "The Marginality of Regulating Marginal Investments." Forthcoming in *Energy Policy* 6/11/91.
13. MIT-AGREA-Energy Lab. "Background Information for the New England Project: Analyzing Regional Electricity Alternatives." April, 1991.
14. S. R. Connors. "Externality Valuation Versus Electric Power Systems Analysis: Identifying Emission Reduction Strategies For the Integrated Resource Management Process." MIT Energy Laboratories, AGREA, October 4, 1991.
15. J. Edmonds, J. M. Reilly. *Global energy: Assessing the future.* Oxford University Press, 1985.
16. A. S. Manne, R. C. Richels. "Global 2100: Model Formation The Economic Costs of CO_2 Emission Limits."
17. D. W. Jorgansen, P. J. Wilcoxen. "The Cost of Controlling U.S. Carbon Dioxide Emissions." Workshop on Economic/Energy/Environmental Modeling Climate Policy Analysis, Washington, D.C., October 22, 23, 1990.
18. W. D. Nordhaus. "The Economics of the Greenhouse Effect." Center for Energy Policy Research, MIT July 1989.
19. T. S. Lee (Chairman). Policy Implementation of Greenhouse Warming: Report of the Mitigation Panel Committee on Sciences, Engineering and Public Policy NRC, National Academy Press, 1991.
20. R. U. Ayres. "Energy Conservation in the Industrial Sector." *Energy and the environment in the 21st century*, edited by J.W. Tester, D.O. Wood, N.A. Ferrari, MIT Press 1991 pp. 357-370.
21. D. C. White, Z. S. Gata. "Energy, Environmental and Economic Development Initiatives: Comparison Among Nations in the form of Global Change." Regional Energy Forum for East & Southern African Countries, Zimbabwe, November, 1990.
22. D. Golomb, H. Herzog, J. Tester, D. White and S. Zemba. "Feasibility, Modeling and Economics of Sequestering Power Plant CO_2 in the Deep Oceans," MIT Energy Lab Report # MIT-EL-84-003 (1989).

ADAPTING TO AGRICULTURAL HAZARDS CREATED BY CLIMATE CHANGE

Paul E. Waggoner
The Connecticut Agricultural Experiment Station
New Haven, CT 06504

ABSTRACT

Farmers will adapt to climate change amidst other changes, especially changing technology and rising demand by a doubled world population. Although moral imperatives may hold some people back from contemplating adaptation, the likelihood of climate change will cause others to strive to adapt. Water supply, already short in the West, is a place to start.

INTRODUCTION

The first challenge in foreseeing adaptations to climate change during 50 to 80 years is to compound the outcome of a changing climate with other changes that will occur during the coming decades. These changes range from technological innovations to social changes and from more people to new landscapes. We can grasp some of the kinds of changes that might happen at the same time as a climate change by looking back eight decades. In 1910 the Ottoman, Austro-Hungarian, British, and Russian empires ruled much of the world. In America my parents attended one-room country schools, which I too attended 25 years later. Crop yields had scarcely changed since George Washington farmed, and about 1920 when the hybrid seed corn invented at the Connecticut station was first produced, oxen pulled the plow.

Although the crystal ball for seeing future life and technology is cloudy, the great changes of the past 80 years teach us that visualizing the effect of future warming on today's farming is hazardous to our minds.

In Figure 1 you can detect two hazards of ignoring changes. The figure shows 30 years of world production of grain and oilseed. Grain encompasses corn and wheat, and oilseeds encompasses cottonseed and soybeans, for example. In 30 years production rose 130%. And, of course, consumption matched the rise of production. Because population only rose 80% while production rose 130%, nutrition improved.

Extension of the curve in the figure reveals the first hazard: minimizing the job farmers face. While climate may change during the 21st century, population will surely double from today's global 5 billion to a future 10 billion. The world wants a farmer scenario that keeps the production curve rising as long as population rises and then keeps it up beyond to improve diets, regardless of climate scenario.

The slope of the curve brings to mind the second hazard: assuming farmers are dumb. Try to foresee how farmers and their allies in science and business will adapt to changing environmental conditions amidst an army of technological, social and economic changes. If we ignore these adjustments, which I call adaptations, we shall write a "dumb farmer scenario." Imposing the climate near the middle of the next century on the activities of 1990 implies that farmers will dumbly suffer new

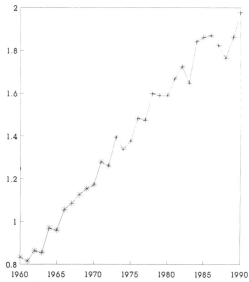

Figure 1. Grain and oilseed production.

circumstances for decades and behave as they do today. The rising production curve in the figure surely shows that farmers will not dumbly suffer.

WHAT ADAPTATION MEANS

In its rising curves, Figure 2 resembles Figure 1. It envisions four paths of change in "yield" of a crop during the time of a climate change. The top curve represents the changes in yield when climate is "unchanged," and the bottom curve when the crop "suffers" a harmful change of climate. One intermediate curve depicts the outcome of measures to "mitigate" the climate change, and the other intermediate curve when the crop is "adapted" to the climate change.

Putting the curve for unchanged climate at the top assumes the present climate is a good one. As time passes in this best climate, the impacts of other factors raise yield. The impact of all factors but climate is the difference, shown as arrow T up from the "yield" at the earliest time to the top of the "unchanged" curve at the latest time. The mnemonic for impact T is "technology."

Assuming climate change is harmful, I place "suffer" at the bottom. On the right at the latest "time," the difference in "yield" between the top and bottom curves shows integrated impact of climate change over the years. This impact, C, of climate change is shown as an arrow at the right of the figure. Sensitivity is the change-ability T/ha per °C of a crop, and impact is the integral of the sensitivity times the climate change (in

256 Adapting to Agricultural Hazards

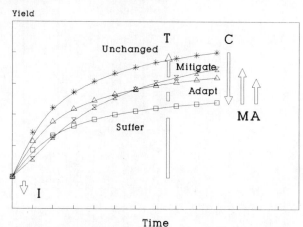

Figure 2. Envisioned courses of yield.

°C per year), year by year. The nature of the sensitivities and relative changes in all other factors will make the impact C of climate change more or less important.

Mitigation is simply stopping the emission of greenhouse gases. To depict the hobbling of farmers in, say, restricting fertilizer to slow emission of the greenhouse gas nitrous oxide, I drew the "mitigate" curve rising slowly at first. Because I assumed that, in the end and in net, mitigation of climate change helps, I drew the "mitigate" curve rising until the impact M of mitigation measured, from the base of the "suffer" curve is positive. If, however, I set aside the hobble by mitigation, I can draw "mitigate" coinciding with "unchanged," making the impact M exactly as large and positive as impact C is negative. If instead mitigation measures do hobble yield and climate change were not to occur and mitigation was not needed, impact M would be measured down from "unchanged" and be negative.

Adaptation is changing sensitivity. In my example, I visualize adaptation as modifying the T/ha per °C during the half century while climate is changing °C per year, keeping curve "adapt" between "unchanged" and "suffer." I visualize that by the middle of the next century, adaptation will have made the T/ha impact on yield by the °C climate change either a smaller loss or a bigger gain. For the harmful change of climate depicted in Figure 2, the integral of the changes in sensitivity integrated over the entire time is the impact A of adaptation.

Finally, the impact, "I" in the lower left, which is measured down from the present "yield," is the impact calculated from the dumb farmer scenario. Impact "I" matters if the climate factor limits "yield" because all other factors, such as fertility and genetics, are now ideal. Or, it matters if we will hold the world steady for a laboratory experiment with climate changing while all other factors are constant. In the unlikely event that climate alone limits "yield" or the unlikely event we devote the planet not just to an experiment but to a controlled experiment about the impact of climate, impact

"I" would represent the blow of climate change. Impact I is the reduction in yield from imposing harmful future climate on dumb farmers. Figure 1 showing production more than doubling over the last 30 years, proves farmers are not dumb and impact "I" is unrealistic.

Before leaving Figure 2, I should mention beneficial climate change. The "suffer" curve would in that case be above "unchanged." And profitable adaptation to an unmitigated change to a salutary climate would put the "adapt" curve topmost.

Saying "an unmitigated change" to a better condition is paradoxical because mitigate connotes moderating the severity of that which is distressing. So, saying mitigation of climate change implies that climate change will be harmful when, in fact, it may be favorable. Nevertheless implying by mitigation that climate change will be harmful and drawing "unchanged" at the top of the figure are usual behavior of risk-averse people.

OBSTACLE TO THINKING ABOUT ADAPTATION

Although farmers are smart and Figure 2 makes adaptation attractive, reluctance to contemplate it obstructs the path to adaptation. Rayner[1] pointed out that from the standpoint of those who advocate mitigation, "Discussion of adaptation to climate change is viewed with the same distaste that the religious right reserves for sex education in schools. Both (adaptation to climate change and sex education) are seen as ethical compromises that will in any case only encourage dangerous experimentation with the undesired behavior."

Bolstered by "moral imperatives," the objections will not be surrendered easily. But if stanching the flow of greenhouse gases proves impractical and climate change grows more likely or is even felt, the hurdle of these imperatives will surely lower. In the meantime, the likelihood and importance of climate change will cause some people to search for adaptations despite the moral imperatives of mitigation.

PARTICIPANT SPORT OF ENGINEERS AND AGRONOMISTS

Like sport, people can be divided into spectators and participants. Spectators are safe in front of the TV and participants are bruised on the field. But engineers and agronomists are naturally participants, and as participants, they are bruised but affect outcomes.

To get on to adaptation, these players don't ask for precise climate forecasts. These players only need a glimmer of the new realm of climate.

Three realms are enough to decide whether farmers should ignore climate change, think of adaptation, or abandon hope. One realm is inconsequential change, at the other end is cataclysmic change, and the intermediate realm is a degree or two warming with accompanying changes in precipitation.

The realm of inconsequential change seems unlikely. The inexorable rise of greenhouse gases and their sure absorption of infrared radiation stand against uncertainties about GCMs.

Cataclysm is possible—melting ice beneath the tundra, released methane, switched ocean currents, melting Antarctica. Despite this apocalyptic litany, a panel of the National Academy of Sciences[2] nevertheless recently decided, "No credible claim can be made that any of these events is imminent." Mixing the merely possible with the probable cheapens the probable because if all things are possible nothing is highly probable. I pass over the realm of cataclysm as I do the realm of inconsequential change.

The realm of a degree or two warming with accompanying changes in precipitation remains. It lies within the range of familiar scenarios of 1 to 5°C warming. Even if more warming followed, a degree or two would be reached on the way.

To consider impact and adaptation, participants can sidestep scenarios and concentrate on a degree or two warming, a realm that the planet is likely to enter, even though it may eventually pass on to another, warmer or cooler.

BIG QUESTION

After removing both the inhibition of thinking about adaptation and the shackle of scenarios, how can a participant proceed with benefits? A panel commissioned by the Council on Agricultural Science and Technology[3] focussed on adaptation by asking the following question:

> For a warmer planet with more people, more trade, and more carbon dioxide in the air, can US farming and forestry prepare within a few decades to sustain more production while emitting less and stashing away more greenhouse gases?

The first phrase begins with warming. If it were more detailed, it would add "for a warmer planet with more but rearranged precipitation." Also, the first phrase emphasizes simultaneous changes in other factors: more people and more trade. The world population will likely double during the half century envisioned for climate change. International trade negotiators are trying to accelerate trade of grain. The first phrase ends with, "more carbon dioxide in the air" because the greenhouse gas CO_2 is the stuff of photosynthesis and crop yields.

The next phrase, "can US farming and forestry prepare within a few decades to sustain more production" focuses on preparation and sets a deadline of the same few decades as climate change.

The phrase "to sustain more production" breaks with dumb farmer scenarios. By the time climate changes in several decades, 1991 production will be irrelevant. True, with no farm animals eating, the present crop production could feed 10 billion people of the 21st Century about 2000 calories per day. But to maintain present diets and improve them, the farmers of 2050 AD must grow twice the present crop—and keep it up.

The final phrase is "while emitting less and stashing away more greenhouse gases." Farming does emit the greenhouse gases methane and nitrous oxide. Although burning fossil fuels dwarfs the greenhouse forcing by farming, farmers might nevertheless be ordered to curb their cows and spare the fertilizer. And, foresters might be commanded to stash away in wood carbon assimilated by photosynthesis from atmospheric CO_2.

The question asked by the Council on Agricultural Science and Technology illustrates that adapting to climate change will be far harder than suggested by studies of the impact of climate alone. The search for answers to the above questions will be amidst other great changes and hobbles and never end. But these hardships are reasons to begin the search now rather than being dismayed later.

A PLACE TO START

A good beginning is on water. Although warming naturally dominates publicity about greenhouse warming, water—not warming—will move farming more. West of Omaha, water is already short, and affairs will be moved for better by a wetter and for worse by a drier climate.

Even though valuable crops grow on irrigated land, farmers do not compete well for water when it is costly. The area irrigated has shrunken for a decade in four out of five regions of the US, Figure 3.

A wetter climate in the west might slow this shrinkage. But if climate does not change, or if it dries, the shrinkage will continue or speed up.

So, a "participant sport" for climate change is growing more food per gallon of water. Figure 4 shows evaporation ET, the "yield" T/ha of grain, and WUE or kg of yield per volume of ET. An engineer naturally thinks of raising WUE by conserving water, i.e., with small hoses trickling water at the bases of tens of trees per acre.

Figure 4, however, represents thousands of cereal plants per acre. It shows how to improve WUE for grain where a hose per plant is impractical. From the 1950s to 1980s the open bars show that ET from Texas grain continued about 0.7 m per year. But yields shown by the central, screened bars, rose helped by breeding, by fertilizing and by controlling pests. Because raising yields raises ET little, the bars shaded with diagonal lines show the water use efficiency, WUE, rising.

The simple, great principle is: remedying one limiting factor raises the efficiency of use of all others.

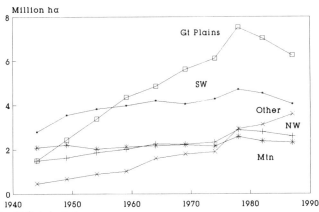

Figure 3. U.S. irrigated area.

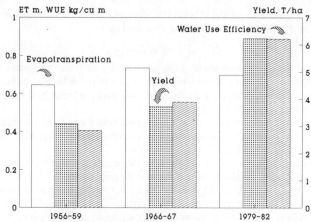

Figure 4. Evaporation, yield and efficiency.

WHAT TO DO?

So, adaptation challenges engineers and agronomists. It challenges engineers to cut useless evaporation. And it urges agronomists to raise yields from each gallon of evaporation.

Foreseeing hard work to keep production rising, the Council on Agricultural Science and Technology asked the question I quoted above. Giving high priority to that hard work, a U.S. National Academy of Sciences[2] study devoted its first recommendation on adaptation to global warming to the farming sector.

> Maintain basic, applied, and experimental agricultural research to help farmers and commerce adapt to climate change and thus assure ample food. ... As climate changes, adapted varieties, species, and husbandry must be more promptly sought and then proven in the reality of fields and commerce. Special challenges are (1) while adapting, to sustain the natural resources of land, water, and genetic diversity that underlie farming; (2) to be productive during extreme weather conditions; (3) to manage irrigation to produce more food with less water; and (4) to exploit the opportunity of increased fertilization provided by more CO_2 in the air.[2]

In that thorough recommendation, the actionable words are: proving new technologies and concepts in the reality of fields and commerce; sustaining the foundation of natural resources; production—not just survival—during extreme weather; producing more with less water; and exploiting the richer CO_2 in the air. These are not selfish pleas for money and scientists. They tell where to find adaptation to climate change.

Hopes for an improved human condition on a warmer planet with more people, more trade, and more carbon dioxide in the air rest on more than continued work by agronomists and grants for universities. Hope rests on an expectation of revitalized agricultural research.

REFERENCES

1. Rayner, S. In M. Grubb et al. (ed.) *Energy policies and the greenhouse effect, II.* Roy. Inst. Intern. Affairs, Dartmouth, 1991.
2. National Academy of Sciences. Report of the Synthesis Panel on Policy Implications of Greenhouse Warming, April 11, 1990.
3. The study is financed by the U.S. Department of Agriculture and being performed under the aegis of the Council on Agricultural Science and Technology, Ames, Iowa.
4. P. E. Waggoner. *Ariz. J. International Law*, in press (1992).

GLOBAL TRENDS IN MOTOR VEHICLES AND THEIR IMPLICATIONS FOR CLIMATE CHANGE

James J. MacKenzie
World Resources Institute
Washington, DC 20006

ABSTRACT

The world motor vehicle fleet has grown eight-fold since 1950 and now numbers over 550 million. While most vehicles today are registered in the industrialized countries, the greatest growth in registration is occurring in the developing world. Motor vehicles are major contributors to a number of important international trends including growth in urban air pollution and increasing atmospheric concentrations of carbon dioxide. Despite a 20 percent reduction in fuel use per motor vehicle between 1973 and 1987, total fuel consumption by vehicles—along with corresponding carbon dioxide emissions—increased by 40 percent. Growth in the motor vehicle fleet is projected to overwhelm future efforts to improve vehicle fuel efficiency. Alternative carbon-based fuels such as ethanol, methanol, or compressed natural gas will not appreciably change these trends. It is argued that electric and hydrogen vehicles—ultimately charged by non-fossil electricity sources—are the only long-term technical options that can reverse these trends.

INTRODUCTION

The World Resources Institute is a non-profit policy research center focusing on problems of sustainability. This paper describes trends in motor-vehicle use both in the U.S. and worldwide, and their implications for several national problems including air pollution, congestion, climate change, and national security. All these problems are linked together through our use of motor vehicles, principally cars, trucks, and buses.

The worldwide growth of motor vehicles contributes to a host of health, social, and environmental problems. Figure 1 shows that in 1950, worldwide there were only 70 million cars, trucks, and buses on the road. In 1990 there were about eight times that number—550 million. The factor of eight in growth is the product of a doubling of the world population and a quadrupling of per capita motor vehicle registrations.

Figure 2 shows—not surprisingly—that most of the motor vehicles in the world are in the U.S. and Canada, western Europe, and Japan. In short, in the wealthy, industrialized nations. In the U.S. there are 190 million motor vehicles but only 168 million drivers to operate them.

The U.S. is the most motor-vehicle dependent country in the world (Figure 3). For every 500 Americans there are almost 400 motor vehicles, far more than in any other country. U.S. motor vehicle use per person is about twice what it is in Europe and four times that of Japan. Worldwide there are only 50 vehicles for every 500 people. In other words, worldwide, there is an enormous potential for growth in motor vehicle use and in the problems they give rise to.

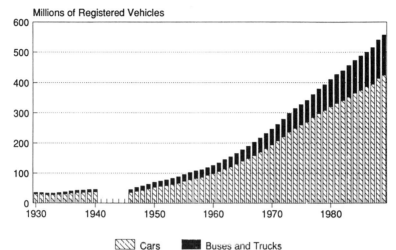

Figure 1. Global trends in motor vehicle registrations.

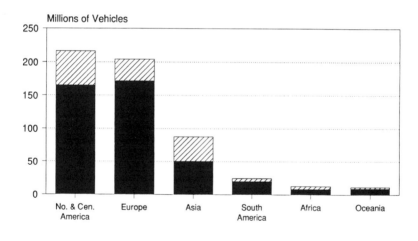

Figure 2. Worldwide motor vehicle registrations (1989).

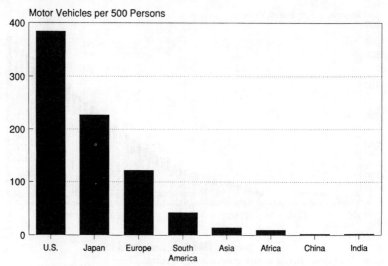

Figure 3. Regional patterns in per capita motor vehicle registrations (1989).

The expanding use of motor vehicles is bringing with it many problems both in the United States and elsewhere. In the U.S., cars and trucks are major contributors to air pollution, especially carbon monoxide, nitrogen oxides, and volatile organic compounds, and hence to our smog and acid rain problems (Figure 4).

Vehicles are also large consumers of oil. Globally, motor vehicles powered almost universally by oil account for a third of world oil consumption. In the western industrialized (OECD) countries, they represent over 40 percent of demand and in the United States over 50 percent (Figure 5).

With U.S. domestic oil production continuing to decline (Figure 6), oil imports and transportation, by far the largest consumer of oil, are playing an ever more important role in the policy debate over national and economic security. In the winter of 1991, the U.S. fought a war in Kuwait primarily over oil. Iraq has about 10 percent of the world's oil, as does Kuwait; Saudi Arabia has about 25 percent. If Iraq had taken over Kuwait and then Saudi Arabia, it would have controlled about 45 percent of the world's proven oil reserves and would have had a major influence—some might say a stranglehold—over the economies and foreign policies of most of the world.

GLOBAL WARMING

Lastly, and most importantly, motor vehicles are major sources of direct and indirect greenhouse gases including carbon dioxide, CFCs, nitrous oxide, tropospheric ozone, and carbon monoxide. The scientific case for the existence of the greenhouse effect is indisputable. The major issues of uncertainty today relate to its augmentation:

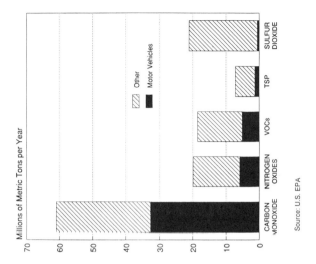

Figure 4. Motor vehicle contribution to U.S. air pollution emissions (1989).

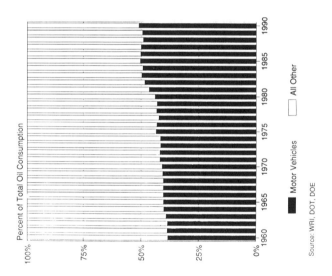

Figure 5. U.S. oil consumption.

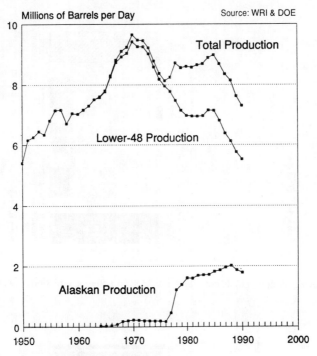

Figure 6. Trends in U.S. Oil production.

where and how fast climate changes will occur as a result of increased concentrations of greenhouse gases.

Motor vehicles currently emit close to 900 million tons of carbon each year, in the form of carbon dioxide, about 14 percent of global fossil fuel CO_2 emissions (Figure 7). These emissions have been growing more or less linearly for the last twenty years. In the United States, motor vehicles account for about 25 percent of total CO_2 emissions.

Measured by percentage growth, emissions in the developing world are growing faster than those of the industrialized countries. However, overall emissions from the industrialized countries still account for the great bulk (over two-thirds) of global motor-vehicle carbon dioxide emissions. The U.S. alone accounts for about 38 percent of global motor vehicle CO_2 emissions.

In the United States, the growth in carbon dioxide emissions has been only slightly interrupted by the three oil disruptions over the past 20 years (Figure 8). Indeed, carbon dioxide emissions from motor vehicles are second only to those of fossil fuel power plants.

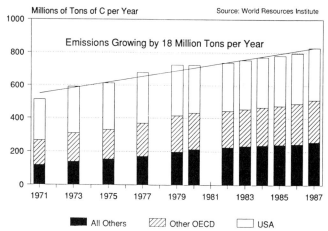

Figure 7. Global motor vehicle CO_2 emissions.

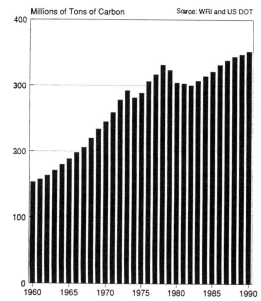

Figure 8. Annual CO_2 emissions from U.S. motor vehicles.

OTHER GREENHOUSE GASES

Motor vehicles are also important sources of other greenhouse gases, including carbon monoxide, which indirectly contributes to global warming. Carbon monoxide is removed from the atmosphere through processes that slow the removal of methane, a potent greenhouse gas. Hence, CO indirectly contributes to the buildup of methane; it also contributes to smog formation.

In the U.S., motor vehicle air conditioners are the largest source of CFCs. These gases are best known for their destruction of the ozone layer. This stratospheric ozone protects life on earth from the damaging effects of high-energy ultraviolet radiation. Before the ozone layer formed billions of years ago, life on land, as we know it today, was not possible.

Global climate change is destined to be one of the most important factors—perhaps the most important—affecting energy planning over the coming decades. And its implications for fossil fuel energy use and motor vehicle use, in particular, are sobering. According to EPA, to stabilize atmospheric carbon dioxide levels, the world will have to cut carbon dioxide emissions by 60 to 80 percent below existing levels.

IMPLICATIONS FOR TRANSPORTATION

We take it for granted that reducing the threat of global warming will require major reductions in worldwide greenhouse gas emissions. Yet, as Figure 9 shows, total CO_2 emissions continue to grow worldwide. In other words, nations have not yet confronted the growth in this, the most important, greenhouse gas.

Environmentalists and others are placing heavy bets that improved new-vehicle fuel efficiency (induced, e.g., by tighter corporate average fuel economy [CAFE] standards) can help control motor-vehicle carbon dioxide emissions. Yet, the history of the past 20 years provides little encouragement that improved new-vehicle fuel efficiency, by itself, can solve our oil-security or climate problems.

Consider that in the U.S. the average amount of fuel used per motor vehicle (cars, trucks, and buses) went down by almost 20 percent between 1970 and 1989. This takes into account all the relevant factors: changes in fuel prices, CAFE standards, changes in driving habits, changes in the mix of vehicles, increased congestion, etc.

Yet the growth in the number of registered vehicles (up 75 percent) and the number of miles they were driven (up 10 percent) overwhelmed these improvements. As a result, total fuel consumption in the U.S. by motor vehicles rose by over 40 percent (Figure 10). Interestingly, virtually the same trends were observed internationally. Between 1973 and 1987, the average quantity of fuel consumed per vehicle in the global fleet decreased by about 20 percent. Yet, total motor-vehicle fuel consumption rose by about the same 40 percent, paced by more than a 70-percent rise in the number of vehicles.

Present trends in motor vehicle growth suggest that even with continued improvements in new-vehicle fuel efficiency, carbon dioxide emissions from motor vehicles worldwide will increase between 20 and 50 percent over the next twenty years (Figure 11). Such increases will make it exceedingly difficult to curb global warming. Thus, it does not look as though improved new-vehicle fuel efficiency, through CAFE

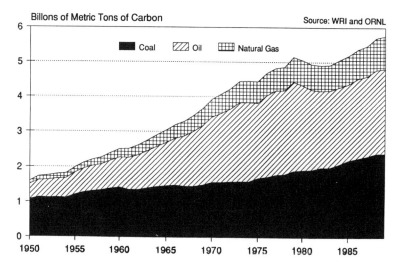

Figure 9. Global fossil-fuel CO_2 emissions.

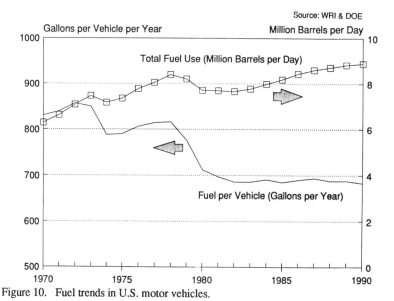

Figure 10. Fuel trends in U.S. motor vehicles.

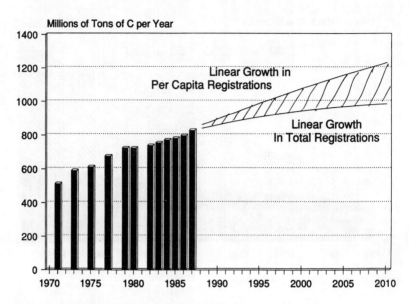

Figure 11. Global CO_2 emissions from motor vehicles.

for instance, can by itself stabilize motor-vehicle carbon dioxide emissions, let alone reduce them by 50 percent or more as long as there is continued growth in the number and use of vehicles worldwide. In short: we cannot improve our new-vehicle fleet efficiency fast enough to overcome the increased emissions from growth in vehicle use.

This conclusion is reinforced by examination of the legislation that has been pending before Congress—the so-called Bryan bill. Although this legislation would require increased new-vehicle fuel efficiencies—it would still lead to increased fuel use over today's levels in the coming years. This is so because CAFE affects only new-vehicle efficiency, not the way people drive. CAFE provides no motivation to ride share, take public transit, live closer to work, drive the speed limit, etc. In fact—all other things being equal—higher new-car efficiencies make driving cheaper and so encourage more driving that offsets much of the benefits.

POLICY RESPONSES

There has been essentially no response to this conundrum on the part of the U.S. Congress or the Administration. Neither will consider seriously economic policies that could be taken to reduce the growth in VMT or encourage more efficient or alternatively fueled vehicles that could help deal with the problem. The Congress won't touch a sizable gas tax or other economic reforms, even though such taxes could be made progressive and offset by reductions in payroll or other taxes.

For its part, recent Administrations have opposed both CAFE and realistic pricing of motor fuels. The position on the motor-vehicle/climate problem is to have no position. The newly passed surface transportation bill will help somewhat by providing more funds for public transport and other alternatives to building more roads. But, at heart, the law only hints at the breadth of changes that need to occur.

There are two prongs to the response that is needed to cope with the global problems spawned by motor vehicles. The first is technological, the second is economic and institutional.

NEW TRANSPORTATION TECHNOLOGIES

Today, motor vehicles depend totally on oil. Until economically and technologically attractive alternatives are widely available, it is unlikely that the pollution, security, and climate problems posed by motor vehicles can be solved. People—in the U.S. and elsewhere—will have little alternative than to continue buying oil-powered cars and trucks. As a result, air pollution will continue, and oil imports will rise as will carbon dioxide emissions.

We need alternatives to oil-powered cars and trucks. And the burden for developing the technological alternatives rests squarely on the industrial countries that make the world's motor vehicles: 80 percent are made in the U.S., Japan, and Europe. There is no one else that can do it.

Alternative forms of carbon-based fuels—methanol, ethanol, and compressed natural gas (CNG)—are being pushed by some as answers even though they are clearly not. Methanol and CNG are based on fossil fuels. Their use is not sustainable over the long haul. They offer little if any improvement over reformulated gasoline in solving our air pollution, security, or global warming problems. It would be folly to place a major emphasis on expanding the use of either.

Ethanol, at least as we now make it from corn, is little more than a farm-support program. In the United States, ethanol is blended with gasoline in a one-to-nine ratio to make "gasohol." All told, about 8 percent of the automobile fuel pumped at service stations in the United States is gasohol. Gasohol is heavily promoted by the federal government and various agricultural states.

In terms of energy, ethanol by itself meets only a tiny portion (about 0.5 percent) of the country's gasoline needs. Over 95 percent of U.S. fuel ethanol is made from corn, requiring about 4 percent of the nation's corn crop. The production of ethanol entails huge amounts of land. To displace the 133 billion gallons of gasoline and diesel fuel consumed in 1990 would require about 670 million acres of cropland—well over twice the total cropland available in the United States! Moreover, ethanol produced from crops will always be at the whim of weather and climate change.

The impacts of ethanol blends on air pollution are marginal. While ethanol blends may reduce carbon monoxide emissions slightly, their overall impacts on smog levels are expected to be negligible and nitrogen oxide emissions could well be higher.

Estimates of the net greenhouse impacts of ethanol are subject to great uncertainty. According to the Office of Technology Assessment, existing ethanol technology offers no significant greenhouse benefit.

ELECTRIC VEHICLES

This leaves only electric vehicles powered by either batteries or fuel cells running from hydrogen. Electric vehicles in large numbers will soon make their debut in California. The Los Angeles Department of Water and Power along with Southern California Edison have ordered 10,000 EVs to be on the road by 1995. Prototypes have been built. These vehicles will be hybrids and have batteries that will power them for about 60 miles at low speeds and gasoline engines to charge the batteries at high speeds.

In addition to the Los Angeles initiative, the California Air Resources Board has adopted regulations requiring that by the year 1998 two percent of all new vehicles sold in the state must have zero emissions and that by 2003 ten percent, a requirement currently met only by electric vehicles. Adoption of these regulations is being considered by other states as well.

How much electric vehicles can cut carbon dioxide emissions depends mostly on two factors: the electric efficiency of the vehicles and the emissions from the power plants that produce the electricity used to charge them. If EVs are charged by electricity from the expected mix of fossil, nuclear, and hydro plants in the year 2000, greenhouse gas emissions for comparable vehicles would fall by 25 percent. If the electricity were produced by power plants that burn natural gas, the reduction would be 30 percent. Charging them with electricity made from coal, in contrast, could increase greenhouse gas emissions slightly compared with gasoline-powered vehicles.

In the longer term, emissions could be eliminated entirely by charging the batteries using either renewable or perhaps nuclear power plants, though the prospects for a revival of nuclear are, in my opinion, not bright. Regardless of which fuel is used to generate the electricity, EVs could be powered strictly from domestic sources, helping both national security and the nation's balance of trade.

Switching to electric vehicles would significantly improve urban air quality. EVs emit no street-level pollutants, and if the batteries were charged at night—as they ought to be initially—no new power plants would have to be built and smog (which cannot form without sunlight) would thin out dramatically. On the other hand, charging the vehicles with electricity from coal plants could slightly increase acid-rain emissions. (Total sulfur emissions, though, are capped by the 1990 amendments to the clean air act.) If renewable energy sources were used to charge the batteries, pollution and greenhouse gas emissions would be very small indeed.

HYDROGEN-POWERED VEHICLES

Interest in hydrogen-powered vehicles has increased over the past decade as well, particularly in Japan and Europe. Most hydrogen today is obtained from natural gas though this cannot be a long-term source. Production of hydrogen from water—either by electrolysis of photochemical reactions—is the most likely long-term source of hydrogen with the primary energy supplied by renewable or nuclear technologies. In this case virtually no carbon dioxide emissions would result.

Although hydrogen vehicles share many of the advantages of EVs, their widespread commercial use is further down the road because of the lack of a supporting infrastructure. Still, like electric vehicles, they would form a natural link in developing a sustainable energy system, and their use would reduce oil imports, alleviate the trade deficit, and cut air pollution and greenhouse gas emissions.

THE NEED FOR ECONOMIC REFORMS

How do we encourage more efficient conventional vehicles and the introduction of zero emission vehicles? And how do we control the growth in motor vehicle use that is strangling our major metropolitan areas? An important part to the response to both questions is that we must adopt more realistic pricing of our transportation services.

The runaway growth in motor vehicle use in this country is largely the result of numerous public policies that encourage and subsidize their use. Underpriced driving is probably the single most important factor in our transportation-related problems. Undoubtedly, similar situations exist in other countries where vehicle use is growing rapidly.

At WRI we have a report now in external review in which we have estimated the various costs of driving that are not borne by motor vehicle drivers. Examples include the costs of road construction and repair, the provision of highway-related public services, and the costs of parking, especially for commuters. In all these cases, motorists pay only a fraction of the costs they should be bearing. Over reliance on motor vehicles results.

Other costs of driving fall outside of normal economic transactions—so-called "externalities"—and are partially borne by non-users of motor vehicles. Examples of externalities are climate change, air pollution, congestion, accidents, noise, and threats to our national security from importing oil.

Estimating these unborne costs is difficult and fraught with uncertainties. But there can be little question the numbers are huge, in the range of hundreds of billions of dollars per year. If internalized, e.g., through a fuel tax, these costs would easily add several dollars to the cost of a gallon of motor fuel (at present consumption rates), raising fuel prices to levels comparable to those of our trading partners.

CONCLUSIONS

Unless, and until, we begin to make drivers pay more of the costs they impose, we will have a hard time introducing badly needed transportation reforms: more efficient conventionally powered vehicles, more attractive public transit, and the introduction of climate-friendly vehicles for the next century. Very few available or emerging options can compete with gasoline priced at a dollar per gallon.

It is clear that the most important set of policies to address the greenhouse/transportation problem is not technological but economic. We have got to get the ground rules into some rational order. Motorists must pay more of the costs for the construction and maintenance of the roads they use and for the parking they enjoy, especially at work. Heavy trucks must pay for the damages they inflict on roadways. And oil consumers must pay for the national security, pollution, and climate risks they impose through their gasoline use.

Such pricing reform must be accompanied with land-use and zoning reforms that will encourage more dense patterns of mixed commercial and residential development that lend themselves to less need for travel and more use of alternative forms of transportation including public transport, bicycling, and walking.

Pricing reforms need not occur all at once. They can be phased in as quickly as is politically feasible. The price increases on motor vehicle services (fuels, road construction, highway services, parking, pollution, etc.) could be offset by reductions in other taxes, such as payroll or social security taxes. Total revenues could be made neutral and reforms could be structured so that they are progressive so as not to adversely impact low-income persons.

To sum up, unless we get the economic signals straight, it will not be possible—short of total regulation of our society—to solve the profoundly threatening problems related to oil use in transportation. Conversely, adopting such pricing and land-use planning reforms would be the single most effective measure we could take to stimulating both more rational transportation planning and the needed technological revolution that will reduce the risks of global climate change.

GLOBAL WARMING, NATURAL HAZARDS, AND TRANSPORTATION IN EUROPE

Allen Perry
University College of Swansea, U.K.

ABSTRACT

Hazardous events disrupt transportation systems in many parts of Europe and on occasions cause considerable numbers of deaths and impose widespread disruption. Projected climate warming is likely to alter the frequency and disruption of climate hazards. The likely effect on different modes of transport will vary and is being investigated.

INTRODUCTION

Efficient, rapid, dependable transport is a prerequisite of an advanced economy; and interruptions, disruptions and dislocations to any mode of transport will quickly have "knock-on" effects across a very wide range of industrial and commercial activity. A clear division can be recognized in the nature of likely impacts of climatic change into:

a) changes of short-term extremes of climate many of which can cause dangerous travelling conditions or impose additional costs on transport operations;
b) changes of the frequency of conditions, which although not extreme, can be hazardous. A good example might be changes in frost and icy road frequencies.

The significant impacts are manifested through the supply and utilization of infrastructure, plant and equipment[1,2].

THE SENSITIVITY OF TRANSPORT TO CLIMATE CHANGE

Sensitivity to weather and climate change is high for all forms of transport and especially for road and air communications[3]. A rank order of impacts can be recognized ranging from least impact, sea transport, to greatest impact, air transport. On the land surface rail transport is probably the mode of transport most tolerant of adverse weather. The main road hazards caused by adverse weather conditions are ice-related slipperiness, wind and fog. In many areas local fog turns out to be the most dangerous of these events. In France fog caused 182 fatalities in 1986, corresponding to 2.6 percent of all traffic accidents. Climatic change would have implications for weather-related road accidents. In France it emerges from statistics that adverse meteorological factors are present in 20 percent of accidents. However, the relationship between bad weather and the number and severity of road accidents is not a simple linear one and less frequent adverse weather could actually increase accidents since less journeys would be postponed or cancelled. Projected winter increases in precipitation could exacerbate problems of road flooding, landslips, breaking up of the road surface and corrosion of steelwork on bridges. At high altitudes greater amounts of snow could occur.

Severe gales can cause damage, disruption and death, as was shown over a large part of western Europe in October 1987 and January 1990. During the exceptionally stormy start to 1990 there were 47 deaths on the roads in England and Wales and a further 39 in the Netherlands, Belgium, France and West Germany. Hundreds more people were injured in road accidents. Although the level of fatalities was very high in these storms a survey has shown that over 100 people were killed and nearly 700 injured on British roads between 1962 and 1980 as a result of strong winds[4]. Although the weather seemed exceptional in 1990 there is no firm evidence of an increase in the frequency of intense winter storms over recent years, and opinion remains divided as to whether global warming will increase or reduce gale frequencies experienced in mid-latitudes of Europe. Measures to reduce the effects of wind include constructing artificial windbreaks, warning drivers to take care on hazardous stretches of road and perhaps introducing short-term closure programs of roads and bridges.

Inland waterway transport would be less affected by the occasional freezing of canals and waterways but perhaps more affected by low water levels as a result of evaporation during hotter summers when navigation would be difficult. Increases in temperature could require increased use of refrigerators and produce changes in the way cargoes are managed in transit.

No systematic analysis exists of the likely aggregate impacts on railways of increases in temperature. At present snow and ice present the most difficult weather-related problem and operating difficulties and revenue losses from this cause could be expected to fall[5]. Increases in flood frequencies as a result of projected increased winter precipitation, if realized, could weaken and wash away track bed more frequently and cause bridge failure due to scour. Sea level rise and more frequent coastal flooding could cause structural damage to both road and rail transport infrastructure in low-lying coastal areas.

Critical thresholds apply to many forms of transport. For aircraft lower pay loads are necessary above 25°C as lower air density reduces the aircraft's pay load capacity. Such thresholds need to be clearly identified in respect of other modes of transport.

MANAGEMENT AND POLICY IMPLICATIONS

There are management implications for winter maintenance activities on both roads and railways. Amongst the beneficial effects of climatic warming would be reduced expenditure. Some 6 million tons of salt was spread on highways in Europe in 1988, representing an expenditure of approximately $US250 million a year[6]. Anti-icing precautions are especially important in areas where the winter overnight air temperature regularly cycles through zero. At present the 0° mean minimum January air temperature isotherm runs from South West Norway–central France–Northern Italy (see Figure 1). As this isotherm is displaced eastwards with climate warming, so Germany, eastern France and parts of Switzerland and Austria will experience winter night-time temperatures above the zero threshold. In several EC countries new technology for ice detection and the forecasting of minimum road surface temperature has been introduced. In some areas mild winters may extend the "pay-back" period of such investment and make its further deployment a less attractive proposition. However, the prediction system could still be expected to contribute to winter

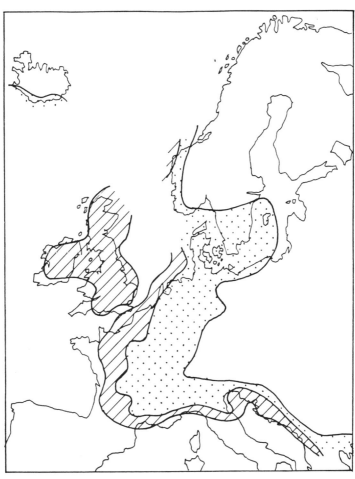

Present zone where ante-icing procedures are most frequently required in winter

Displaced zone where ante-icing procedures are likely to be most frequently used by 2030

Figure 1. Air temperature isotherms over Europe

maintenance decision making, since milder winters would still include some frost and ice occurrences. Savings should also be possible in capital expenditure programs of snow clearing equipment. Decreases in the number of freeze-thaw cycles and of the severity of frost should reduce structural damage to roads and bridges in many places.

Policy options will need to be reviewed in the light of more frequent and severe episodes of low level pollution occurring during hot sunny weather in summer. The efficiency of measure such as low speed limits to discourage motorists from undertaking some journeys by private car, or restrictions on the use of vehicles in city centers on particular days, as in Athens, will have to be estimated. The influence of meteorological stresses on the incidence of traffic accidents requires more intensive investigation. This should include studies of the physical stress to the driver from the ingestion of toxic substances into the vehicle.

Policies for the monitoring of bridges and other transport infrastructure may need to be reviewed. Impairment of port facilities could result from both sea level rise and any increase in storminess, with concomitant implications for shipping.

RESEARCH NEEDS

Comprehensive studies of the likely impacts of climatic change on transport have not been carried out and only fragmentary, small-scale studies exist[7]. Both broad-scale and specific research is needed, for example:

a) Broad-scale assessments of likely climatic induced demands for transport as life styles, residential and migration patterns change. More specifically many southern European coastal areas may be perceived as less ideal than at present for summer holidays, but become more attractive at other times of the year with consequences for seasonal traffic flows;

b) Specific analysis of the level of savings that can be expected in winter maintenance of highways in different areas is required. Detailed cost-benefit figures need to be collected to examine whether further investment in costly technology like ice detection systems, can be justified.

c) Summer low level ozone pollution episodes are likely to become more frequent, especially in and around cities. Research is needed into the efficiency of legislative options available to control and constrain traffic during such episodes and to investigate to what extent transport "substitutability" is possible.

d) Increased damage to bitumen surfaces in hot spells could occur. Research is needed into what increased costs might occur for highway managers, especially in Northern Europe and what changes in road surface materials should be introduced.

A number of semi-official bodies already exist with interests in aspects of European transport, for example the Standing European Road Weather Conference (SERWEC) and they should be encouraged to promote research on the implications of climate change on transport amongst their members. In addition bodies such as COST (European Co-operation in the Field of Scientific and Technical Research) have expertise in transport studies.

REFERENCES

1. A. H. Perry. 1981. *Environmental hazards in the British Isles.* George Allen and Unwin Londen.
2. M. L. Parry and N. J. Read. 1988. AIR Report No. 1, University of Birmingham, U.K.
3. A. H. Perry. 1991. Transport. In *The potential effects on climate change in the U.K.* Department of the Environment, HMSO London.
4. A. H. Perry and L. Symons. 1991. *The Geographical Magazine* **63**:46–42.
5. K. Smith. 1990. Weather sensitivity of rail transport. In W.M.O. Economic and Social Benefits at Meteorological and Hydrological Services No. 733 Geneva Switzerland.
6. R. Pettifer and H. Melama. 1990. *Vaisala News*, pp. 4–11.
7. A. H. Perry and L. Symons. 1991. *Highway meteorology.* Spon Ltd London.

POTENTIAL IMPACT OF CLIMATE-INDUCED NATURAL DISASTERS ON THE CONSTRUCTION INDUSTRY

Ahsan Kareem
University of Notre Dame
Notre Dame, IN 46556-0767

ABSTRACT

This paper discusses the potential impact of climate-induced natural disasters on the construction industry. Following a general overview of climate change and its influence on natural disasters, the question of how present construction practice will be influenced by climate change is addressed. The need for risk assessment as an efficient tool for planning and tailoring adaptive measures, e.g., retrofitting, coastal management for communities at risk is examined.

BACKGROUND

Damage to constructed facilities due to extreme winds and storm surge currently is in the billions of dollars annually in the United States, and it is expected to increase every year due to the rapidly accelerating growth of coastal developments and movement of population to these communities[1]. Such disasters, besides causing the region devastating economic losses and hardship, also disrupt the community lifelines, i.e., the communication links, transportation, and supply. The situation will be further exacerbated by global warming and the resulting rise in sea level. Coupled with the climatic changes, coastal hazards will further increase as a result of accelerated erosion, higher sea levels, transformation of wetlands, and intrusion of salinity. The frequency of hurricanes may not increase, but their intensity and associated wind speeds are speculated to increase[2,3]. Furthermore, those regions that escaped hurricane landfalls in the past may become more likely candidates as warmer waters will further assist in steering storms further north; e.g., California, New England, and Nova Scotia/Newfoundland. The major threat will be posed by a possible increase in the storm surge level. Unless significant measures are taken to lessen the impact of wind/wave/surge catastrophe coupled with climate change effects on the built environment, the cost of replacement or repairs and loss of lives is likely to escalate.

GLOBAL WARMING

In the past few decades, attention has been focused increasingly on global climate warming. This is primarily resulting from the heating caused by emissions of greenhouse gases that are trapping outgoing infrared radiation. The worldwide burning of fossil fuels is a primary source of these gases. The impending threat of global warming and associated sea level rise are gaining fast recognition as a new natural hazard in the making that also has the potential of intensifying known hazards, such as hurricanes. There is still a considerable controversy over this; nevertheless, should the warming trend continue, the implications are staggering. The nature and magnitude of weather conditions and events that might accompany greenhouse warming at a location are extremely uncertain. The reason for the uncertainty stems from a specific deficiency in the general circulation models employed for climate simulation. These models represent climatic forcing on two different spatial scales, namely, large scale—1000 Km to global—and mesoscale that encompasses a few kilometers to several hundred kilometers. The larger scales are responsible for determining sequences of weather events which characterize climate change, whereas mesoscales regulate the regional distribution of climatic variables. Current models are too coarse to adequately describe mesoscale forcing for a more accurate prediction of regional climatic details[4]. Furthermore, the clouds resulting from initial warming subsequently end up offsetting the effect of warming by blocking the sun. This cycle is not well understood. The role of oceans in the climate dynamics is an important one. Despite the level of overall uncertainty in data stemming from the types of thermometers used, reading practice, location of measurements, and a possible urban bias, some believe that climatic warming has a definite identity and it is distinguishable from the climatic noise[5]. Recent developments in the theory of dynamical systems will make it possible to learn more about the underlying dynamics of weather and climate and to find out, independently of any modeling, to what extent these are predictable[6].

SEA LEVEL RISE SCENARIO

Unlike the quantitative assessment of global warming, the historical records of sea level changes taken from tide gauges around the world are less ambiguous. The compiled data indicates a century-long rise from 10–30 centimeters suggesting a global warming trend. Atmospheric warming would cause melting of glaciers and ice sheets that will ultimately travel down to the ocean. Also, the warming of ocean waters results in an expansion of the water volume. Both scenarios would lead to a sea level rise and a confirmation of the global warming. Nevertheless, questions remain regarding the role of uneven distribution of tide gages, land subsidence, and crustal motion on the sea level rise scenario.

In addition to sea level rise by global warming there are climate change-induced changes in waves, winds, and tides density of water that cause spatio-temporal variations. IPCC[7] Business-as-Usual Scenario in the year 2030, projects global-mean sea level 8–29 cm higher than today, with a best estimate of 18 cm. That implied rate of rise for the best-estimate projection corresponding to the IPCC Business-as-Usual Scenario is about three to six times faster than that over the last century. It is noted that even if the emission of greenhouse gases is reduced, the sea level will continue to rise due to time lag in the climate system. Other sea level rise scenarios and their impact may be found in the literature[7,8,9,10,11,12].

The primary impact of sea level rise entails the direct physical impact of increased sea level on various components of coastal resources system, whereas the secondary impact concerns socio-economic effects on human activities or interests. The primary impact will be exacerbated by the impending threat of increase in intensity, frequency, duration, and region of influence of tropical storms. The warmer seas provide essential ingredients critical to the developments of these storms. It is noted that upwelling of cooler subsurface water due to stronger winds would offset any increase in intensities. However, there are regions in the Gulf of Mexico where loop current and its eddies bring warm Caribbean water to depths of 60 meters which would negate adverse effects of upwelling[12]. While the debate over the linkage between the frequency of storms and global warming is still on, there is some evidence that more intense storms may result from global warming. Even in the absence of an increase in the intensity and frequency of storms, the sea level rise threatens the coastal construction due to higher storm surge. The frequency of extreme events, e.g., wind gusts and storm surge, varies with changes in the mean and standard deviation of these processes. These changes are possible and are being speculated as a consequence of projected global climate change. An increase in the standard deviation is reflected in an increase in the probability of both high and low extremes. Due to an increase in the mean value with constant variability, as expected with global warming, the probability of high extremes rises. Therefore, a change in the mean of only one standard deviation would increase the frequency of occurrence of an event with MRI of twenty years by many fold. Other issues of concern are in the context of two successive extreme events that have significant influence on damage due to their cumulative impact. If the MRI of one hurricane is reduced by a factor of 2, there is a corresponding reduction of the MRI of two hurricanes in successive years by a factor of 4. Accordingly, the projected global warming threatens to increase the likelihood of damaging double events from being unlikely to happen within an average lifetime to a hazard which we can expect to experience. The prospect of sea level rise coupled with the potential of super-storms threaten the low-lying coastal areas with flooding under high tide, storm surge and wave action, beach erosion, and damage to coastal construction around the world. An increase in sea level would also threaten earlier flooding of some major hurricane evacuation routes. Accordingly, the present margins of safety may become inadequate, thus necessitating the implementation of adaptive measures. For the existing construction, this will entail retrofitting structural components and the system, and construction of levees and seawalls. There is a concern, however, that despite protecting the coastal buildings from being swallowed by ocean pounding, seawalls have led to the destruction of beaches[13].

IMPACT OF GLOBAL WARMING

Despite the controversy over the several global warming aspects stated above, it is essential that the question concerning the impact of these climate-induced changes be addressed. Any anticipatory adaptation measures on the one hand will certainly be very effective and timely in lessening the impact of the impending climate change as feared by many, and on the other hand should the projected climate change not materialize these measures will pay for their cost in improved performance under hazards precipitated by natural climatic variability.

Should the climate change start having an impact, most cities would opt for adaptation rather than abandoning their present sites. In most cases this cost will be much lower than the cost of moving an entire city. In the coastal communities, the cost will be much higher due to added storm protection needed from rising sea and protection from coastal erosion. In some cases of less populated regions, protective zoning or retreating may be more suitable. The construction industry will face many challenges in such undertakings that will require innovative techniques, such as submerged barriers in lieu of sea walls to protect beaches, the use of special equipment, and construction activity that promises not to disturb the remaining environmentally critical wetlands[14].

The coastal communities will have to make major adjustments in construction practice to cope with such changes. Places like Manhattan, Miami, New Orleans, Galveston, Charleston, and surrounding barrier islands will need to address serious and costly issues like elevated future construction, raising existing structures, and the building of dikes, levees, berms, and pumping facilities[15]. For example, the cost of building a seawall in Charleston is estimated to be $6,000/linear meter. Once the structures are elevated these are further exposed to the enhanced wind gusts which needs to be addressed in future design codes. Estimated cost of raising a house by 3 ft. is 10 to 40 thousand dollars. Strengthening the U.S. coastal property for wind resistance may cost between 30 to 90 billion dollars. Besides new construction practice, the transportation network will need to be elevated to avoid flooding during storms as these serve vital evacuation routes for coastal communities in the event of a hurricane landfall threat. Coastal erosion will result in shifting sand dunes and eroding recreational beaches that may require dredging and replacing sand at strategic locations. The cost of this effort may vary between 15 to 60 billion dollars for the recreational beaches based on an estimated sea level rise of 50 cm by year 2100[16].

The design practice will have to take into account possible climate change for structures with long life, e.g., bridges, levees, coastal structures, and offshore installations. The safety margins for such structures are generally computed based on the historical frequency of extremes, like storms. The possibility of greenhouse warming may now be considered in evaluating these safety margins. Regarding future construction, attention must be paid to account for climate change when long-lived structures are built or renovated through enhanced building codes and land-use planning. In this regard, different issues need to be addressed, e.g., economic tradeoff where the damage to or loss of a structure is gambled against the cost of incorporating extra safety margins, justification of investment in a wider safety margin in view of upfront cost or additional cost of retrofitting in light of the probability that the alteration will in fact be needed. An exercise in utility theory may help to systematically sort out these questions.

The construction industry can also offer additional measures to help mitigate the effects of global warming. The first step should be the adoption of nationwide energy efficient building codes that include efficient lighting systems to reduce energy consumption; more efficient building envelopes, claddings and curtain walls to cut down heating and cooling demands; the development of white roofs and pavement surfaces that reflect sunlight; the development of a new glazing to retain heat in winter and deflect in summer; and the utilization of passive solar techniques in locating windows on the south side of buildings and overhangs and other architectural features for shade in warmer months.

It is recommended that in order to minimize uncertainty in the effects of global changes one should institute careful monitoring and cataloging of the climate, sea surface level, stream runoff, occurrence, and tracks of hurricanes and tornadic events and their relative intensities. This information will eventually provide a rational basis for code modifications and selection of specific adaptation measures. Another important initiative in the wake of uncertainty regarding the climatic change is to improve our present ability to build structures to resist current climatic extremes. Such a construction practice will not only pay for its cost in the event no climate change occurs, but it will certainly be an effective adaptation measure to lessen the impact of an extreme climate change as feared by many.

RISK ASSESSMENT OF CONSTRUCTED FACILITIES

Risk assessment for tropical storm winds, storm surge, and wave action and consequences of projected global climate change on coastal built infrastructure is an essential prerequisite for instituting measures to lessen catastrophic damage potential. Despite a spectre of staggering economic losses from tropical storms, there is only fragmentary information on the relationship between the characteristics of the geophysical event and its damaging effects. Several studies in earthquake engineering have addressed this issue in relation to earthquake risk, however, little or no work appears to have been conducted in relation to wind/surge action[17]. The need for risk assessment is further emphasized by the impending threat of global climate change to better plan and tailor adaptive measures to confront the projected magnification of tropical storm catastrophe by climate change. Of critical importance is the development of a framework for risk assessment for identifying facilities at risk in any given geographic area.

The resulting analysis framework will permit integration of future climate change scenarios in the overall tropical storm hazard and the evaluation of its impact on the risk to coastal zones. One of the outcomes of global warming is sea level rise. The prospect of a sea level rise coupled with the potential of super-storms threaten the low-lying coasts worldwide with flooding under high tide, storm surge, wave action, beach erosion, and damage to coastal construction. Accordingly, the present margins of safety may become inadequate, thus necessitating the implementation of adaptive measures.

Risk is referred to here as the chance of property damage, impairment of intended function, and lost opportunities. The determination of hazard, vulnerability, and significance of the facilities, and analysis of damage potential constitute essential prerequisites for risk assessment. The assessment of hazard entails determination of a probabilistic model that best describes the occurrence of extreme events at a site and their severity. The vulnerability and significance of a facility encompass structural characteristics, e.g., construction system and material, sensitivity to hazard, and utility. For damage analysis, a relationship is sought that relates expected damage, as a percentage, to the exposed assets, due to the occurrence of a specific event. A synthesis of these attributes leads to the risk assessment analysis framework.

Phenomenological models combined with historical records, and the Monte Carlo simulation procedure provide a reliable means of predicting tropical storms[18]. Further improvements are needed in modeling the hurricane wind field, modeling of swirling rainbands, and filling rate (degradation of tropical storms strength after land fall). Improved modeling of storm surge and wave fields driven by storm wind field is needed to better describe these hazards in light of regional topography. The projected influence of climate change in terms of modification to storm intensity, frequency of occurrence, and sea level rise need better predictive models.

The assessment of damage potential is the key element of risk assessment exercise, but it poses a very difficult challenge due to a lack of a well-founded basis for evaluation. To date only very simplistic approaches have been utilized to assess damage potential. This area needs better quantitative procedures to evaluate damage potential in light of a host of factors ranging from imprecise nature of loading, variability in structural strength, and complexity of possible load paths. Ideally, the damage potential can be derived from a series of tests conducted under varying loads applied incrementally to obtain total collapse. In addition to the difficulty in the realization of a test of this scope for different classes of structures, it is economically prohibitive. Alternatively, reliability based mathematical modelling of load effects and resistance, and expert knowledge gleaned from the exposure of similar structures to extreme events offer algorithmic and inferential means of ascertaining damage potential. These areas need focused attention to develop techniques that provide realistic measures of damage to a wide range of constructed facilities. A multi-hazard approach should be considered, i.e., damage due to wind and storm surge/wave action. This effort should cross-cut similar activities in earthquake damage assessment. One should not only concentrate on structural damage, but also rain penetrating into building enclosures subsequent to minor damage of roofing, siding, or window which can result in staggering losses (e.g., wind-related damage in Hugo was nine times the flood losses).

There is a need for demonstration projects to illustrate the effectiveness of the risk assessment scheme and its utilization as a tool for planning adaptive measures, e.g., retrofitting and coastal management for communities at risk. Galveston and Charleston present ideal sites for the demonstration study in view of their vulnerability and recent storm histories.

Future utilization of GIS (Geographical Information Systems) for cataloging inventories of built facilities in different communities will promote use of the risk assessment tools not only in the U.S., but also overseas.

CONCLUDING REMARKS

Proactive hazard management strategies that involve assessment of risk and mitigating effects to reduce the impact of hazards rather than continuing the current practice of only responding to natural hazards by disaster relief efforts need to be developed. The impending threat of an increase in exposure to hazards by climate change places a growing need to assess the risks for constructed facilities in tropical storm-prone regions. This will serve as an efficient tool for better planning and tailoring adaptive measures, e.g., retrofitting, coastal management, and construction of dikes for communities at risk.

ACKNOWLEDGMENT

Financial assistance is provided in part by NSF Grant No. BCS-90-96274. The author would like to thank Dr. J. H. Golden from NOAA for providing many useful references.

REFERENCES

1. G. A. Berz. Global warming and the insurance industry, *Proceedings: The world at risk: Natural hazards and climate change*. American Institute of Physics, 1993, this issue.
2. R. A. Anthes and S. W. Chang. Response of the hurricane boundary layer to changes of sea surface temperature in a numerical model. *J. Atm. Sciences* **35** (1978).
3. K. A. Emanuel. The dependence of hurricane intensity on climate. *Nature* **326** (1987).
4. F. Giorgi and L. O. Mearns. Approaches to the simulation of regional climate changes: A review. *Rev. of Geophysics* **29**(2) (1991).
5. M. L. Brecque. Detecting climate change. *MOSAIC, NSF* **20**(4) (1989).
6. C. L. Keppenne and C. Nicolis. Global properties and local structure of the weather attractor over western Europe. *J. Atm. Sciences* **46**(15) (1989).
7. *Climate change: The IPCC Scientific Association WMP/UN environmental programme*. Cambridge University Press. 1990.
8. National Research Council. *Responding to changes in sea level: Engineering implications*. D.C. National Academy Press. 1987.
9. D. A. Moser and Z. Eugene Stakhiv. Risk-cost evaluation of coastal protection projects with sea level rise and climate change: Risk analysis and management of natural and man-made hazards. ASCE. 1989.
10. Climate change: Sciences, impacts and policy, Proc. Second World Climate Conf., WMO. Cambridge University Press. 1991.
11. F. Rijsberman. Potential costs of adapting to sea level rise in OECD countries. In *Responding to climate change: Selected economic issues*. OCDE Publications. 1991.
12. C. Cooper. A preliminary case for the existence of hurricane alleys in the Gulf of Mexico. OTC#6831, Proc. Offshore Tech. Conf. Houston, Texas. 1992.
13. O. H. Pilkey and W. J. Neal. Save beaches not buildings. *Issues in Science and Tech.* **8**(3) NAS/NAE/IM. 1992.
14. A. Kareem. Adaptability of the construction industry to predict stresses of climate changes. Symp. sum., The world at risk: Natural hazards and climate changes. MIT Center for Global Change Science. 1992.
15. S. T. Schneider. *Global warming*. Sierra Club Books. 1989.
16. Policy implications of greenhouse warming. Policy implications of greenhouse warming—Synthesis panel, NRC. National Academy Press. 1991.
17. Nat'l. Res. Council. *Estimating losses from future earthquakes*. National Academy Press, Washington, D.C. 1989.
18. P. N. Georgiou. Simulation of hurricane wind speeds. Proc. of Hurricane Alicia: One year later. (A. Kareem, Editor.) ASCE, New York. 1985.

SOCIETAL RESPONSES TO GLOBAL CLIMATE AND NATURAL HAZARDS

SOCIETAL RESPONSE TO GLOBAL CLIMATE CHANGE: PROSPECTS FOR NATURAL HAZARD REDUCTION

Joanne M. Nigg
Disaster Research Center
University of Delaware
Newark, Delaware 19716

ABSTRACT

This paper provides a sociological perspective on societal responses to disasters (extreme natural hazard events). A definition is provided for a "social," as opposed to a physical, disaster event. Several factors are identified which affect the ways societies adapt to hazard agents—societal resources, realistic expectations, vulnerability, and past experience. The ability of societies to anticipate future intensified or new disaster events under conditions of uncertainty is also addressed.

THE SOCIOLOGICAL QUESTION

The prospects for reducing the consequences of extreme natural hazard events—that is, of natural disasters—that may result from the global climate change phenomenon provide an interesting question for social scientists: How might communities and societies respond to unknown future hazards such as intensified droughts, riverine floods, tidal surges, hurricanes, and coastal water level rise?

Social scientists, particularly sociologists and social geographers, have investigated societal responses to natural and technological disaster agents for over 50 years[1] and have identified several factors associated with the types of hazard reduction techniques societies employ to decrease their exposure to potential danger and loss from *known* hazard agents. The prospect of a society's ability to anticipate and respond to *possible* future disaster events due to a changing, but still not characterized, physical environment may not be so different from the other natural hazards studies of decision making under conditions of uncertainty.

For example, recent scientifically credible earthquake predictions for the New Madrid Fault system in the Central United States forecast a potentially damaging earthquake somewhere in the region during the next 20–40 years; an event which has not been experienced in that part of the country in modern times. Many of the characteristics of such a forecast—a broad geographic area; a long time window covering decades; an unusually large event magnitude which has potentially severe social consequences; and scientific projections based on limited empirical data—are also present in global climate change scenarios. What, then, can be learned from past social science research on societal hazard reduction efforts under conditions of uncertainty?

Before addressing this question, it is important to begin with some common definition of what a "disaster" is and a general understanding of why social scientists have been concerned with this topic from a theoretical perspective.

THE DEFINITION OF DISASTER

Since the late 1940s, social scientists have studied disasters as a way of investigating issues of hazard perception and social organization. Disaster situations are the "natural laboratory" within which social scientists develop generalizations about human nature and the basic processes of social stability and social change.

A *true* disaster must be distinguished from a routine emergency or accident.[2] A true disaster threatens four vital functions of a society:

1. Its biological survival—that is, the ways a society provides subsistence and shelter for its citizens
2. The social order—its division of labor, cultural norms, authority structures, social roles and general social organization.
3. Its meanings—the definitions of reality and values shared by its members
4. The motivations that guide human and organizational behavior within its social systems.

In a true disaster, then, there is widespread, systemic disruption of social life due to human and physical losses that prevent the society from carrying out its essential functions. However, a true (or *social*) disaster is *not* synonymous with an extreme physical hazard event, unless that event substantially disrupts the social life of a major segment of the community or society.

Yet the physical characteristics of the disaster agent should not be overlooked in terms of their importance for disaster consequences. "Event variability"—that is, the physical characteristics of the hazard agent— will have consequences for the degree to which an extreme event will impact a society.[3] Some of the significant features of event variability include:

1. the predictability of the event's occurrence
2. the probability that the event could occur
3. the controllability of the agent
4. the type of agent (technological or natural)
5. the speed of onset
6. the scope of the impact (focused or diffused)
7. the destructive potential of the event.

In addition to the differences in the physical nature of the disaster agent, it must also be recognized that the occurrence of a natural hazard event will affect different social systems differently. Social systems differ in terms of both their vulnerability and in terms of their ability to adapt to hazard events.

SOCIETAL ADAPTATION

The popular notion, so often portrayed in "disaster movies,"[4] that societies disintegrate when confronted with natural disaster events and that individuals engage in irrational or antisocial behavior[5] has long been rejected by social scientists. Instead, when communities and societies are confronted with direct challenges to their continued existence, they tend to resolve these problems through collective approaches, adopting changes that allow them to rebound with greater resilience when faced with these problems again.

This tendency has been referred to as *adaptation*; that is, the development of options to help human and social systems adjust or adapt to new climate conditions or events.[6]

There are five general ways in which social systems (communities or societies) engage in adaptation with respect to extreme natural hazard events. They can:

1. Modify the hazard itself (Example—channelization and the construction of dams for flood hazards)
2. Prevent or limit impacts (Example—the use of building codes, disaster response planning, warning systems)
3. Move or avoid losses (Example—zoning or land use management)
4. Share the loss (Example—disaster relief assistance and insurance)
5. Bear the loss (that is, do nothing)

HAZARD EXPECTATIONS AND PAST EXPERIENCE

The degree of disaster-induced societal disruption is determined by several factors, but principal among these factors is the type of adaptation(s) a community or society has undertaken. Depending on the availability of societal resources (financial, human, and governmental), not all of these adaptation strategies are feasible for some countries to undertake.[7] Changing the uses of land in order to avoid the exposure of the population to an extreme event is almost impossible in areas that are already densely populated (for example, in Mexico City), densely developed (for example, Tokyo), or where people are dependent on the land for their subsistence (for example, in coastal Bangladesh).

Where resources are not so scarce, however, disaster-induced societal disruption is closely related to the extent to which a society or community has developed *realistic expectations* about, and preparations for, extreme events.

The effects of Hurricane Hugo on the lifeline infrastructures of two communities on the southeastern coast of the United States illustrates this last point.[8] When Hurricane Hugo made landfall at Charleston, South Carolina at midnight on September 22, 1989, the hurricane did significant damage to the electrical power and telephone distribution systems of that community. However, because the community had been tracking the developing storm for the preceding three days, had taken preparedness measures to deal with the post-hurricane problems, and had "hardened" their lifeline systems through structural mitigation measures in anticipation of such storms, the time it took to restore utility services was relatively quite short (18 days).

Charlotte, North Carolina, however, was another matter. Hugo passed over Charlotte four hours after hitting Charleston; and although Charlotte is 200 miles inland, it sustained winds of only slightly less intensity than those that struck Charleston. In contrast to Charleston, Charlotte was caught largely unprepared.

After the eye of the hurricane passed over Charleston, the National Weather Service called the "warning point"—a communications center for severe weather warnings—in

Mecklenburg County (where Charlotte is located) just after midnight to warn local officials of the approaching hurricane. Because of the early hour and because the storm had not been expected to come that far inland, delays in mobilizing the county's emergency management personnel and in contacting representatives of the local utility companies occurred. No preparatory activity took place—either on the part of the utility companies or by local government; neither had their infrastructural systems been "hardened" because of a lack of perception that hurricane winds could have such damage potential so far from the ocean. This lack of preparedness and mitigation action resulted in 98% of electricity customers losing power immediately throughout the metropolitan area. Because of widespread damage to the transmission and distribution systems and to a central power station, some areas in and around Charlotte were without power for up to six weeks.

In response to the same disaster event, these two communities had two very different reactions. Charleston adapted to the potential for such an event through both structural (the hardening of lifeline systems) and non-structural (disaster response planning and warning system development) measures; while Charlotte, seemingly, opted to "bear the loss."

Why would such a difference occur when these two communities are in the same region of the country, in adjacent states? The issue was *not* a difference in resource availability. Charlotte was, in fact, economically "healthier" than Charleston with a growing population, higher income families, and relatively newer building stock and infrastructural systems. The answer to this question principally lies in the two communities' different perceptions of the nature of the hazard and their expectations of loss. "Coastal" Charleston was known to be at risk from high intensity hurricanes and planned on ways to reduce the impacts of such storms; while "inland" Charlotte assumed that distance would protect the community from the damage and loss that accompany "coastal storms."

This example also illustrates a second point about adaptation strategies; namely, that societies develop protective measures on the basis of their *past experience*. Social systems (as well as individuals) usually prepare to deal with problems of the last worst disaster that effected them. In other words, their *future* hazard reduction strategies are developed to handle *past* problems that were experienced.

Although the implementation of "lessons learned" from past disasters is beneficial and does provide communities some protection from future disaster events, it often does not allow them to anticipate the types of problems that might arise from larger, more catastrophic events. Under what conditions, then, are social systems likely to consider adaptation strategies *prior* to a "worst-case" event?

SCIENTIFIC UNCERTAINTY: A QUESTION OF WHEN TO ACT

With respect to the societal implications of global climate change research, the situation currently faced by atmospheric scientists is, in fact, quite similar to that of geoscientists involved in earthquake prediction. The science upon which potential, catastrophic events or physical changes are based is in its infancy; the models are crude; and the available empirical data are quite limited.[9] But the potential consequences can not, or should not, be ignored.

In the global climate change field, there appears to be a high degree of scientific consensus that global temperatures are rising; however, the consequences of that temperature increase are still being debated.

The challenge, it seems, is "when do we know enough to begin to do something," despite the scientific uncertainty that exists. I would like to suggest that merely by raising the possible linkages between climate change and increased disaster consequences, the scientific community has begun "to do something."

It should be remembered that knowledge dissemination and utilization is *not* an outcome; rather, it is a process whereby research findings and conclusions come to change the way the general public, governmental decision makers, and emergency management practitioners come to define a problem and its many solutions.[10] From the "enlightenment" model of knowledge utilization,[11] then, research doesn't directly solve problems. Instead, it provides an intellectual set of concepts, propositions, orientations, and generalizations that can be used by those who are distant from the research process itself to define their problems and evaluate the options for coping with them. The imagery is that scientific generalizations "percolate" through informed publics to shape ways in which they come to think about the issue.

This symposium provided a first step in the process whereby scientific knowledge is beginning to define a set of environmental problems that are related to global climate change, although the implications of those changes may still be somewhat fuzzy. The significant function of such an effort is its *sensitization* of different potential user communities—political leaders, emergency managers, engineers, water resource managers, land use planners, to name a few—that would allow them to consider how the possibility of change (for example, of excessive or sparse rainfall) would effect current structural mitigation and emergency management practices.

However, it must also be noted that this is *only* a first step.[12,13] In order to keep this issue on the public agenda, the physical science communities must continue to raise these potential consequences in appropriate forums, not just among their scientific peers. Continuing interaction with potential user communities is imperative in order to provide an enlightened policy environment within which action can take place. Although there are a number of factors that constrain the development of policy, policies to address these changing environmental conditions—that could result in disaster events—will not occur without this continuing dialogue that stresses the linkages between the science of global climate change and disaster possibilities.

REFERENCES

1. E.L. Quarantelli. 1987. *International Journal of Mass Emergencies and Disasters* **5**:285–310.
2. C. Fritz. 1961. In R. K. Merton and R. A. Nisbet, Eds., *Contemporary social problems*, New York: Harcourt, Brace, and World, pp. 651–694.
3. R.R. Dynes. 1970. *Organized behavior in disaster.* Lexington, Mass.: Heath-Lexington Books.
4. E.L. Quarantelli.. 1985. *Communications* **11**:31–44.

5. J.M. Nigg. 1989. Unravelling the myths: How people really respond to natural hazard threats and events. Paper presented at the annual meetings of the American Association for the Advancement of Science, San Francisco, January 14, 1989.
6. National Academy of Sciences, National Research Council. 1991. *Policy implications of greenhouse warming*. Washington, D.C.: National Academy Press.
7. W.E. Reibsame. 1990. In Alcira Kreimer and Mohan Mumasinghe, Eds., *Managing natural disasters and the environment*, Washington: Resources for the Future, pp. 9–16
8. J.M. Nigg. 1990. Lifeline disruption in two communities: Charleston, South Carolina and Charlotte, North Carolina. Paper presented at the University of Puerto Rico Conference, Six Months After Hurricane Hugo, Mayaguez, March 12–14, 1990.
9. R. Krasnow. 1986. *Policy aspects of climate forecasting*. Washington: Resources for the Future.
10. J.M. Nigg. 1988. In Walter W. Hays, Ed. *A review of earthquake research applications in the National Earthquake Hazards Reduction Program 1977–1987*, Reston, Va.: USGS pp. 13–33.
11. C.H. Weiss. 1978. In L. E. Lynn, Jr., Ed. *Knowledge and policy: The uncertain connection*, Washington: National Academy Press, pp. 23–81.
12. J.W. Jacobs and W. E. Reibsame. 1989. The greenhouse effect: Recent research and some implications for water resource management. Boulder: University of Colorado.
13. J.G. Titus. 1987. *Greenhouse effect, sea level rise and coastal wetlands*. Washington: EPA.

SOCIETAL RESPONSE TO CHRONIC ENVIRONMENTAL CHANGE:
THE ROLE OF EVOLVING SCIENTIFIC AND TECHNICAL
INFORMATION IN STATE COASTAL EROSION MANAGEMENT

Mark Meo
Thomas E. James
Science and Public Policy Program
University of Oklahoma, Norman, OK 73019

Robert E. Deyle
Department of Urban and Regional Planning
Florida State University, Tallahassee, FL 32306

ABSTRACT

Natural hazards, such as acute storm events, often exacerbate chronic change processes and heighten public awareness of overt and latent risks to society. This paper reports the results of a study of the development of coastal erosion management policies from 1960 to 1990 in Florida, Massachusetts, and North Carolina, all of which have been subject to marked chronic and acute environmental changes. The case studies were guided by a conceptual framework that links acute and chronic environmental changes with state and federal policy initiatives as well as the scientific community. We traced the evolution of policies and programs and the policy innovation process in these states that were designed to mitigate the risks associated with coastal erosion, and documented the role that scientific and technical information played. From information drawn from published and unpublished literature as well as personal interviews, the case studies reveal a wide variation in policy responses to coastal erosion phenomena and exhibit differing degrees of program success. Although the case studies preclude broad generalizations, they illustrate clearly the importance of institutions to enable policy entrepreneurs and other key individuals a base for acquiring and using scientific and technical information for policy development and implementation.

INTRODUCTION

Critical scientific uncertainties associated with global environmental change in general and climate change in particular avail little guidance for policy makers seeking to design cost-effective responses to the range of societal risks that might ensue.[1,2] Recently, the Committee on Earth and Environmental Sciences initiated a comprehensive research program to improve understanding of global change phenomena and provide a credible source of knowledge to guide the formulation of appropriate public policies.[3] An assumption implicit to this task is that better scientific understanding of global change processes will enhance society's ability to reduce associated risks.[4,5] Thus, a salient aspect of formulating societal responses to present and projected global changes is appraising the degree to which improved understanding of severe or acute environmental events and slower, more chronic long-term changes has influenced environmental policy in the past.[6] In this respect, prior experiences with state coastal erosion management can be studied to glean useful insights about the role of scientific and technical information (STI) in the policy change process.

Acute events in the context of policymaking are inherently political phenomena; the definition of what constitutes an acute event is in the eye of the policy maker. Nonetheless, it is useful to differentiate between acute events that are linked to phenomena in the physical environment and those that are purely political in origin. A catastrophic storm may be perceived by policymakers as an acute event because it generates expectations of prompt and effective response by the electorate of the affected jurisdiction. A purely political acute event arises from action taken by one or more stakeholders in the policymaking arena, e.g., members of the public, elected officials, etc.

Acute events can influence policy innovation in one of two ways. First, an acute event can be viewed by policymakers as a political crisis; an unanticipated event that results in a crisis reaction and what Polsby[7] has characterized as acute policy innovation. On the other hand, an acute event may provide a "policy window"[8] for policy entrepreneurs to advance an incubated policy innovation.[7] In this case, the policy entrepreneur may play a role in defining the event as a problem or political crisis, but has a well-developed solution at hand. This is the process we have called "strategic policy innovation."[9] The acute event that precipitates an acute policy innovation or provides the policy window for a strategic policy innovation can be purely political or can originate in the physical environment.

We have hypothesized that acute events originating in the physical environment might have provided the impetus for policy innovations dealing with both acute erosion and chronic erosion of ocean front shorelines. Acute erosion is what results from severe storms and tends to be manifest in dramatic changes in beach profile, loss of dunes, and accompanying overwash, flooding, threats to public safety, and damage to property. Chronic erosion, on the other hand, is the cumulative result of the dynamic processes that operate on sedimentary shorelines over the long term. The impacts are typically less dramatic, although the ultimate effect on property can still be substantial.

We have also hypothesized that access to and use of STI about erosion processes would influence policy innovations concerning erosion management. Extension of Polsby's dichotomy between acute and incubated policy innovation would imply that STI should be most influential in cases of incubated or strategic policy innovation and should be less likely to influence acute policy innovations.

In our research, detailed case studies of three coastal states, Florida, Massachusetts, and North Carolina, were carried out through in-person and telephone interviews of knowledgeable individuals and reviews of the pertinent literature. The state case studies were identified based on the results of a questionnaire sent to 44 individuals in Atlantic and Gulf coast states who are expert in either coastal science or coastal resource management. In this portion of the research, states with a generally proactive stance toward coastal erosion management were selected in favor of those more inclined to be reactive or inactive. A total of 44 personal interviews were conducted during week-long visits to each state. In addition, numerous telephone interviews were conducted with other knowledgeable individuals to gather information about or clarify specific points of interest. Prior to visiting the states, background papers were developed that chronicled the evolution of federal coastal management policy; discussed the range of state and local coastal management initiatives for mitigation of coastal erosion; and provided information about each state concerning the geophysical setting (storm events, geological structure, erosion trends), socio-economic description of the coastal areas, and the historical development of the policy setting and initiatives during the 1960 to 1990 interval. The state case studies were conducted to develop in-depth information to explore the conceptual model of strategic policy change and hypotheses discussed above (see Figure 1). The model includes linkages between acute events and chronic change, the expert community, and the federal government with the state and local government institutional settings. The results of the case studies are intended to help understand the operation of the relationships and the relative importance of the various factors (e.g., acute events, availability of STI and the expert community) for influencing the development of policy initiatives.

The three states included in this paper share some common traits, although the relative importance of the institutional elements external to the state and local policy areas varied across the states. For example, all of the states had formal or informal communities of experts with which state agencies or individuals within agencies could communicate in order to initiate and sustain policy innovations that addressed coastal erosion. The expert communities tended to be institutionally distinct from state

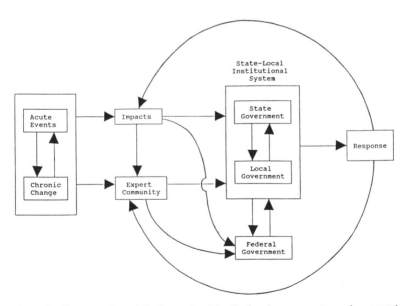

Figure 1. Conceptual model of state-local institutional response to environmental change.

government, but individuals often became closely involved with state programs and were important to their success. Although their total numbers were usually small, the ability of experts to communicate research findings and explain their implications was valued highly by state agency staff and the public. The most effective communication of STI was often interpersonal, and scientific credibility and impartiality were essential.

Policy entrepreneurs were clearly present in the three proactive states and generally sought to use acute events as opportunities for motivating strategic policy changes in coastal erosion management. Often acute events were extremely important to the development of policy windows and the political will necessary for the initiation of policy innovations. Not all acute events, however, were geo-physical. In some instances they were related to social or political changes and activities. Proactive states also tended to use federal programs, either directly or indirectly, as opportunities to promote coastal erosion management. However, more often the states saw themselves as setting examples in coastal barrier management for the federal government to emulate. Although federal programs do not appear to have had much influence on state policy responses, the results of federal studies do appear to have been important sources of STI for the states. Support for coastal research by the federal government was indispensable for acquiring a data base, building scientific understanding, conducting demonstrations, and generating public awareness about erosion.

In general, the recognition of the impacts of chronic erosion in concert with acute events, the evolution of scientific understanding, and credible representatives from the expert community were important factors influencing the process and nature of state institutional response to coastal erosion. In addition, each state had only a few individuals within state government who were instrumental in taking a leadership position to take advantage of the opportunities presented by the external influences. Generally, more than one factor was present that interacted to create the environment within which policy innovation could take place. Which factors were important over time and how they interacted varied somewhat across the states. Following are very brief discussions of the findings from the individual states.

FLORIDA

Florida's Coastal Zone Management Program received federal approval in 1984 and is based on 27 state laws and their implementing regulations, which are administered by 16 agencies.[10] The Department of Community Affairs is responsible for overall administration of the program. Key coastal management programs, particularly those addressing coastal erosion, also are administered by the Departments of Environmental Regulation and Natural Resources. The Governor and the Cabinet are the ultimate authority for issues related to coastal barrier island erosion.

The evolution of coastal erosion management policies was heavily influenced by key individuals in both the political and science and engineering communities, the availability of STI, chronic and acute events, and the activities of other states. Each of these influences may have been necessary but not sufficient factors in guiding the development of Florida's coastal erosion policies. In many instances, two or more of these factors combined to create the environment within which policy debate could take place or actually lead to the development of policy initiatives.

The mid-1950s marked the beginning of a concerted effort to develop an organized program to address beach erosion at the state level.[11] The impetus for this effort was largely outside state government and is an example of the confluence of multiple influencing factors. In 1955, the Florida legislature provided a biennial budget of $25,000 to the Engineering and Industrial Experiment Station of the Department of Engineering Mechanics at the University of Florida to conduct a study to provide recommendations on the control of beach erosion. The same year, Per Bruun, an expert on coastal processes and model laboratory study procedures, arrived at the University of Florida from Denmark. In 1957, Bruun and his colleagues released a report entitled "Studies and Recommendations for the Control of Beach Erosion in Florida."[12]

Recognizing the growing interest in Florida's coastal problems related to chronic erosion (for example, the erosion of Miami's beaches and the potential economic consequences), Bruun and others at the University seized the opportunity to promote the development of a coastal engineering laboratory at the University and the state's capacity to manage the coasts. In 1957, Bruun organized a meeting of individuals from around the state and the group became the Florida Shore and Beach Preservation Society. The Society was instrumental in helping to develop legislation that provided funding to create the Coastal Engineering Laboratory at the University of Florida and which authorized the Government and Cabinet to establish and maintain a Department of Beach and Shore Erosion, the conceptual ancestor to the current Division of Beaches and Shores in the Department of Natural Resources.

STI and experts from the science and engineering communities have played important roles throughout the evolution of Florida's policies. For example, in 1963 the legislature provided money to the Division of Beaches and Shores to conduct research and restoration projects.[13] Coastal engineers at the University of Florida (Bruun, James Purpura, and T. Y. Chiu) became the main source of scientific data on beach dynamics for the Division. Chiu, Purpura, and Robert Dean, also from the University of Florida, developed the methodology used to establish the Coastal Construction Control Lines. Location of the line is based on storm surge, erosion, existing topography, the dune and vegetation line, and upland development and is intended to predict the zone of severe impact of the 100-year storm event. Construction and excavation seaward of the line requires a permit from the state. Also, Robert Dean's studies in the early 1980s concluded that 80 percent of the erosion on the east coast of Florida is due to navigation inlets cut through barrier islands. This led to a policy stating that improved inlets must provide for by-pass of 100 percent of the sand.[14]

Acute storm events have been very influential in the development of erosion management policies. In 1971, a storm with unusually high tides caused damage to buildings in the Panama City area. This reinforced the need to move from the politically based 50-foot setback lines established in 1970 to the more science-based Coastal Construction Control Line. The combination of an acute event and the influence of another state was evident in 1984. In that year, Florida's Governor Graham and Debbie Flack, Director of the Division of Beaches and Shores, surveyed the damage from a hurricane that hit North Carolina and noticed the effect of that state's 30-year erosion zone policy. This policy prohibits construction that would be underwater in 30 years due to erosion. They returned home determined to develop such a policy for Florida. Later that year, a Thanksgiving Day storm ravaged the east coast of Florida, tearing down piers and washing away hundreds of feet of beaches. Flack took legislators to view the damage and told them about North Carolina's 30-year erosion zone policy and the administration's claim that the policy did not have a negative economic impact on the state. North Carolina's experience and the Thanksgiving Day storm were important factors leading to the initiation of a task force to develop recommendations for beach restoration and nourishment programs and the inclusion of the 30-year erosion line as part of the 1985 Omnibus Growth Management Act.[15]

Federal coastal management programs are not perceived by Florida interviewees to have had much direct impact on the evolution of the state's policies to manage coastal erosion. Data from federal studies are used and the Corps of Engineers are recognized by local officials as playing an important role at their level. However, for the most part, state officials believe that Florida is ahead of the federal initiatives, and they do not believe that federal programs have directly influenced Florida's policies and programs.

MASSACHUSETTS

Massachusetts became actively involved in state coastal erosion policy in 1978 when its coastal management program was formally approved by NOAA.[16] At that time, the state declared the use of nonstructural measures to be a policy goal. With an explicit networking arrangement for interagency cooperation, the Massachusetts Coastal Zone Management (MCZM) program addressed coastal erosion policy by building specific prohibitions into the Wetlands Protection Act which are enforced by the Department of Environmental Protection (DEP). The state, which has a tradition of strong local government and strong central leadership, has relied on local conservation commissions with technical assistance provided by MCZM and DEP to manage development on coastal barriers.

The scientific basis for the state erosion policy was written by an MCZM task force that included reputable scientists who understood well the erosion-prone character of the coast. With the earlier creation of the Cape Cod National Seashore in 1961 and the research-oriented presence of the National Park Service, the state was able to benefit not only from research on coastal erosion, which was directly helpful to those towns near the national seashore, but also through its public education efforts and cooperative university research program which were highly regarded.

Acute storm events as well as mounting public distress over growth-related impacts figured prominently in the policy change process. After the devastating blizzard of 1978, for example, MCZM sought to acquire damaged coastal structures with National Flood Insurance Program (NFIP) funds and eventually succeeded over local opposition when federal funds became available. Further, MCZM leaders convinced the governor to issue an executive order in 1980 (EO 181) that prohibited the use of state funds and federal grants on developed and undeveloped coastal barriers.[17] Over the years, the state has strengthened its position regarding erosion control structures, holding to it even when coastal properties in Chatham were lost to the sea following a 1987 Nor'easter that permanently breached the protective barrier beach.

For their part, acute events of a political nature have been important as well. In 1974, faced with mounting public concern over growth-induced environmental impacts, the state created the innovative Martha's Vineyard Commission (MVC) to regulate land use on the island and to avoid federal initiatives similar to the National Seashore concept.[18] The MVC was modeled on the American Law Institute's model land use code which addressed large scale developments and critical environmental areas. The coastline was one of the first Areas of Critical Planning Concern designated by the MVC. On Cape Cod, concern about environmental quality, particularly groundwater quality during the 1980s, led to increased reliance on staff for technical expertise. A policy window, which opened up when a moratorium on growth was approved by public referendum, provided for the creation of the Cape Cod Commission, a strategic innovation that was built on the earlier success of the MVC, and that provided an organizational structure that better enables technical knowledge to be applied more readily to erosion management.[19]

Scientific and technical expertise, available at area universities and research centers, has been used frequently for informal counsel and through formal task forces, as occurred in 1978 and 1990, to set erosion management and development guidelines. In recent years, MCZM staff have tapped that expertise to create state maps that display long-term erosion rates for the entire coastline, to examine the implications of sea-level rise for coastal regulations, and to write local town by-laws that promote natural sand sharing processes. For the most part, federal expertise has been used indirectly more than directly. Although the National Park Service (NPS) supported frontier research in the Cape Cod National Seashore, almost no state or local persons acknowledged its importance to state policy development.

NORTH CAROLINA

North Carolina's barrier-island erosion management policies have evolved largely under its Coastal Area Management Act (CAMA) of 1974.[20] The Division of Coastal Management (DCM), within the Department of Environment, Health and Natural Resources, administers CAMA and provides staff support to the Coastal Resources Commission (CRC—the primary policymaking and regulatory body for CAMA) and to the Coastal Resources Advisory Commission (CRAC—a citizens' advisory body to the CRC). Under CAMA, local land use planning in compliance with CAMA guidelines is mandatory, and environmentally sensitive lands and waters are regulated directly by the state through the Areas of Environmental Concern (AEC) designation and permitting process.

The major erosion management policies of CAMA are ocean erodible and inlet hazard AEC regulation, annual erosion rate-based setbacks in those AECs, and a prohibition of shoreline hardening erosion control structures. These policies can all be defined as strategic innovations, implemented after thorough searches for technically and legally defensible policy solutions. When implementation has been linked to an acute event, the event has been of a political, rather than a physical nature.

The evolution of STI about coastal barrier systems has been a major force behind development of erosion management in North Carolina. As understanding of coastal geology evolved in the 1970s, for instance, the technical staff of DCM realized that the emphasis of erosion policies should be not on dunes but on chronic and storm-induced erosion processes and took on the task of convincing the CRC and CRAC of this view. At the same time, the staff realized that theoretical understanding of long-term erosion was less advanced than that of acute erosion and subsequently contracted with researchers at North Carolina State University to study long-term erosion.[21] Thus, when ocean erodible AECs were initially designated, in 1976, their boundary was defined by the 25-year storm surge, because that was the defensible methodology available. Later, in 1979, when AECs were revised, the methodologies were available to define the ocean erodible AEC based on average annual long-term erosion rates and on the 100-year storm erosion hazard. A setback based on annual erosion rates was also implemented.[22]

The ban on hardening structures is an example of policy implementation awaiting a policy window. The notion of a prohibition on hard structures had existed for some time among staff and some CRC members. When several hotels essential to the tax base in Dare County were threatened by erosion and the county considered no longer enforcing CAMA regulations, policy makers responded by forming an erosion task force, which they then used as a vehicle to coalesce support for greater restrictions on hard structures.

The information network among DCM staff, the CRC, and the CRAC is centralized, and, at least until recently, the program has been characterized by a lack of polarization, a high level of understanding of coastal processes, and a general consensus on goals for coastal development. Principal sources of STI were the DCM staff and members of the CRC and CRAC. DCM staff made a concerted effort to educate CRC members about coastal geology, engineering, and related phenomena through staff reports and presentations.

Social impacts have also been considered in the policymaking process. For instance, a study of the economic impact of the rate-based setbacks was part of the consideration process for that policy, and the detrimental effect of hard structures on the public beach and use thereof was instrumental in motivating the ban on such structures.

According to interviewees, federal policies and programs have had little impact on the development of state policy.[23] The imminent passage of the CZMA, however, was part of the political climate that finally saw passage of CAMA. Surprisingly, the Coastal Barrier Resources Act, the NFIP, and NPS and Army Corps policies appear to have had little explicit influence on policies in North Carolina. State policy decisions regarding coastal barriers in North Carolina have obviously influenced local governments, since the state controls permitting in AECs and mandates local planning consistent with state-promulgated guidelines.

Our observations from North Carolina suggest that both types of acute events have stimulated both kinds of policy innovation in the arena of erosion management policy. We have also identified examples of strategic policy innovation that are not linked to acute events of either kind. Severe storms have been associated primarily with policy innovations at the local level, some of which are evidently acute innovations while others may qualify as strategic. Erosion management policy innovations at the state level have been predominantly strategic and are linked more to political acute events than to severe storms or other acute events originating in the physical environment, the one exception being the Sea Ranch Motel which was sited in an area with an unusually high range of erosion and accretion. STI appears to have been more influential in the state policy innovations than in those initiated by local governments, but our research on the local policies is not yet complete.

CONCLUSIONS

Although the case studies provide a limited basis for making generalizations, they illustrate clearly the importance of institutions to enable policy entrepreneurs and other key individuals a base for acquiring and using STI for policy development and implementation. Should the potential acute and chronic impacts induced by global environmental change impinge upon society in a distributed and nonuniform manner, care should be taken to enable affected institutions to respond in a timely manner. If the current investment in global change research is to be used widely for policy and planning purposes, greater attention should be afforded the political, social, and related factors that help to shape policy change in environmental management.

ACKNOWLEDGMENTS

This material is based upon work supported by the National Science Foundation under Grant No. SES 90-12445. The Government has certain rights in this material. We thank D. G. Aubrey, P. Dornbusch, and N. Hanks for their assistance.

REFERENCES

1. J. H. Ausubel. 1991. *Am. Sci.* **79**:210.
2. Committee on Science, Engineering, and Public Policy. 1991. *Policy implications of greenhouse warming.* National Academy Press, Wash., D.C.
3. Committee on Earth and Environmental Sciences. 1991. *Our changing planet: The FY 1992 U.S. global change research program.* Office of Science and Technology Policy, Wash., D.C.
4. E. S. Rubin, L. B. Lave, M. G. Morgan. 1992. *Iss. Sci. Tech.* **8**(2):47.
5. G. M. Hidy, S. C. Peck. 1991. *J. Air Waste Manage. Assoc.* **41**:1570.
6. P. C. Stern, O. R. Young, D. Druckman, eds. 1992. *Global environmental change: Understanding the human dimension.* National Academy Press, Wash., D.C.
7. N. W. Polsby. 1984. *Political innovation in America.* Yale University Press, New Haven, Conn.,
8. J. W. Kingdon. 1984. *Agendas, alternatives, and public policies.* Scott, Foresman and Co., Glenview, Ill.
9. R. E. Deyle, M. Meo, L. A. Wilson. 1989. *Am. Wat. Res. Assoc.*
10. U.S. Department of Commerce. 1981. Final environmental impact statement of the proposed coastal management program for the State of Florida. Wash., D.C.
11. J. H. Balsillie. 1991. *Florida's history of beach and coast preservation.* Nat. Res. Dept., Tallahassee, Fla.
12. P. Bruun et al. 1957. *Studies and recommendations for the control of beach erosion in Florida.* Univ. Fla., Gainesville, Fla.

13. J. H. Balsillie. 1991. *Florida's history of beach and coast preservation*. Nat. Res. Dept., Tallahassee, Fla.
14. H. Bean. 1991. Pers. comm.
15. D. Flack. 1991. Pers. comm.
16. U.S. Department of Commerce. 1978. Final environmental impact statement of the proposed coastal management program for the State of Massachusetts. Wash., D.C.
17. L. Smith. 1991. Pers. comm.
18. L. Susskind, ed. 1975. *The land use controversy in Massachusetts*. MIT Press, Cambridge, Mass.
19. P. Herr. 1991. Pers. comm.
20. J. M. DeGrove. 1984. *Land growth and politics*. Planners Press, Wash., D.C.
21. A. Cooper. 1991. Pers. comm.
22. Coastal Resources Commission. 1979. Raleigh, N.Car.
23. S. Benton. 1991. Pers. comm.

A CONSORTIUM APPROACH FOR DISASTER RELIEF AND TECHNOLOGY RESEARCH AND DEVELOPMENT: FIRE STATION EARTH

Douglas C. Ling*
Massachusetts Institute of Technology, Cambridge, Mass. 02139

ABSTRACT

A new paradigm is proposed for alleviating the chronic problem of inadequate response to natural and man-made disasters. Fundamental flaws and weaknesses in the current disaster mitigation system point to the need for an international consortium involving governments, academia, industry, and businesses. Recent changes in social and political framework offer a unique opportunity to rethink and reform the existing disaster response mechanism. Benefits of a collaborative consortium approach may include commercial incentives, improved cost effectiveness, coherence in research and development efforts, conduciveness for long-term planning, and improved deployment of technology for disaster mitigation.

INTRODUCTION

It is estimated that natural disasters have claimed over three million victims in the past two decades alone. Human suffering and strain on governments wrought by disasters can be unnecessarily prolonged and magnified due to delays in disaster mitigation process. International relief efforts are often compromised due to political haggling, logistics, and management and communication breakdowns. Lack of local preparedness leads to an ad hoc, low-tech, and fatally slow path to recovery.

Several major disasters in the early '70s (e.g., the Peruvian earthquake) exposed the weakness of the world's ability to respond to major disasters. Subsequent reviews culminated in several major disaster studies by the National Academy of Science[1,2]. While specific recommendations and criticisms were not lacking, agreement and concrete follow-up action were. The early '80s saw sporadic activities in disaster research, while disaster kept on its usually menacing pace.

Based on recommendations in a 1987 National Research Council report, *Confronting Natural Disasters*[3], the United Nations (UN) finally moved in 1989 to proclaim the 1990s the "International Decade of Natural Disaster Reduction" (IDNDR or the "Decade"). The office of the Decade, based in Geneva, is charged with the responsibility to guide, facilitate and coordinate global dialogues and national programs to design the framework for eventual reforms.

While much attention is being paid to natural disasters, more preventable types of disaster—environmental and technological or industrial disasters—are being ignored. Unchecked economic activities, and our careless and callous mentality against our living environment, have increased dramatically the threat of large-scale environmental disaster. Nuclear accidents the magnitude of Chernobyl (1986), severe industrial

---

*Current address: MIT Branch P.O. Box 156, Cambridge, MA 02139-0902

accidents as in Bophal, India (1984), and environmental disasters the scale of the Exxon Valdez oil spill (1990) should have served ample warning to the global community. The threat of man-made disaster and even more capricious response system present another major challenge.

In May, 1991, Bangladesh was hit by cyclones packing 145-mph wind and 20-foot waves. The afflicted area had over ten million inhabitants. Eight days after the initial hit an estimated of 125,000 were killed and 10 million more were left homeless; local response was still limited to 141 tons of air-dropped relief goods. An estimated forty percent of the material sank in flood waters. While Oxfam and other private-organization personnel were on the scene quickly, they had few resources at their disposal. Two hundred Japanese-donated surface vessels were reported by government officials as "misplaced." French-donated medical supplies were allowed to sit idle for days at Dhaka airport for lack of customs documentation. Nearly two weeks had passed before a U.S. navy amphibious group, led by the assault vessel Tarawa, en route from the Persian Gulf, arrived at the area to relieve victims stranded on remote islands in the Bay of Bengal. In fact, they were the first relief group to land on those embattled islands. The global disaster response community, only recently taxed to the limit by events in the Persian Gulf and in Kurdistan, was slow to provide adequate relief.

The above description is not unique. Millions of dollars are donated and spent every time there is a major disaster, only to have the same calamity, same destruction, same delays, repeated again and again. There must be a better way.

THE FRAMEWORK

Disaster studies have for many years recommended reform of the existing disaster response and relief mechanism. A window of opportunity is opening with the fast-changing global political composition:

1) IDNDR is designed to increase global activities in disaster research and resource allocation. Although the outline for the Decade is sparse, it leaves much room for innovation, flexibility, and creative planning and organizing for a new global cooperative projects. The initiative of technology cooperation must now be taken back into the hands of the private sector, where better efficiency and richer experiences can be taken advantage of.

2) Breakdown of communism and the end of Cold War means a potential new world order of cooperation. A de-militarized era forces us to rethink the role for the military, and the future of the surplus research capacity hitherto dedicated to military technologies. A world more receptive to trans-national cooperation allows us to redefine the role man and technology can play in mitigating natural and man-made disasters.

3) In November 1991, the UN General Assembly adopted a resolution to appoint a high-level humanitarian aid coordinator with authority to deal with governments that deny assistance to suffering people. The agreement was reached after months of negotiation between the industrialized countries (the "North")—the U.S., Canada, and the European Community (EC)—and the less developed countries

(LDCs, or the "South"). It marks another small but significant step toward establishing a right of humanitarian intervention in international law that would empower relief organizations to assist the afflicted wherever they are. Such a convention has been called for recently by Congressman Tony Hall[4] (D-Ohio). Establishment of a "common responsibility of all people and government to provide protection and relief to the victims of natural disasters" that makes "the denial or delay of needed protection/relief a violation of the basic human right to life" will open up new channels of operation for the disaster response mechanism.
4) Cooperative effort in global change science research provides a good model of large-scale collaboration among scientists, industry, commerce, and academia. The Symposium, "World at Risk: Natural Hazards and Climate Change," sponsored by the Center for Global Change Science at Massachusetts Institute of Technology, is a demonstration of receptivity of the scientific community to multi-disciplinary approach to the problem of disaster.
5) Public outcry invoked by vivid media coverage of calamities can lead to, at least in spurts, support for relief programs. Television imageries spectacular as those from the Bangladesh cyclone disaster, the Kuwaiti oilwell fires, or the Ethiopian famine, will always command public attention. As the world becomes more and more linked electronically, and information transfer becomes almost instantaneous in most parts of the world, the public cannot help but become increasingly intolerant of the delays in disaster relief.

CHARACTERISTICS OF CURRENT DISASTER RESPONSE SYSTEM

Existing disaster mitigation mechanisms have largely been centered around the United Nations, namely UN Disaster Relief Organization (UNDRO). Chartered in 1971, UNDRO has been given the mandate to coordinate international disaster relief operations. Schmitz[5] provided a comprehensive inventory of the UN disaster response capabilities and remained critical and skeptical about the long-term usefulness of UNDRO. He found that internally, UNDRO felt hand-tied with inadequate funding; while external opinions criticize its remoteness, over-centralized operations, and lack of management or technological expertise.

Cuny[6] also doubts the need for a central coordinating entity (as an extension of UNDRO), but he conceded that existing response system is unsatisfactory whether it is based on accountability (i.e. national government must provide indigenous solution), humanitarian considerations, social-economics, or developmental policies. He challenges the disaster community to shift from disaster response to prevention and mitigation; abandon the current short-term "program approach"; and develop new institutional forms for disaster intervention.

Military units remain an effective intervening force. Trained soldiers are well disciplined to provide orderly distribution of aid in chaotic post-disaster environment[7]. Military communication, transportation, and medical equipment are well suited for rugged field use and can be deployed rapidly. In an operation where expediency can mean life or death to disaster victims, the military is the only quick-strike response available. The Japanese Ground Self-Defence Force, the Swedish Standby Force, and the Swiss Technical Unit are some of the military-based disaster expert-teams which are fully equipped for disasters.

However, the appearance of foreign military forces, even non-combat units, is not easily palatable to many national governments. Lengthy negotiations, protocol exchange, and command and control issues can mean loss in lives. Moreover, the high cost of military-type intervention would only allow quick-fix responses that leave little long-term effect.

Government agencies like US Agency for International Development (USAID) and its Office for Foreign Disaster Assistance (OFDA) provide much funding for disaster relief and economic development annually. However, bureaucracy and political agenda often limit the return on the spending or scope of their involvements.

Private non-governmental organizations (NGOs) like Oxfam and Interaction have been indispensable in bridging the lapses in official disaster relief operations. Their resolve and efficiency often put them first on-site (as in the case of Bangladesh). However, without consistent, long-term funding for their own development, these NGOs are often limited to stop-gap operations. Some are criticized for being more responsive to donors' rather than their constituents' needs. Private-sector donations, for example, medical supplies by major drug companies, are often sporadic and dependent on the public-relations value available.

The characteristics of the existing disaster response system can be summarized in Table I. While each institution can make some positive impacts on disasters, the negative constraints hinder full utilization of its capability. One may consider the

Table I. Comparison of existing disaster response systems

	Advantage	Disadvantage
United Nations	impoving image trans-national	inter-agency competition bureaucracy
Military	rapid response inject discipline and order well-equipped	short-term relief top-down management political barriers sovereignty issues
Government agency	commands national resources tie-in with long-term development	political agenda/constraint request-response mode
Non-governmental organizations	seek indigenous solution highly appropriate	resources limited lack of sustaining support inter-group competition more responsive to donors techno-phobic
Private sector	motivated	random

advantages and summarize the final desirable traits for a global disaster mitigation mechanism as: "a financially stable, well coordinated, multi-national, multi-disciplinary, rapid deployment force armed with appropriate technologies and resources that can catalyze long-term, as well as indigenous, solutions."

Kent[8] further suggests that one of the great impediment to disaster relief has been the lack of definition and recognition of the disaster professionals. He focuses on the human dimension of disaster operations and concludes that the breakdown in continual professional development for disaster experts is partially to be blamed for lack of improvement, over time, of the system.

Stephenson[9] recounted one experienced disaster worker's comment on the current state of disaster operations: "... one would never run a MacDonalds franchise that way!" Standard tools of management like critical path analysis, budgeting, competitive and cost-effective, analysis, or business planning are not commonly found utilized in disaster organizations. A fresh business and management approach to the disaster problem may offer new insights and justification to some reform ideas. Privatizing allows for more competitive structuring of the response system and better allocation of funding.

ROLE OF TECHNOLOGY IN DISASTER MITIGATION

Disaster research, for example, those conducted at the Disaster Research Center at University of Delaware, has been directed mainly at organizational and sociological aspects. Application of technology in disasters (in particular, natural disasters) has sometimes been described as "overkill" and "inappropriate." It is, however, difficult to argue that most disasters will not benefit from some form of technology. A more accurate criticism is that poorly planned and coordinated deployment of technology diverts and wastes resources much needed elsewhere. Mitchell[10] called the problem a lack of reconciliation between sophisticated technologies, which tend to receive funding from ill-advised donor-nations, and low or intermediate technologies, which are under-funded but have higher success in disasters. Poorly integrated technological solutions compound confusion and frustration at the disaster site, and are likely to be shunted.

Aside from monitoring and predictive technologies, however, studies have pointed to technologies that can have positive impact on disaster mitigation[11]. Dynes et al.[12] identified remote sensing, data collection and information dissemination as the areas of greatest impact to disasters. The importance of satellite imaging technology, communications, or food packaging, storage, and transportation technology cannot be passed over as luxuries. It was further suggested that training and management techniques must be co-developed to complete an integrated disaster mitigation system.

Some encouraging signs of acceptance of technology have emerged recently. Microcomputers, having permeated into all facets of contemporary life, have at last found their ways into disaster mitigation. Although not perfect, the microcomputer is slowly gaining support as one of the most significant innovation for disaster during the past decade[14].

The American Red Cross is developing a two-year pilot program that will link all its U.S. field operations with a central computer via modem linkup. If successful, the program will be adopted for use around the world.

Volunteers for International Technical Assistance (VITA) has also implemented a computer-based database and communication system which will offer African farmers the latest famine warning, weather projections, and agricultural technologies.

There has also been proposal for the establishment of a International Disaster Coordination Center (IDCC) which will act as an information clearinghouse and tracking center for global disaster response deployment[15].

In January 1992, the Public Satellite Service Consortium (PSSC) announced a partnership program with Iridium, Inc., a subsidiary of the electronics firm Motorola, to develop an affordable, reliable global personal communication network. The proposed low-earth-orbit-satellite-based network, scheduled to go on-line in 1996, will afford disaster workers to communicate effectively via voice, data, and fax, from disaster sites. Although the project was first put forth by studies done more than fifteen years ago, it is not until recently when some commercial benefits are evident that the industry responded with a commitment.

TECHNOLOGY: TRANSFER, DEVELOPMENT, INTEGRATING, TESTING, AND DEPLOYMENT

Scholars of innovation research and practitioners of technology and R&D management have now fully embraced the notion that "over-the-wall" management style, in which R&D, design, and marketing functional groups do not interact, is a sure way to failure. In the business world, as well as in the disaster relief community, where resources are scarce, technology, if managed correctly, can improve efficiency and productivity. Well conceived and design technological solutions need not be forbiddingly expensive to produce and deploy. The key to effective and cost-effective industrial R&D is a fundamental change into a participatory style of management in which users, designers, marketers, and researchers collaborate. R&D of disaster technology likewise can no longer be treated as an isolated, independent activity. Participation from users, planners, and technologist must be integrated to produce real usable disaster technologies like robots for excavation, optical seeker for open-sea search and rescue, or unmanned remote sensing platforms.

Technology transfer is another mechanism to harvest benefits from investment in R&D activities. The newly established National Technology Initiative provides incentives for the private sector to transfer technology from U.S. government laboratories. The National Aeronautics and Space Administration (NASA) has just opened six Centers for Technology Commercialization (CTC) around the country in an effort to increase the flow of technology into the private sector. Cooperative Research and Development Agreements (CRADA) provide tax-deferment as incentive. A unique opportunity now exists for innovative entrepreneurs, socially responsible investors, and technologists to work together with a win-win financial proposition.

For example, hazardous waste cleanup technologies, widely used for refurbishment of contaminated domestic military bases, should be transferred from the Departments of Energy and Defense to the private sector for export to Eastern Europe, where the market for environmental cleanup is estimated to be close to $500 billion. Licensing fees generated can subsidize the massive cleanup bill the military is to face. The technology transfer can continue horizontally in a East-East fashion, adapting to the local particularities and needs.

Call for long-term disaster prevention measures inevitably leads to discussions of economic sustainability and stability. Relief and development are now considered interlinked[15]. Here is perhaps where technology can play a major role. Alternative indigenous renewable energy, computer-enhanced education, and information systems have the potential of breaking that vicious cycle of poor economic and human conditions leading to unsustainable rate of resource depletion, and eventually, high risk to disasters. Technology transfer from industrialized countries to LDCs ("North-South") has almost been synonymous with "industrial colonization"—dumping of undesirable, hazardous industrial operations into those countries which cannot afford to refuse, and are least likely to complain. Modern thinking of "technology cooperation" discards the force-feeding connotation of "transfer," and takes into account indigenous resources, needs, and potential for growth. The direction of transfer can be North-South, West-East, or East-East. Environmental technologies, in particular, have tremendous suitability and market potential from this standpoint.

A NEW CONSORTIUM PARADIGM: FIRE STATION EARTH

An industrial consortium is typically an umbrella organization created by potential industry competitors to cooperate in R&D programs for pre-competitive technologies or to provide complementary expertise and resources for the sake of improving competitiveness. It allows companies to venture into non-core-business areas or new technological frontiers with shared risk, and perhaps lower cost. Eventually, rapid advancement and diffusion of technology will benefit all participants and the industry as a whole.

Academia-government-industry consortia like the Waste-management Education and Research Consortium (WERC) represents another potentially fruitful collaboration between multiple participants—from scientists, students, industrialists, to policy makers. WERC was founded in 1990 as a joint venture between University of New Mexico, New Mexico State University, New Mexico Institute of Mining and Technology, Sandia and Los Alamos National Laboratories. Administered by the Department of Energy's Albuquerque Office, the annual grant of $5-million is spent on academic and research programs at the three institutions, research and testing laboratories, technology transfer, and education and outreach programs. The consortium brings together a vast array of experts from environmental professionals in a wide array of disciplines, as well as industry partners like Dupont. Commercializable technology is expected to be spun off as rapidly as possible through the industrial partners or entrepreneurial startups.

A consortium that develops appropriate disaster technology, both for transfer to LDCs can be invaluable. Small-size technology developers can take advantage of lowered entry barriers, rapid dissemination of information, and diffusion of technology. Larger partners can benefit from lowered risk and faster transfer of non-core technology. Shorter time-to-market, particularly external markets, is the goal of the consortium for large or small companies. Collectively, the consortium will be able to negotiate, relate, and cooperate with national governments in developing appropriate programs for disasters.

Analogous to a fire department, Fire Station EARTH (FSE)[16] is a consortium envisioned to provide functions of research and development, rapid response, training, education, and information management. This hybrid institutional mechanism will encompass characteristics of the public sector (non-profit), as well as the private sector (commercial). The goal is to provide a self-sufficient mechanism that can augment existing disaster response mechanisms, generate appropriate technological innovation and operating funds, maintain lean management style, and induce multi-disciplinary and long-term participation. The modular design shall include:

R&D consortium for disaster and environmental technologies

Academia plays an important role as the major source of fundamental knowledge. Scientific knowledge is allowed to be disseminated and diffused rapidly throughout the organization. A centralized integration facility will bring users, researchers, and developers of technology together. Feedback and initiative from users can guide the R&D and design processes to ensure relevance and effectiveness. Product development and design, including testing and modification, can be cycled through more rapidly. Development and demonstration of preventive alternative technologies, particularly environmentally "clean" technologies, must also be explored.

Incubator for emerging technologies

Small entrepreneurial companies arguably represent some of the best innovations in the U.S. Their spirited drive and flexibility mean that they can react to new situations quickly. Their smaller size and limited resource, however, often preclude them from tackling the bureaucratic jungle associated with international government agencies and markets. As part of a large and established consortium, these smaller enterprises will be able to focus on the innovations instead.

Transferring of high-profit-potential innovations to the commercial market provide incentives for industries and investors to participate in the consortium. Potential of dual-usage in disaster and commercial market is stressed in selecting technology for transfer and development. Co-development projects can be done via startup, or corporate partnering. Income generated from business operation can be partially retained, as royalty or fee, to support base operation of FSE.

Planning, training, and deployment center for rapid-response teams

As many disaster researchers have recommended, experienced professionals must be retained and be allowed to transfer their experiences and knowledge through post-operation debriefing. Their involvement in developing long-term disaster preparedness planning is invaluable.

Stand-by rapid-response teams, drawn from the international community, can be trained on-site for various disaster scenarios. Standard protocol and command and control mechanisms can be established to allow rapid deployment. Equipment and material can be placed on modular platforms for easier airlift or other transportation modes.

International governments can provide funding, personnel, or logistical support, in exchange for technologies transferred, and response coverage by FSE. This enables the countries of lesser wealth to participate fully as contributors in the consortium.

Center for information management and education

As a focus of global disaster response, FSE must serve as a center for integration of knowledge, information, resources, and technology. At the same time, it must also maximize the use of information in education, outreach, and provide up-to-date information in all aspects of disaster.

CONCLUSION

The new consortium paradigm, or Fire Station EARTH, is meant to provoke new thinking, and to suggest a new perspective on the persistent problem of disaster relief and development. Introduction of the commercial element and business management experiences may be just the right prescription. Surely, time is right for unifying our knowledge of disaster, in sociological, technological, managerial, and humanitarian terms, into one integrated discipline. (Lexicologists may cringe at attempts to coin the study of disaster "Disastology.") But as sure as natural and man-made environmental disasters are inevitable, so must we garner our collective will and knowledge to foster new mechanisms and innovations to combat them. Our better understanding of disasters must be translated into better preparedness, reduction of risk, more-rapid response, minimized destruction, more expedient recovery, and, ultimately, less human suffering.

REFERENCES

1. R. Dynes, ed. 1979. *Assessing international disaster needs*. Committee for International Disaster Assistance Report, National Academy of Science, Washington, D.C., 1978. National Academy of Science, Washington.
2. S. Green. 1977. *International disaster relief: Towards a response system*. McGraw-Hill, New York.
3. National Research Council. 1987. *Confronting natural disasters*. National Academy Press, Washington, D.C.
4. T. Hall. 1991. *Roll Call* June 24, 1991.
5. C. Schmitz. 1986. *Disaster! United Nations and international relief management*. Council for Foreign Relations, New York.
6. F. Cuny. 1985. What has to be done to increase the effectiveness of disaster interventions. *Disasters/Harvard Supplement* pp. 27–28.
7. L. Gordenker and T. G. Weiss. 1989. *Disasters* **13**(2):118–133.
8. R. Kent. 1983. Reflecting upon a decade of disasters: The evolving response of the international community. *International Affairs* **59**:693–711.
9. R. S. Stephenson. 1986. *Disasters* **10**(4):242–246.
10. J. K. Mitchell. 1988. Confronting natural disaster: The IDNDR. *Environment* **30**(2):25–29.

11. B. Cowlen and L. Love. 1979. Application of technology to disaster relief,. In S. Green and L. Stephens, eds., *Disaster assistance: Appraisal, reform, and new approaches*. New York University Press, New York.
12. R. Dynes, ed. 1977. The role of technology in international disaster assistance. Proceedings of the Committee on International Disaster Assistance Workshop, Washington, D.C., March 1977. National Academy of Science, Washington.
13. T. Drabek. 1990. Microcomputers and disaster responses. *Disasters* **15**(2):186–192.
14. L. Kan and C. Givens. 1989. In *Communication when it's needed most*, Annenberg Washington Program in Communication Policy Studies of Northwestern University, Washington, pp. 110–115.
15. F. Cuny. 1983. *Disaster and development*, Oxford University Press, New York.
16. D. Ling. 1991. Fire Station EARTH: On call for disasters. *Environmental Impact Assessment Review* **11**(4):287–296.

NATIONAL RESEARCH ACTIVITIES

ENGINEERING RESPONSE TO THE DUAL RISK OF NATURAL HAZARDS AND GLOBAL CHANGE

Joseph Bordogna
J. Eleonora Sabadell
National Science Foundation, Washington, D.C. 20550

ABSTRACT

The research programs on natural hazards mitigation and in global change supported by the National Science Foundation are briefly described. The potential relationships between the two types of phenomena, natural and man made, are discussed together with a summary of possible future directions in engineering research.

INTRODUCTION

The causes, effects, prediction and mitigation of natural hazards are research areas funded by the Directorate for Engineering (ENG) at the National Science Foundation (NSF). The reason for this support is immediate since every year in the United States floods, hurricanes, earthquakes, volcanic eruptions and landslides, among other hazards, result in loss of life and injuries, loss of property and disruption of the social system, and damage to the infrastructure as well as to the natural environment. Many other parts of the world are also afflicted by these kind of disasters where millions of people are affected and billions of dollars are lost annually.

NSF is interested in fostering, through research and by facilitating the use of research results, the following concepts: first, that a natural hazard should not inevitably become a disaster; and second, that prevention and mitigation procedures can decrease substantially all types of losses.

The goals of the two ENG natural hazards mitigation research programs are: to strengthen and disseminate the fundamental engineering and scientific knowledge about these geophysical events and through it increase public safety; to reduce economic losses and dislocations; to reduce the structural and environmental damage; and to enlarge the number of professionals dedicated to research and application in this area, and the scope of their expertise.

The NSF, together with eight other Federal agencies, is conducting research on the causes and some of the effects of a possible global climate change (GC) due to increasing concentrations of long-lived gases in the atmosphere. These "greenhouse" gases are: carbon dioxide, methane, nitrous oxide and chlorofluorocarbons or CFCs.

Twenty programs in the Directorate for Geosciences (GEO), two programs in the Directorate for Biological Sciences (BIO) and two programs in the Directorate for Social, Behavioral and Economic Sciences (SBE) support research projects on the acquisition of data; on the basic physical, biological and social processes by which the climate could change or is changing; and on the development of predictive global climate models (GCMs).

The main purpose of this research is to better understand the process of change and to monitor it. The gathered knowledge will be used to validate the thesis of change, and if true, to better understand the possible impacts of GC on the world's ecology, agriculture, forestry, hydrology, water resources, oceans and coastal areas, inland areas, snow cover, ice and permafrost, human settlements and human activities.

POTENTIAL IMPACTS OF GLOBAL CHANGE ON NATURAL HAZARDS

There are hypotheses about the effects of "greenhouse" gases on the global climate and thus on the environmental conditions of continents and regions of the world. The reality of the change itself and its consequences, their timing, intensity and physical distribution are some of the subjects of intense debate within the scientific community.

On the other hand, based on records of past climatic variations, there is substantial agreement on the kind of potential consequences of a future climate change. Some of these consequences could be: a global-mean surface warming; a global-mean precipitation increase; a reduction of sea ice with a rise in global-mean sea level; and a rise in climate variability with increases in climatic extreme events.

The possibility also exists that if GC is a reality its impact will affect more severely regions of the world which are already under environmental stress and areas which today are at risk from natural hazards. For example, hurricanes, typhoons and floods could become more violent than today's, and droughts more intense in areas where these are now seasonal occurrences. The geographic and areal distribution of hydrologic and atmospheric extremes could also change as well as their frequency and duration.

It has been postulated also that the probability is high for climate variability and climatic extremes to become manifested quite sooner than any evidence of the gradual change in climate. If such is the case, a significant number of people, infrastructures and economies in the United States and elsewhere could be under risk conditions in the near future for which they are not prepared.

To date, the natural hazards research supported by NSF includes examination of issues such as: the underlying geophysical processes; the interactions between extreme loads and physical structures; the uncertainty in predicting and forecasting floods, droughts, hurricanes and tornadoes; the synergism between diverse land uses and recurrent natural hazards; the development and improvement of methods for the assessment of structural, economic and social damages; the development, implementation measures and evaluation of natural hazards mitigation. All these issues are highly pertinent to many of the possible effects of global change.

It should be noted that, in contrast to the ongoing GC research, the natural hazards research is highly interdisciplinary in nature because it deals with problems such as the capability of physical and social systems to cope with changed and changing geophysical conditions; the development of a systems approach to the mitigation of and the adaptation to natural hazards; methods for dealing with uncertainty and risk; and the design and management of systems under stress.

The issues addressed and the approaches developed by the natural hazards research community in the last two decades should be used for developing the needed public policies for a sustainable economic development of the countries around the world being affected by changing environments. These changes, which are the result of undue anthropogenic stresses and natural forces, will necessitate both short term and long term policies. This means that all available and forthcoming knowledge and capabilities should be used for designing and implementing these policies.

The ENG response to the opening of these novel areas of concern is to develop new programmatic lines in existing programs, set comparative research activities with other Federal agencies, and support joint international research projects.

ITALIAN NATIONAL RESPONSE TO THE INTERNATIONAL DECADE FOR NATURAL DISASTER REDUCTION

Lucio Ubertini
Director, Research Institute for Hydrogeological Protection in Central Italy
National Research Council
Via Madonna Alta 126
Perugia, Italy

ABSTRACT

The Italian National Group for the Prevention of Hydrogeological Disasters has been created under both multidisciplinary and interdisciplinary frameworks, necessary for advancing research activities in the sphere of hydrogeological hazards. The Group collaborates with relevant institutions abroad not only in its research activities but also in organizing seminars and summer schools. In this paper, the activities of the Group carried out in preparation towards the United Nations declaration of the IDNDR and subsequent ones under the Decade are presented. The paper concludes with proposals for integrated data collection activities on hydrogeological hazards and a drought monitoring observatory within the context of an impending climate change.

INTRODUCTION

In recent years the Italian scientific community has been giving increased attention, under the auspices of the National Research Council, to the question of natural hazards. It is a well known fact that the Mediterranean environment, and for that matter Italy, is highly vulnerable to frequent occurrences of natural disasters triggered by large scale atmospheric perturbations, such as floods and landslides, earthquakes and volcanoes.

This growing awareness of the vagaries of natural hazards has not only been confined within the walls of research institutions but also a concerted effort has been created to coordinate the activities of prediction, mitigation and monitoring at the governmental level. The Department for Civil Protection, under its Commission on High Risks, groups together scientific and professional leaders in the various aspects of natural and man-made hazards such as hydro-geological, seismic, volcanic, nuclear, transport and chemical. The main activities of the scientific communities are coordinated under competent National Groups for the Prevention of Disasters. These Groups in turn operate a number of operational units within universities, research institutions and also some relevant agencies at regional and local levels. The main source of their finances comes from the Department for Civil Protection, which aims to create a national scientific network useful for the promotion and realization of scientific initiatives and for providing consultancy in emergency situations needed for prompt interventions during extreme atmospheric and geologic perturbations, such as floods, landslides, earthquakes and volcanoes. This arrangement offers the opportunity to establish both horizontal and vertical interactions between research needs and demands of operation and vice versa.

As a step towards a more organic multi-disciplinary collaboration between the various sectors involved in natural disaster prevention, a Natural Disaster Finalized Project (still under consideration for financing) was advanced with the following objectives:
- a concerted information management;
- development of strategies for the diffusion of warning communications;
- the adoption of emergency measures (e.g., evacuation) in response to natural disasters.

Further discussion of the response of the Italian scientific community to the Decade will be centered around the activities of the National Group for the Prevention of Hydrogeological Disasters for which the author is responsible[3].

THE GNDCI AND THE DECADE

The National Group for the Prevention of Hydro-Geological Disasters (GNDCI) was instituted, under the auspices of the National Research Council (CNR), by a governmental decree on Dec. 12, 1984. It is sponsored by the Ministry of Scientific Research and Technology in concert with the Ministry for Civil Protection and the Ministry for Public Works. As stated above, the finances for the activities of the Group are provided by the Ministry of Civil Protection.

The statute setting up the GNDCI assigns the underlisted objectives to the group:
1) to promote and develop coordinated interdisciplinary research, directed towards the acquisition and improvement of scientific knowledge necessary for the control of floods and landslides;
2) to provide scientific and technical consultancy in the sector of hydrogeological hazards to interested ministries, regional authorities and other local departments, with particular reference to civil protection problems and the education of populations exposed to the dangers of inundations and landslides;
3) to assure the coordination of actions of scientific interventions in the occasion of floods and landslides;
4) to formulate proposals of specific research programs;
5) to formulate proposals for guides and provisions appropriate for forecasting and prevention;
6) to maintain a liaison with public organizations charged with territorial development and, in particular, to assume the coordination of activities of scientific character related to the planning of river basins;
7) to maintain relations with analogous research initiatives of other countries, promoting the exchange of international experience in this sphere.

Coordination of Scientific Research

The scientific activity of the group is directed towards two complementary objectives:[3,1]
1) to realize a framework of knowledge, homogeneous within the national territory, on the basis of which the planning of structural interventions could be carried out. The framework should be compatible with the available resources, so as to optimize the benefits achievable while reducing damages;

2) to organize an inventory of vulnerable areas, and a system of forecasting which allows the quantification of the probability of occurrence of an event. The objective is to permit the adoption of measures of intervention and a system of communications with the residents during the event, which could allow for a reduction in anticipated damages.

These activities are subdivided according to four general lines of research as follows:
1) Forecasting and prevention of hydrological extremes and their control.

The accurate evaluation of flood risk is the key point for a national program of reducing flood damages. Three general objectives are:
   a) Evaluation of the instantaneous maximum flood of assigned duration which has a given risk of exceedance;
   b) Evaluation of the effect of floods triggered by structural interventions;
   c) Evaluation of the possibility of flood forecasting.

Under the auspices of the project, which is methodological in character, a special operational program for the evaluation of flood flows corresponding to specified return periods for water courses in Italy (VAPI program) was initiated. The program involves studies based on the statistical analysis of the frequency of annual maxima of extreme rainfall and observed discharges, as documented by the Italian Hydrographic Services (S.I.I).

2) Forecasting and prevention of events of high risk landslides.
   The activities of this line of research are programmed as follows:
   a) a systematic collection of data related to the phenomena of landslides and other mass movement to obtain an initial picture of the location and recurrence of the events, together with the approximate dimension of the phenomena and the extent and magnitude of damages produced;
   b) study of landsliding phenomena for integration and interpretation of the available data, including the delineation of unstable zones, topographic definition of the movements, geologic-structural grouping in the local context and geotechnical characterization of soils.

3) Evaluation of hydrogeological risk, zoning and intervention strategies for the mitigation of the effects of extreme events.

The third area of research has as a general objective the evaluation and use of structural and nonstructural alternatives to flood control. It involves information for the zoning of areas subject to the risk of inundation. The expected outcome of the activities of this effort is the preparation of a scientific and technical reports on areas vulnerable to flood risk, comprising in detail:
   a) the description of the cartographical, topographical, hydrographical, hydrological, hydraulic, meteorological, geologic and geotechnical data related to flow channels, and other information of an urban and socio-economic nature used in the evaluation of risk;
   b) the description of the hydrological and hydraulic procedures used in defining risk maps;

c) the presentation of the results, in a consistent, common format: the hydrological and hydraulic evaluations, their return periods and associated confidence intervals.

The research activities under the Arno project, which fall under the umbrella of this research line, led to collaboration with the Massachusetts Institute of Technology (MIT). This international collaboration concentrates on rainfall monitoring and forecasting, the development of rainfall-runoff models and the employment of meteorological radar for rainfall measurement.

4) Evaluation of the vulnerability of aquifers.

This line of research has as an objective the study in space and time of the phenomena of pollution (effective and potential) of underground water bodies. The objectives are to:
a) provide an information base for each regional situation;
b) provide a document for the rapid evaluation of contamination potential;
c) evaluate the possibility of installing monitoring systems for the assessment of the effectiveness of interventions intended to safeguard the population subject to risk.

Organizational activities

The organizational structure of the Group ensures both multidisciplinary and interdisciplinary approaches in scientific and technical collaboration not only within academic and research institutions but also within relevant ministries, regional and local authorities. The Group, right from its inception, recognized that the immense complexity of atmospheric and geologic extreme perturbations underlying hydrogeological hazards calls for an active collaboration and exchange of experiences at the international level. It was precisely with this view that it convened a series of seminars and workshops with participation of scientists and engineers from all over the world.

As a result of this the Group developed an active network of collaboration with institutions and agencies abroad such as the U.S. Geological Survey, the French Delegation on High Risks, the Massachusetts Institute of Technology, Pennsylvania State University, and the Institute of Hydrology in Wallingford, just to mention a few.

As mentioned earlier, the Group's activities were initiated well ahead of the proclamation of the Decade. The main organizational activities of the Group can be divided into two broad areas, i.e., scientific meetings and educational programs[2].

Scientific Meetings and Educational Programs

In the period 1985–86 two international scientific seminars were sponsored by the Group on "Combined Efficiency of Direct and Indirect Estimation for Point and Regional Flood Prediction" and "Drought Analysis." The Italian National Research Council Group has co-sponsored with the U.S. National Science Foundation two international workshops of a multidisciplinary nature. The first was on "Natural Disasters in European Mediterranean Countries" in 1988, convened jointly by the Group and MIT[4]. Since this meeting was held during the preparation for the Decade, the participants came out with specific proposals and recommendations relevant to the activities being envisaged under the Decade. The other was the first international scientific meeting held after the U.N. proclamation of the Decade with active participation of the Secretariat of the IDNDR. It was organized on the theme "Prediction and Perception of Natural Disasters," convened by the Group in collaboration with the Disaster Research Center of the University of Delaware. The novelty of this workshop was the fact that the deliberations were shared by both physical and social scientists which gave rise to active exchange of ideas on the common linkages within the state of the art of prediction, monitoring and mitigation of natural disasters with special emphasis on uncertainties.

Additionally, the Group has been organizing a one-month Summer School since 1988 under a standing agreement between the Italian National Research Council and the U.S. Geological Survey as "Hydrogeological Hazards Studies." In September 1990, a second school was organized on "Stability of River and Coastal Forms." The above-mentioned summer schools attract participants from Italian research and academic institutions as well as from the international community.

SPECIAL PROGRAMS

Two of the most important issues facing the world scientific community are natural disaster reduction and the prospect of climate change[7]. Even though scientific evidence is not yet conclusive on the pace of the impending climate change, it is advisable to institute measures to prevent the negative effects on the hydrological cycle and most importantly any effects on the frequency and magnitude of extreme hydrogeological events. It is evident that there cannot be any sustainable program for hydrogeological disaster reduction if the ideas of possible climate change are not incorporated into it. Hydrogeological hazards and climate change are phenomena which know no geographical boundaries but rather respond to the geology and climate of a region.

Focal Center for IDNDR:

One of the salient conclusions of the International Seminar on Natural Hazards in European Mediterranean Countries was a recommendation advanced to the Water Resources Research and Documentation Centre (WARREDOC) of the Italian University for Foreigners, Perugia, to act as a focal point for the activities envisaged under the IDNDR for the European Mediterranean countries. It must be emphasized that WARREDOC has successfully served as the organizational seat of the Group for most of scientific meetings and workshops. This recommendation became a reality through a convention signed between the Centre and the National Research Council under the scientific patronage of the Group. The provisions under the convention amplifies the sphere of activities to cover the entire international arena. The following programmes are planned for the coming years:
1. Fifth Summer School on Hydrogeological Hazards;
2. International Workshop on Artistic Perception of Natural Hazards;
3. International Workshop on Geographical Information Systems and Hydrogeological Hazards;

4. International Seminar on Global Circulation Models and Climate Change.

In coming years we also expect to organize short courses on relevant issues on natural hazards and climate change and also to offer opportunities to interested researchers to use the facilities of the Centre in carrying out research activities for some short periods of time.

Project on Interactions of Climate Change with Hydrogeological Hazards in the Mediterranean Basin

It is apparent that one of the most important limitations to understanding natural disasters, including climate change, is our inability to sample important hydrologic, meteorologic variables at global scales. Examples are our inability to sample rainfall, soil moisture and atmospheric humidity over large regions. Furthermore it is also clear that large-scale phenomena require coordination and integration of data from many parts of the world. Rainfall and soil moisture are some of the variables most closely associated with such natural disasters as floods, droughts, landslides.

One of the most serious problems facing the international scientific community is the lack of extensive climatic and hydrometeorological data in time and space. Attempts at data analysis for quantitative evidence on the possible scenarios of an impending climate change confirm this situation. The transnational nature of the physical phenomena in question calls for strong cooperation at the international level. Scientists at different institutions in many countries should work in a coordinated way to have a global view of the phenomena and to better understand causes and effects of natural disasters[5]. Italy and the United States have already started defining the necessary global and interdisciplinary agenda that will be required to handle the activities of the next ten years. One such idea is based on integrated hydrometeorological monitoring to cover sampled areas of the globe. The following are two derivatives of the project, supplementary in nature but separated for convenience of possible funding.

Climatological Studies: The main objective of this project is to study and analyze the interactions and possible influence of climate change on extreme meteorological and hydrological events over the Mediterranean Basin. Towards this goal, the project is intended to:
- analyze extreme disturbances and storms in the Mediterranean area;
- determine possible changes in space and time of the extreme disturbances;
- detect, on the basis of the aforesaid changes, possible ongoing climate change;
- analyze the fields associated with extreme disturbances;
- detect relationships among extreme disturbances and mean fields;
- correlate extreme disturbances with extreme events studied by other climatic teams;
- define the impacts of extreme events on the mechanisms of floods and desertification;

- determine possible future variations in space and time of the extreme disturbances in the light of scenarios developed with GCMs or other methods of climatic predictions.
- determine the need for an integrated data collection over the area.

Drought Monitoring Observatory

An extensive plan for mitigation of drought damages should at least include the development of a drought forecasting and monitoring system, e.g., of a "Drought Monitoring Observatory." This could be a system operating at a regional (or national) level, able to follow in real time the evolution of the drought phenomenon in all its aspects (depth, time, and areal coverage for several hydrological variables) and to evaluate the related risk of water shortages.

The aim of such an Observatory would therefore be to provide decision-makers with valuable indications and practical tools to be used in evaluating drought effects on society and the environment and in selecting the most appropriate drought control measures. This could lead to the development of emergency plans for water supply systems, based on forecasted or ongoing droughts, to face water shortages for municipal, agricultural and industrial uses.

A research program should include:
- application of Global Circulation Models and climate scenarios to the Mediterranean part of Europe, which previous studies indicate as the most endangered. It is necessary to confirm the hypothesis that climate change at the middle latitudes will lead mainly to increases in the dry season length, as well as of the soil moisture deficit.
- necessary investigations toward the design of the most appropriate structure for the Observatory, including the choice of significant parameters and indicators, and the design of a monitoring network.
- preparation of emergency programs, including the design of models aimed at forecasting water shortages and the development of operating rules to manage water supply systems in order to reduce the drought damage risk.

CONCLUDING REMARKS

The creation of the National Group for the Prevention of Hydrogeological Disasters has led to a variety of research activities within the sphere of hydrogeological disaster monitoring. The main achievements of the Group involve multidisciplinary collaboration.

The task facing the international scientific community in the face of growing vulnerability of society to natural disasters and the impending climate change is immense while the resources are systematically decreasing. The experience of the Group in collaborating with other institutions abroad indicates that without pooling resources there cannot be any meaningful progress towards natural disaster reduction. Moreover, the study of potential climate change calls for an international concerted effort at creating transnational projects which will ensure homogeneity in data collection and methods of analysis.

REFERENCES

1. K. Andah. 1990. Report on the scientific activities of GNDCI/CNR. Italian National Research Council Internal Report, Rome.
2. K. Andah. 1991. Activities of the Italian national group for the prevention of hydrogeological disasters towards and within the IDNDR Programme of the United Nations. Italian National Research Council Internal Report, Rome.
3. Annual reports of the national group for the prevention of hydrogeological hazards. 1987–90. National Research Council, Rome.
4. R. Bras, F. Siccardi. 1988. Natural disasters in European Mediterranean countries. Selected papers from international workshop, organized by U.S. National Science Foundation and the Italian National Research Council, Perugia.
5. M. Sanderson, ed. 1990. UNESCO sourcebook in climatology for hydrologists and water resource engineers. UNESCO.
6. The WMO hydrology and water resources programme 1988–1997. 1988. Second WMO Long-Term Plan, Part II, Vol. 5. WMO publ. No. 695.
7. World climate programme: Water. 1988. Fourth Planning Meeting, Paris, Sept. 12–16, WMO/TD-No. 271.

UNITED STATES RESPONSE TO THE INTERNATIONAL DECADE FOR NATURAL DISASTER REDUCTION

J. Eleonora Sabadell
National Science Foundation, Washington, D.C. 20550

ABSTRACT

The activities developed by the Federal Government of the United States in response to the International Decade for Natural Disaster Reduction, which began in 1990, are described in this paper. The main goal of both endeavors, the international and the domestic, to maximize the opportunities for reducing the losses created by natural hazards is discussed. The objectives of the program developed by the Federal Subcommittee for Natural Disaster Reduction, its framework and priorities will be presented as well as some examples of specific activities.

INTRODUCTION

Every year natural hazards such as floods, earthquakes, hurricanes, tornadoes, landslides, and wild fires claim hundreds of lives and at least two billion dollars in property losses and damages in the United States alone. These estimates do not include the cost of major local disasters like Hurricane Hugo or the Loma Prieta earthquake, with each of them reaching a ten billion dollars loss. The impact of a single extreme event on some countries can be devastating not only because of the very high number of deaths and injuries, but also because the economic losses are of inordinate proportions compared with the country's resources, national budgets or annual gross national product (GNP).

The trend in economic losses is increasing in the United States and all over the world as urban population grows and locates in hazard-prone areas; as the value, complexity and vulnerability of the physical, economical and social infrastructures increase; and as some of humankind's own actions trigger disasters such as wild fires and landslides, and intensify the impact of hazards such as floods, droughts and coastal erosion. It has been found also that fatalities and injuries are generally declining because of improved warning systems and the construction of safer structures but, for developing countries this latter trend may not hold for long.

To date, the national and international response to natural disasters has been mainly remedial, with limited investments in mitigation, with a shortfall of resources for the identification, assessment and management of these kind of risks, and for the development and implementation of preventive measures.

It should be pointed out that in the United States and elsewhere methods for curtailing damages and losses due to natural hazards are available though not always implemented and that, through research, even better procedures are being developed. One of the keys for successfully mitigating natural hazards is to develop improved mechanisms for the transfer of this knowledge.

Moreover, by integrating available knowledge and by shifting emphasis from emergency response to anticipatory activities it should be possible to affect savings far greater than those resulting from the current remedial and fragmented approach. These are some of the tenets of the International Decade for Natural Disaster Reduction (IDNDR) proclaimed by the United Nations in December 1989 with the endorsement of the United States.

OBJECTIVES OF THE IDNDR

The main objective of the Decade is to reduce, through concerted international efforts, the probability for a natural hazard to become a disaster. The goals of the IDNDR to be fulfilled during the decade of the 1990s are:
- To improve the capabilities of nations to mitigate the effects of natural disasters and to assess disaster damage potential;
- To improve the dissemination and use of available knowledge;
- To foster scientific and engineering research aimed at closing critical gaps in knowledge;
- To disseminate existing and new scientific and technical information related to measures for assessing, predicting and mitigating natural disaster impacts;
- To develop mechanisms for fostering technical assistance, technology transfer, implementation of demonstration projects, education and training, and evaluation of the effectiveness of existing and forthcoming mitigation measures.

FEDERAL ROLE IN NATURAL DISASTER REDUCTION

Natural hazards, which do not recognize political boundaries, may cause the individual states, localities, to the public and private sectors problems far greater than they can handle with the resources they have available. It is under these circumstances, that the United States federal government can improve significantly the ability of local and regional entities to recover from such events.

The role of the federal government in the area of natural disasters is quite varied, spanning research to operation, providing important services as, for example, furnishing information and warnings; offering guidelines, regulations and incentives for risk reduction; supplying engineered and structural protective measures; and designing and implementing emergency response plans. In each case federal activities complement and supplement the local, state, private and public resources, by undertaking programs and projects that are most efficiently handled on a national scale.

To date, federal agencies with responsibilities for understanding, predicting and coping with natural hazards and disasters have each quite specific missions. The communication, cooperation, and coordination efforts among these agencies are far from satisfactory. This fragmentary approach is not as effective nor economical as it should be. During the last two decades some model programs, where agencies worked together for a common goal, were developed and implemented. One example is the National Earthquake Hazards Reduction Program (NEHRP) which has produced encouraging results especially in the development of preventive measures. The intent of the pertinent Federal agencies is to foster the setting up many other cooperative activities.

The United States, in supporting the IDNDR's goals, has recognized the opportunities that the Decade provides the nation for improving the coordination and cooperation between the public and the private sectors, for developing a stronger support for mitigation and prevention policies and measures, and for fostering its leadership posture in assisting other countries, especially developing nations, in their pursuit for a sustainable development.

UNITED STATES RESPONSE TO THE IDNDR

The idea of dedicating the 1990s to the reduction of natural hazards' impacts originated in the United States in 1984 and was adopted by the world community in 1989. In 1990 the Subcommittee on Natural Disasters Reduction (SNDR), of the Committee on Earth and Environmental Sciences (CEES) within the Federal Coordination Council for Science, Engineering and Technology (FCCSET), was constituted with the charge of designing a plan for domestic and international activities in compliance with the objectives of the IDNDR.

The SNDR, which has representatives from more than a dozen federal agencies, set out to identify the national goals for the Decade, to develop strategic and integrating priorities, to catalog research needs and application opportunities, to identify demonstration projects, and to develop an implementation plan for the federal government.

To date, the SNDR has accomplished almost all these tasks, with the implementation plan now in progress. What follows is a summary of the recommendations for the action program presented by the Subcommittee to the Office of Science and Technology Policy serving the President.

The main goal of the program, as established by the SNDR, for "hazard-proofing" the Nation is similar to that of the IDNDR:

To reduce fatalities, human suffering, environmental damage, and economic losses caused by natural disasters.

PROGRAM FRAMEWORK

A framework for the program was developed early on, recognizing that priorities should be established for any of the activities to be undertaken during the Decade. In the program framework set by the SNDR the strategic and integrating priorities to be used by the cooperating agencies were identified as follows:

The Strategic Priorities, which define the overall approaches and provide guidance for the evaluation of the program, or parts of it, include:
- Supporting the activities of the IDNDR and increasing the cooperation with international programs and organizations;
- Anticipating and preventing the impacts of natural hazards rather than reacting to disasters;
- Advancing the scientific and technological knowledge related to natural hazards and applying the results of research;
- Improving the efficiency of and coordination among federal, state and local programs and organizations; and
- Developing and strengthening the cooperation and coordination between the private and public sectors.

The Integrating Priorities, which define the specific type of activities within the program where all interested federal agencies, private and other public organizations, and experts should participate include:
- Documenting, characterizing and predicting single and interacting multiple natural hazards;
- Assessing the various types of damage caused by natural hazards to the constructed and the natural environments;
- Assessing actual and acceptable natural hazards risks;
- Developing options for risk and loss reduction;
- Implementing risk and loss reduction measures; and
- Upgrading, expanding and integrating the knowledge acquired after a disaster episode, incorporating these experiences into mitigation measures.

RESEARCH AND APPLICATION ELEMENTS

Specific areas where gaps in knowledge exist, or where available knowledge is not applied were identified by the SNDR. In realizing these research and application elements the strategic and integrating priorities described above should be observed as to maximize the benefits resulting from these activities.

The Research Elements belong to three distinct topical areas:
- The Physical and Biological Nature of Natural Hazards
- The Managed Systems
- The Human Interactions

Research on the Physical and Biological Nature of Natural Hazards addresses the need to improve the capabilities for predicting time, frequency, duration, place, and intensity of extreme events by better understanding the underlying processes. These processes pertain to the:
- Climate, Weather and Hydrologic Systems
- Solid Earth Systems
- Ecological Systems

Examples of extreme processes for each one of these three systems are: hurricanes, floods and droughts; volcanic eruptions, landslides and earthquakes; desertification, infestation and wildfire.

Research on Managed Systems addresses the methodologies and procedures by which the impacts of natural hazards can be minimized or prevented. The two general areas belong to the:
- Engineering Systems
- Environmental Systems

Some examples of areas of research for each of these two systems are: response of structures to extreme loads, design of hazard-resistant structures, and assessment of damage to buildings, lifelines and facilities; drought-resistant vegetation, erosion and sedimentation control, and biological controls.

Research on Natural Hazards–Human Interactions addresses the social, health, institutional and economic processes affecting the impacts of natural hazards. The processes of interest appertain to the:
- Behavioral, Health and Communication Systems
- Institutional Systems
- Economic Systems

Examples of research topics within these three general systems are: incentives for safe behavior, social assessment for diverse hazards, health impact assessment, disaster epidemiology, communication program design and delivery; role of insurance and construction industries in mitigation, role of financial institutions, policy adoption and regulation; economic damage assessment, cost/benefit of mitigation, economic incentives for mitigation and prevention.

The Application Elements place a major emphasis on the dissemination and application of available and new knowledge in the development of techniques, measures and procedures by which the destructive consequences of natural hazards can be reduced drastically. The logical progression in time of actions to be taken is:
- Before a Natural Hazard hits
- During the time the Natural Hazard strikes
- After the Natural Hazard has taken place

Before a Natural Hazard hits preventive and mitigation measures should be taken to reduce losses and to facilitate recovery. In addition, accurate, reliable and timely prediction and warning systems are needed to protect the population at risk. The types of actions to be taken are:
- Preparation
- Prediction and Warning

Examples of these actions are: risk assessments, and risk and emergency management plans; public education and awareness programs; implementation of novel monitoring and warning systems; data analysis and modeling.

During the time the Natural Hazard strikes plans should be made for coordinating the actions of federal, state and local organizations in their response to and the control of the impact of natural hazards. These types of actions fall under the rubric of:
- Intervention

Examples of intervention are: coordination of resources; tactical and logistical response; on-the-spot training.

After the Natural Hazard has taken place, both immediate and long-term measures should be taken to revitalize the stricken community and to maximize its resistance to future disasters. These types of actions are:
- Emergency Assistance
- Recovery

Some examples of after-the-fact actions are: search and rescue; delivery of emergency services; deployment of emergency communication; provision of financial assistance; damage assessment; restoration, retrofitting and reconstruction; evaluation of recovery effectiveness.

Although all the described actions are needed in order to minimize losses due to Natural Hazards, the emphasis, as stated before, of the proposed program for the United States and other countries is on what can and should be done *before* these kind of extreme events occur. In time, consideration should be given also to the technological disasters triggered by Natural Hazards.

The implementation plan is now being developed by the SNDR in which several domestic and international demonstration projects will be recommended for implementation. It is expected that by the end of this decade mechanisms will be in place for the effective and active cooperation among the different levels of government, agencies, academia, professional associations, and the private sector for the protection and safety of populations at risk in the United States and the world.

RESPONSE OF NATIONS

NATURAL HAZARDS AND THE U.S. GLOBAL CHANGE RESEARCH PROGRAM

Dallas L. Peck
Director, U.S. Geological Survey
and
Chairman, Committee on Earth and Environmental Sciences

ABSTRACT

The goal of the U.S. Global Change Research Program (USGCRP) is to gain an understanding of complex Earth systems so that change can be predicted and a scientific basis for formulating national and international policy can be established. The USGCRP has three objectives: documenting the Earth's system through careful coordinated observations; improving scientific understanding of the processes; and developing models that will allow accurate global and regional predictions. Improvements in computer technology, including the use of coupled models, are making it possible to document how the different components of the Earth system work together. With increased understanding of these interactions, capabilities for predicting global change and associated natural hazards will be greatly increased.

As a geologist who spent much of his career studying volcanoes, I believe that natural hazards merit serious and sustained attention, and I was pleased to see that this Conference focused on translating climate change into Earth systems impacts.

Before I begin to address U.S. policies on global change and natural disasters, I'd like to spend a few moments discussing the relationship between natural disasters and global change. We all saw the dramatic pictures of the eruption of Mount Pinatubo in the Philippines June 1991—the largest volcanic eruption in the world since 1912. I see the Pinatubo eruption not only as a major natural hazard, but also as an outstanding opportunity to establish a better understanding of the influence not of climate change on natural hazards, but of a natural hazard—volcanic eruptions—on climate.

We have substantial historical evidence on the climate effects of major eruptions, such as the well-known years without summer that followed the eruptions of Tambora in 1815 and of Krakatoa in 1883. The mid-latitude, stratospheric winds helped distribute ash from the Krakatoa eruption worldwide, producing spectacular sunsets—just as Pinatubo is doing today. Temperature records from the period show that global temperatures dropped as much as 0.3°C for several years following the eruption.

Today, we are able to explore in much greater detail the reasons for these reactions. Microwave sounding unit data show substantial stratospheric temperature variability coincident with volcanic effects. And, using information collected by the NASA Total Ozone Mapping Spectrometer (TOMS) satellite as well as ground stations, we were able to detect and monitor the amount of sulfur injected into the stratosphere by Pinatubo and the spread of that plume of volcanic material on a global scale. We are already seeing the first signs of the effect of this material in the stratospheric temperature record. Over the next two to four years, the Pinatubo eruption may reduce global surface temperatures somewhere between 0.3 to 0.5 degrees Centigrade, although some estimates range as high as one degree Centigrade.

But Pinatubo is much more than an outstanding research opportunity for earth scientists. The hazard reduction efforts made at Pinatubo exemplify the kinds of efforts that the United Nations is promoting through the International Decade of Natural Hazard Reduction.

At Pinatubo, we were able to predict the eruption with enough accuracy to save thousands of lives and billions of dollars in equipment—much of it the property of the U.S. Air Force. In cooperation with Filipino scientists headed by Raymundo Punongbayan, Chris Newhall and his team of volcanologists from the U.S. Geological Survey (USGS) instrumented the volcano shortly after it stirred back to life. Punongbayan installed the first seismometer at Pinatubo, and Newhall's team tied that seismometer into a network of seven seismometers and two tiltmeters within a month. The USGS team also brought a computer program for acquiring and analyzing seismic data that allowed them to follow magma movements within Pinatubo and correlation spectrometers for analyzing gas emissions.

Using the experience we have gained at the Hawaiian Volcano Observatory and during the eruptions of Mt. St. Helens and Mt. Redoubt, along with the insights developed by the Japanese scientists at Asama Volcano, we were able to use those data to predict Pinatubo's eruption.

The signs of imminent eruption were clear. Seismic signals indicated that magma movements, once widespread, were concentrating beneath the summit. Pinatubo's surface began to bulge. When the seismic signals indicated that the magma beneath the surface had become pressurized and the amount of magmatic gas escaping to the surface dropped, the volcanologists were able to determine that Pinatubo had sealed itself. The subsequent surge in seismicity made it clear that it was time to evacuate the area around Pinatubo.

We were able to get that message, clearly and accurately, to the local population, and, by and large, they evacuated the area around the volcano. While it is true that 500 Filipinos died in the eruption of Mt. Pinatubo, several hundred thousand had been at risk and the death toll could have been much higher than it was.

The United Nation's International Decade of Natural Disaster Reduction is based on the premise that much of the scientific understanding necessary to help prevent natural hazards from turning into natural disasters has been established. As the success at Pinatubo showed, what we have to do is apply that knowledge.

I have been a strong supporter of the Decade since the idea was first proposed by Frank Press, President of the U.S. National Academy of Sciences, in 1984. The USGS, in addition to its volcanoes program, is actively involved in improving scientific understanding of several other hazards—earthquakes, landslides, floods, and droughts. The Decade has had two Executive Secretaries in Geneva, Jim Devine and Bob Hamilton—both employees of the USGS who served with partial support from the National Academy of Sciences and several other Federal agencies.

The Committee on Earth and Environmental Sciences (CEES), which I will discuss in greater detail later, established a Subcommittee on Natural Disaster Reduction in 1989 to work on the Decade under the leadership of Bill Hooke of the

National Oceanic and Atmospheric Administration (NOAA). I serve on a Science and Technology Committee, advisory to the Secretariat for the Decade. The U.N. has also formed a Special High-Level Council for the Decade.

As the Pinatubo example demonstrates, the U.S. hazard reduction efforts have been foreign as well as domestic. The City of Los Angeles is working in partnership with Mexico to apply the Los Angeles earthquake hazard reduction program known as SCAPE in Mexico City. This fall, the Central United States Earthquake Consortium conducted a workshop to promote hazard reduction throughout Latin America. And we are taking a cooperative approach with our colleagues in Japan to improve natural hazard planning in the third world.

Continued efforts to better understand and predict the behavior of volcanoes and earthquakes are needed—but these hazards shouldn't be affected by global change. A range of other natural hazards, however, may well present new problems in the future. Under some global change scenarios, windstorms, rainstorms, landslides, floods, and droughts may be more frequent or more intense—or they may affect entirely different areas.

Predicting these extreme events is one goal of the climate modeling done under the U.S. Global Change Research program. Better understanding of why and how these events occur is closely connected with identifying the forces behind natural variability—which is a central element of the U.S. research effort.

Tropical storms, for example, can pose extreme hazards. The control state of one general circulation model does a good job of reproducing the observed behavior of tropical cyclones today. When that model is run with a scenario that doubles concentrations of atmospheric CO_2, the number of storm cycles increased by 50%. But storm intensity did not change at all.

The storm prediction results of the most widely used general circulation models are mixed. Some have predicted fewer cyclones for the Northern hemisphere under a doubling of atmospheric CO_2, others have predicted an increase. We have even greater difficulty making predictions on a regional scale. Given a doubling of atmospheric CO_2, some of the major global circulation models predict a hotter, drier Midwest and others predict a cooler, wetter Midwest.

Nevertheless, in many cases, the models we have today suggest that natural hazards would be felt most severely in regions that are already under stress. Drought risk appears to be the most serious concern at both global and regional levels. Concern over drought, in fact, helped lead to establishment of the U.S. Global Change Research Program.

Five years ago, the scientific community was already well aware of a number of troubling research findings—many of which were made by the individuals participating in this Conference. We had good paleoclimate information for the past 100,000 years that indicated that global average temperatures and atmospheric carbon dioxide levels both increased and decreased together. Analyses of ice cores from Greenland and Antarctica show that concentrations of carbon dioxide in the atmosphere have increased about 25 percent since the beginning of the Industrial Revolution (about 1760). And, between 1957, when systematic annual measurements were begun at Mauna Loa, Hawaii, by Keeling, and 1989, atmospheric CO_2 increased by about 12 percent. Global temperature records also seemed to have increased by 0.3 to 0.6 degrees centigrade over the last century—although this range falls within the range of the planet's natural temperature variability. When General Circulation Models were applied to predict the impact of a doubling of atmospheric CO_2 concentrations, they calculated anywhere from a 1.5 to 4.5 degree Centigrade increase in global average temperatures. At the same time this information was reaching the popular press, concerns about the ozone hole over Antarctica were being made public. More to the point, the United States was experiencing both a sustained summer heat wave and a drought in the agricultural regions of the West and Midwest. Scientific concerns and public policy questions merged to underline the need for a sustained program to address climate change.

Bill Graham, then the President's Science Advisor, formed the CEES at that time. CEES is made up of representatives from 19 different agencies working to put together a coherent, integrated global change program. Tony Calio, formerly of NOAA, was the first director of CEES, and I've been chairman for the last four years. CEES is part of the Federal Coordinating Council for Science, Engineering, and Technology (FCCSET). FCCSET has been in existence since 1976, but it was raised to prominence by Allan Bromley.

Why is there CEES? The basic point is that we need coordination and partnership. The Federal government needs to identify the knowledge gaps and missing research elements. We have to focus on the interdisciplinary science aspects. We need to relate to international science programs. And we need to coordinate with the policy-makers. I think we've been successful in doing that.

The work really gets done through a whole series of working groups. One of the most active and successful has been the Working Group on Global Change that Bob Corell of the National Science Foundation heads. We have a number of other important subcommittees tackling subjects like water resources, coastal oceans, and atmospheric sciences. We also interact informally with another very successful group, the Interagency Working Group on Data Management for Global Change headed by Tom Pyke of NOAA.

Through all the Federal agencies, we work with scientists at universities and at the Federal laboratories. In consultation with the National Academy of Science's Committee on Global Change, the Federal program has identified the key scientific priorities that need to be addressed. Through the National Academy, we work with the International Geosphere-Biosphere Program of the International Congress of Scientific Unions, as well as the World Climate Research Program, international analogs of our domestic Global Change Program. We are also working in cooperation with the United Nation's Intergovernmental Panel on Climate Change (IPCC), which is assessing the science of climate change and developing response strategies on an international basis. Frederick Bernthal, Deputy Director of the National Science Foundation, who will soon replace me as Chairman of CEES, chaired the working group that drafted the response section of the IPCC report.

To help us coordinate all of these efforts, one of the first tasks of the CEES was developing a research strategy. The CEES Federal research strategy for global change accompanied the President's budget in January 1989. Since then, we've prepared similar documents expanding and adapting the details of the program as our understanding of the data needs has improved.

Our research program goal is to gain an understanding of complex Earth-systems interactions so we can predict changes and establish a scientific basis for formulating national and international policy decisions. Our objectives break down into three parts: documenting the Earth's system through careful, coordinated observations; improving our scientific understanding of the processes; and developing models that will allow us to accurately predict what is going to happen globally and regionally.

The Administration has been consistently supportive of the global change program. I know of no better way of illustrating this point than pointing to our budget. When the program got underway in Fiscal Year 1989, we had a budget of 133.9 million dollars. For Fiscal Year 1992, it is nearly 1.2 billion dollars.

President Bush also sponsored the White House Conference on Science and Economics Research Related to Global Change in April 1990. Clearly, the focus of that conference was not only science policy concerns, but the economic impacts of global change as well. At that conference, President Bush proposed North-South partnerships for global change research. The White House Office of Science and Technology Policy and the CEES are working to establish a Western Hemisphere Global Change Research Institute this year as part of that initiative.

I want to point out, however, that the U.S. global change program is more than a research program. Where there is solid information, where the costs and benefits are clear, the United States has taken action. The U.S. banned chlorofluorocarbons for non-essential purposes years ago and we are moving rapidly towards full implementation of the Montreal Convention and total phase-out of these long-lived, powerful gases that contribute to the greenhouse effect. As part of the Clean Air Act, a number of actions have been taken to cut industrial emissions. And, as new Clean Air Act requirements are implemented, further cuts are on the way. President Bush proposed even further emissions reduction steps in his National Energy Strategy, although this strategy has yet to receive favorable action by the Congress. The combination of measures taken so far should reduce greenhouse gas emissions by 15 percent over the next decade.

But, in many areas, the Administration believes that we simply don't have the information needed to take policy action, and this is particularly the case with CO_2 emissions. Reducing manmade atmospheric CO_2 emissions will be a tremendously costly task for everyone, but the benefits of slashing CO_2 emissions are not yet clear. We are far from being able to develop and document the kind of information that lets us assess exactly what the cost of global change might be internationally, nationally, or regionally.

As I've already pointed out, the global climate models we are working with today aren't as consistent as they need to be, particularly at the regional level, in order to allow us to make the kind of predictions we need to make. And we are faced with scientific puzzles. Global average temperatures actually decreased between 1945 and 1970—a period of rapid growth in industrial CO_2 emissions.

Simply understanding the carbon cycle is one of our most fundamental challenges. The annual flux related to man's activities is about 6 billion tons of CO_2. Now, of that, we don't know where 2-1/2 billion tons goes. It may be in the oceans, or it may be in the mid-latitude biosphere, including the soils. Lacking that knowledge makes crafting effective response policies difficult.

Recent global change research has changed our view of the future. For example, the warming we've observed since the 1970s may be due, at least in part, to an increase of daily minimum—or night-time—temperatures rather than daily maximum temperatures. Since this would tend to lengthen the growing season, such a temperature increase may well be beneficial. Over the last year or so, more sophisticated modeling of the effects of clouds on climate decreased by half the amount of warming predicted by general circulation models. And the recent use of coupled atmosphere-ocean models has shown that the West Antarctic ice sheet isn't nearly as likely to melt—and increase sea levels—as had been predicted earlier.

The findings being generated by global change research today are increasingly being made through the use of coupled models. Improvements in computer technology are allowing this to happen at a time when it is increasingly important that we understand and document how the different components of the Earth system work together.

Understanding how the various Earth systems work together as a whole is a good deal more complex and challenging than understanding how and why volcanoes erupt. Still, as with the our volcano research, the CEES global change program is aimed at providing the predictive power we need to prevent natural hazards from becoming natural disasters. That is anything but a modest goal. Reaching it is critically important. Without solid predictive powers, the prevention steps we take may be ineffectual, or, worse yet, counterproductive.

As we better understand how Earth systems work the way they do, our capabilities for predicting global change—and natural hazards—will be greatly increased. Just as critically, our ability to apply this global change information to prepare for, respond to, and prevent natural disasters will improve. So, regardless of the actual outcome of global change, our integrated approach of looking at Earth systems will provide us with valuable information that will help improve the lot of all people in the future.

VENEZUELAN POLICIES AND RESPONSES ON CLIMATE CHANGE AND NATURAL HAZARDS

Claudio Caponi and Anibal Rosales
Ministerio del Ambiente y
de los Recursos Naturales Renovables
Venezuela

ABSTRACT

Venezuela is an intertropical country which has the fortune not to suffer the severities of natural hazards which are usual in other countries of this region. It is a developing country, whose economy is heavily dependent on oil production and exports. Its greenhouse gas emissions are relatively low, but it is expected that the planned industrialization development will bring an associated increase in emissions.

As a nation, Venezuela has a highly developed environmental consciousness. The Ministry of Environment, the first in Latin America, was created in 1977, and has been the main contributor to the national policy of Disaster Prevention and Reduction. As in many developing countries actions and responses in this regard have been rather limited in scope, and even though legislation has been developed, many problem arise for its enforcement. Several local warning systems, civil defense procedures, and infrastructural protection measures are operational, however they have not been designed, revised, or planned taking into consideration the potential impacts of climate change.

Presently Venezuela is an active participant state in the negotiation for a framework convention on climate change. That is a very difficult negotiation for our country. Here we have to conciliate environmental principles with national economic interests. The elements of our position in this context are presented in this statement.

Venezuela is an intertropical country, with an area of nearly 1 million km^2 and a population of 20 million inhabitants, 85% of which is urban. The population is mainly concentrated in the northern coastal region, with high densities in mountainous valleys extremely vulnerable to natural hazards, especially hydrometeorological and seismic ones.

On the other hand, it has had the fortune to be only marginally influenced by the pass of hurricanes. Furthermore, the same lack of balance in population density results in a relatively small percentage of its total territory being subject to the risks of natural hazards, at least those involving losses of human lives.

The Ministry of Environment, the first of its kind in Latin America was created in 1977 and has been the main contributor to the national policies of Natural Disaster Prevention and Reduction, which mainly deal with flood prevention in urban areas. There has been a change in approach from the period 1950–1976 when the Ministry of Public Works mainly concentrated on structural works for flood prevention: reservoirs, dikes, river channel improvements, derivation channels and such. Emphasis has now been placed on non-structural measures such as laws on regulation of flood plains occupation, establishment of forecasting and warning systems, coordination with local civil defense authorities and watershed conservation as a fundamental element for flood control.

There have been many obstacles in the implementation of these measures:
- Law enforcement has been difficult in regard to flood plain occupation. Occupation of flood plains offers economic advantages to low-income population (easy access to water for supply and waste disposition, lowland plains which implies low development costs, good soils for agriculture). Commonly, occupants with low levels of education ignore risks associated with living in the flood plain, and in some cases a conscious decision of living in a risk area is made to benefit from aid provided by the Government.
- Only a few local and non-integrated forecasting systems (consisting of rainfall, river level and mudflow detecting telemetering stations) have been established, and only in areas where a disaster has actually occurred previously.
- Civil defense authorities are better prepared to deal with the consequences of a disaster than to take preventive measures when one is forecasted.

As a result of the declaration of the IDNDR, the three main meteorological services (Air Force, Navy, Environment) have created a National Center for Hydrometeorological Alert (NCHA), which aims to join their human and technological resources, recognizing that each service on its own has only access to part of the data needed to establish a reliable forecasting Center.

The NCHA has a high resolution GOES receiving station, a weather radar (eight more are planned), direct access to the WMO World Weather Watch (WWW) Programme through the Regional Communication Center run by the Air Force in Maracay, and is planning to link the computer centers of the three services to share their climatic and hydrological data banks.

However, the potential impacts of climate changes in our country have not been taken into consideration in the actions mentioned above. Experts have brought up to the attention of authorities the possible consequences of climate change to the Nation. Items of concern are sea-level rise, the possible increase of intensity and frequency of hurricanes, displacement of cyclogenesis to the south, changes in rainfall patterns affecting the agricultural regions, and other possible impacts. Serious limitations of human, economic, and technological resources as well as lack of long-term reliable data in many areas hinder the development of more detailed research studies.

Presently, Venezuela is an active participant state in the negotiation for a framework Convention on Climate Change, which for a developing country such as ours, with a highly developed environmental conscience, but also with an economy heavily dependent on oil production and export (oil business provides 75% of government revenues, and 81% of foreign exchange receipts), has not been an easy task.

Most of the comments to follow are taken from the document "Energy-Economy and Climate Change Convention" prepared by the Venezuelan Delegation to the United

Nations Intergovernmental Negotiation Committee for a Framework Convention on Climate Change (INC).

For more than half a century oil revenues have brought wealth to Venezuela, creating a significant rate of economic growth and the development of a modern and important infrastructure for the country. However, this same wealth distorted Venezuela's economy and with the sudden fall in oil prices in 1982 and 1986, and the external debt crisis, the affluence reached an end.

Our country's economy is actually facing a new reality, experiencing a recession that reached the lowest point in 1989, when inflationary pressures reached 80%. In response, a radical economic reform and restructuring programme is in progress as an effort to promote a more outward-looking, diversified and market-oriented economy, less dependent on petroleum exports.

Reform measures taken during the last two years include liberalization of exchange rate and trade regimes, price decontrol, liberalization of interest rates to allow for positive real interest rates, measures to reduce the fiscal deficit, and privatization of state-owned industries in the service sector. We can already perceive some results: foreign debt payments have been restructured, annual inflation has been reduced to 30%, and support of the multilateral financial institutions has been reestablished.

However, we paid a high social cost for the severity of the incoming adjustment program: 25% of Venezuelans live today under a commonly accepted line of absolute poverty, and we do not forget than in February 1989, huge crowds, reacting to the sharp increase in bus fares and food prices began burning buses and looting groceries and other businesses . Certain constitutional rights had to be temporarily suspended and military personnel deployed to restore order. Finally, several hundred deaths resulted from these riots that were the only incident of its kind to have occurred under democratic rule in Venezuela (33 years).

In summary, oil exports have permitted us to have access to the hard currency necessary for alleviating poverty and developing our productive infrastructure in the past, and in coming years oil exports revenues should also allow us to transform our productive systems for diversifying and stabilizing our economy making us less vulnerable to the changes of the international petroleum market and more capable of responding to the challenge of protecting the environment.

Of course, this will not only depend on our domestic efforts, but on foreign technical cooperation and on the liberalization of international trade, which do not seem to advance at the same pace as the demands for international environmental protection measures, such as those proposed under the Climate Change Convention.

The stabilization or reduction of carbon dioxide emission in developed countries by the year 2000 at the 1990 level, through the use of taxes, shall represent direct revenue losses for us. It will also lead to economic imbalances as a result of the incremented costs of essential goods and services that we import from industrialized nations. As a consequence, we may face more difficulties in correcting our economic disequilibrium. Increased poverty, health care, and educational needs will necessarily make the allocation of resources for environmental protection far more difficult. These implications of the Climate Change Convention have not been sufficiently discussed, nor a clear compensation mechanism proposed. Under these circumstances, which imply a reduction of our country's income and new difficulties in making the needed investments in our oil industry to guarantee continuity of oil supply to the international market, it is very difficult for us to accept the Convention.

We could summarize the Venezuelan position in this negotiation, by stating that we would support a Convention on Climate Change, but a balanced one that would take fully into consideration the following aspects:

- It should be flexible, to allow for adjustments according to advances in scientific knowledge that would reduce the uncertainties on the magnitude and time period in which the climate changes will occur, the emissions-sinks equilibrium and the social and economic consequences of the proposed preventive measures.
- It should include commitments on emissions, with measures such as improved energy efficiency and conservation, decrease of biomass burning and better control on forest fires.
- It should include commitments on sinks and reservoirs, such as stopping deforestation, increase of protected forest areas, increase of afforestation, and eliminaton of contamination of ocean, seas, and other water bodies.
- It should take into full account the developing countries right to develop themselves, by giving them access to the technology needed for a more efficient energy use and allowing them to increase their energy consumption, using the more convenient and appropriate sources in each case.
- It should not promote alternative sources of energy which are not environmentally safe and sound, such as nuclear energy, which still implies risks which may be worse than those of climate changes.
- It should include special considerations for countries whose economies are highly dependent on fossil fuels production and exportation, with compensation mechanisms accounting both for the direct revenue losses and the incremented costs of imported goods for these countries.

Even though the predicted climate changes would be the result of the development of the industrialized countries, Venezuela recognizes that the solution must be global and come from the joined efforts of all nations.

To conclude, we feel that more research must be undertaken worldwide on the socio-economic impacts of the diverse proposed mitigation measures, and it is in this field where research centers such as the one hosting this symposium may contribute their invaluable assistance to this, our world at risk.

CONCLUDING REMARKS

Peter S. Thacher
Senior Advisor to the Secretary General of the U.N. Conference
on Environment and Development
and
Senior Counselor
World Resources Institute

As a former Program Director for the 1972 Stockholm Conference on the Environment, it is useful to recall that MIT played a key role more than 20 years ago in the preparations for that conference. Dr. Carroll Wilson organized and convened an international scientific meeting in Sweden in 1971 that is reported in the MIT Press book—the 1971 *Study of Man's Impact on Climate* edited by Bill Matthews—that shaped many of the recommendations adopted by governments at the 1972 conference and is still directly relevant to this meeting, and the Earth Summit in Rio in June 1992.

Two pollution problems that were then seen distantly on the horizon as possibly posing *global* risks of highly adverse impacts are now front-page news: stratospheric ozone depletion (then seen in terms of high-flying second-generation SSTs), and greenhouse gases (GHGs) which threaten to commit the planet to rates of warming unprecedented in human experience.

As a result of actions initiated at Stockholm, we now have encouraging evidence in the ozone issue that international cooperation works. Cooperative integrated analysis can lay a basis for international negotiations leading to agreements that are flexible, so as to be strengthened as the need arises, as well as effective, with financial assistance available to ensure follow-up.

Threats to the ozone layer were put on the Stockholm agenda 20 years ago and in 1973 UNEP's first Executive Director, Maurice Strong, got governments to agree to an "Outer Limits" program that would address both stratospheric ozone and climate. Despite its hypothetical nature, the 1974 warning about CFCs by Rowland and Molina triggered a quick response by UNEP which together with WMO, WHO, FAO, and others, in a March 1977 international expert meeting, drew up an Action Plan to Protect the Ozone Layer that governments approved later that year. Among its recommendations was the reactivation of a Dobsen spectrometer in far-away Halley Bay that many years later was to be the instrument first to discover the Antarctic Hole phenomenon.

Also present at this meeting 15 years ago was the International Chamber of Commerce with representation by private sector manufacturers of CFCs. Although they rejected the notion that CFCs posed a threat, corporate scientists were active participants thereafter in the Coordinating Committee on the Ozone Layer (CCOR) and provided key production data that became critical when it came time to unravel the Antarctic Hole mystery. And when the NASA/UNEP Panel found CFCs responsible in 1988, DuPont—the biggest CFC manufacturer—proposed a total shutdown of CFC production and use—a move which led the U.S. and others to tighten up the 1987 Montréal Protocol.

Since then economic studies have led governments to set up a Trust Fund under the Montréal Protocol to provide incentives for countries like China, India, and Brazil to become parties to that treaty, despite the high price involved.

The ozone issue illustrates the important roles played by a wide variety of actors in addition to government representatives. These include international civil servants, uninstructed scientists, corporate leaders, media, academicians, and grassroots NGOs. Looking to the future, and at climate specifically, it's clear we will need this same high level of active participation by many key actors—including, I hope, MIT. Of course the climate problem is far more complex than stratospheric ozone.

As usual, the North must lead, but the South will matter even more, because demographic growth in the so-called developing countries will soon contribute major atmospheric heating commitments, even if there is no increase in their per capita energy consumption figures. This means that no matter what the North does to get more energy for less CO_2, if the greenhouse works as most scientists expect, the South alone can commit the planet to heating at unprecedented rates. Of course the North, especially North America, should increase its energy efficiency for all sorts of reasons that have nothing to do with climate change, e.g., just to be able to compete in international markets.

Both the North and the South are preoccupied with immediate problems (even the rich feel poor) and find it convenient to hide behind uncertainty. The South sees the North as having used up planetary ecological space to their disadvantage—the South produced no CFCs and, up to now, only a small fraction of increased GHGs in today's atmosphere, yet they are expected to forgo industrialization and refrain from burning their coal to save our skins, or so they think. While acknowledging our responsibility as major contributors to global problems, we in the North seem to be having second thoughts about the principle we used to champion, that the Polluter Pays.

Legitimate questions abound about costs of reducing climate risk, and the broader question of how a consumption-oriented economy might be converted to providing goods and services on a *sustainable* basis; living off the income of our resource base rather than the capital, as the Brundtland Commission put it.

In the June "Earth Summit" in Rio, governments agreed on a broad action plan called "Agenda 21" that addresses all the environment-and-development issues including climate and other forms of global risk, such as the fast loss of biological diversity, as well as safe drinking water and other issues we in the North look upon as "environmental" but which for most people are still a "development" problem. Governments also agreed on a statement of principles, on the means of implementation for Agenda 21, and on some treaties, including one on climate.

The "means of implementation" call for additional finances to help those in need to become partners in reducing global risk. President Nixon's proposal 20 years ago for a new fund of $100 million with the U.S. contributing 40% on a matching basis got UNEP started. There is nothing like this in the present U.S. position.

Nonetheless, a cross-cutting theme on which virtually all governments agree is the need to strengthen human and institutional capacity locally, on the ground, where all of the local decisions are made that we now recognize drive the process of global change.

In this regard, MIT has played a leading role before and since the 1972 Conference that I hope will be continued in the years after Rio.

Author Index

A

Avissar, Roni, 156

B

Berz, G. A., 217
Bordogna, Joseph, 317
Bras, Rafael L., ix, 174
Bruce, James P., 3

C

Cahoon, Donald R., Jr., 131
Caponi, C., 343
Clague, John J., 48
Cofer, Wesley R. III, 131

D

Dennis, Karen C., 193
Deyle, Robert E., 295
Dooge, James C. I., 13
Dracup, John A., 86

E

Eagleson, Peter S., 168
Eltahir, Elfatih A. B., 174
Emanuel, Kerry A., 25
Entekhabi, Dara, 168
Evans, Steven G., 48

G

Gregory, James M., 125

H

Heckman, Stan, 77
Hughes, James P., 112

J

James, Thomas E., 295

K

Kahya, Ercan, 86
Kareem, Ahsan, 280
Krajewski, Witold F., 180

L

Leatherman, Stephen P., 193
Lettenmaier, Dennis P., 112
Levine, Joel S., 131
Ling, Douglas C., 305

M

MacKenzie, James J., 262
Marotzke, Jochem, 150
McGeer, Tad, 206
Meo, Mark, 295

N

Nicholls, Robert J., 193
Nigg, Joanne M., 289
Nuttle, William K., 43

P

Peck, Dallas L., 337
Pelto, Mauri S., 61
Perry, Allen, 275
Peterson, Richard E., 34, 125
Price, Colin, 68

R

Rind, David, 68
Rodríguez-Iturbe, Ignacio, 96
Rosales, A., 343

S

Sabadell, J. Eleonora, 317, 328
Solow, Andrew R., 38
Stocks, Brian J., 131
Stone, Peter H., 143

T

Thacher, Peter S., 347

U

Ubertini, Lucio, 320

V

Volonte, Claudio R., 193

W

Waggoner, Paul E., 254
Warner, Thomas E., 34
White, David C., 224
Williams, Earle, 77
Winsted, Edward L., 131
Wood, Eric F., 112

AIP Conference Proceedings

		L.C. Number	ISBN
No. 247	Global Warming: Physics and Facts (Washington, DC, 1991)	91-78423	0-88318-932-1
No. 248	Computer-Aided Statistical Physics (Taipei, Taiwan, 1991)	91-78378	0-88318-942-9
No. 249	The Physics of Particle Accelerators (Upton, NY, 1989, 1990)	92-52843	0-88318-789-2
No. 250	Towards a Unified Picture of Nuclear Dynamics (Nikko, Japan, 1991)	92-70143	0-88318-951-8
No. 251	Superconductivity and its Applications (Buffalo, NY, 1991)	92-52726	1-56396-016-8
No. 252	Accelerator Instrumentation (Newport News, VA, 1991)	92-70356	0-88318-934-8
No. 253	High-Brightness Beams for Advanced Accelerator Applications (College Park, MD, 1991)	92-52705	0-88318-947-X
No. 254	Testing the AGN Paradigm (College Park, MD, 1991)	92-52780	1-56396-009-5
No. 255	Advanced Beam Dynamics Workshop on Effects of Errors in Accelerators, Their Diagnosis and Corrections (Corpus Christi, TX, 1991)	92-52842	1-56396-006-0
No. 256	Slow Dynamics in Condensed Matter (Fukuoka, Japan, 1991)	92-53120	0-88318-938-0
No. 257	Atomic Processes in Plasmas (Portland, ME, 1991)	91-08105	0-88318-939-9
No. 258	Synchrotron Radiation and Dynamic Phenomena (Grenoble, France, 1991)	92-53790	1-56396-008-7
No. 259	Future Directions in Nuclear Physics with 4π Gamma Detection Systems of the New Generation (Strasbourg, France, 1991)	92-53222	0-88318-952-6
No. 260	Computational Quantum Physics (Nashville, TN, 1991)	92-71777	0-88318-933-X
No. 261	Rare and Exclusive B&K Decays and Novel Flavor Factories (Santa Monica, CA, 1991)	92-71873	1-56396-055-9
No. 262	Molecular Electronics—Science and Technology (St. Thomas, Virgin Islands, 1991)	92-72210	1-56396-041-9

No. 263	Stress-Induced Phenomena in Metallization: First International Workshop (Ithaca, NY, 1991)	92-72292	1-56396-082-6
No. 264	Particle Acceleration in Cosmic Plasmas (Newark, DE, 1991)	92-73316	0-88318-948-8
No. 265	Gamma-Ray Bursts (Huntsville, AL, 1991)	92-73456	1-56396-018-4
No. 266	Group Theory in Physics (Cocoyoc, Morelos, Mexico, 1991)	92-73457	1-56396-101-6
No. 267	Electromechanical Coupling of the Solar Atmosphere (Capri, Italy, 1991)	92-82717	1-56396-110-5
No. 268	Photovoltaic Advanced Research & Development Project (Denver, CO, 1992)	92-74159	1-56396-056-7
No. 269	CEBAF 1992 Summer Workshop (Newport News, VA, 1992)	92-75403	1-56396-067-2
No. 270	Time Reversal—The Arthur Rich Memorial Symposium (Ann Arbor, MI, 1991)	92-83852	1-56396-105-9
No. 271	Tenth Symposium Space Nuclear Power and Propulsion (Vols. I–III) (Albuquerque, NM, 1993)	92-75162	1-56396-137-7 (set)
No. 272	Proceedings of the XXVI International Conference on High Energy Physics (Vols. I and II) (Dallas, TX, 1992)	93-70412	1-56396-127-X (set)
No. 273	Superconductivity and Its Applications (Buffalo, NY, 1992)	93-70502	1-56396-189-X
No. 274	VIth International Conference on the Physics of Highly Charged Ions (Manhattan, KS, 1992)	93-70577	1-56396-102-4
No. 275	Atomic Physics 13 (Munich, Germany, 1992)	93-70826	1-56396-057-5
No. 276	Very High Energy Cosmic-Ray Interactions: VIIth International Symposium (Ann Arbor, MI, 1992)	93-71342	1-56396-038-9
No. 277	The World at Risk: Natural Hazards and Climate Change (Cambridge, MA, 1992)	93-71333	1-56396-066-4